Seed Science and Technology

Malavika Dadlani • Devendra K. Yadava
Editors

Seed Science and Technology

Biology, Production, Quality

 Springer

Editors
Malavika Dadlani
Formerly at ICAR-Indian Agricultural
Research Institute
New Delhi, India

Devendra K. Yadava
Indian Council of Agricultural Research
New Delhi, India

This work was supported by ADT Project Consulting GmbH

ISBN 978-981-19-5890-8   ISBN 978-981-19-5888-5   (eBook)
https://doi.org/10.1007/978-981-19-5888-5

This Springer imprint is published by the registered company Springer Nature Singapore Pte Ltd.
The registered company address is: 152 Beach Road, #21-01/04 Gateway East, Singapore 189721, Singapore

त्रिलोचन महापात्र, पीएच.डी.
सचिव, एवं महानिदेशक
TRILOCHAN MOHAPATRA, Ph.D.
SECRETARY & DIRECTOR GENERAL

भारत सरकार
कृषि अनुसंधान और शिक्षा विभाग एवं
भारतीय कृषि अनुसंधान परिषद
कृषि एवं किसान कल्याण मंत्रालय, कृषि भवन, नई दिल्ली 110 001

GOVERNMENT OF INDIA
DEPARTMENT OF AGRICULTURAL RESEARCH & EDUCATION
AND
INDIAN COUNCIL OF AGRICULTURAL RESEARCH
MINISTRY OF AGRICULTURE AND FARMERS WELFARE
KRISHI BHAVAN, NEW DELHI 110 001
Tel.: 23382629; 23386711 Fax: 91-11-23384773
E-mail: dg.icar@nic.in

FOREWORD

Seed science and technology, once considered only an ancillary to variety development, is now recognized as a full-fledged multi-disciplinary subject that is an essential component of the crop improvement programme. Seed security is vital to attain food security. Therefore, a strong seed system is a major instrument to achieve nutritional security in a sustainable manner. Since the era of Green Revolution, systematic crop variety improvement programme and seed system development were given priority in India, which have continued with the changing scenario of plant breeding, variety development, emerging issues relating to IPR and seed quality consciousness. In order to keep pace with the changing philosophies and emerging technologies, capacity building of the scientists, teachers, seed analysts, and all seed professionals engaged in the public and private sector organizations is of utmost importance.

The book "Seed Science and Technology – Biology, Production, Quality" is an advanced textbook and a reference book for the post-graduate students and researchers; plant breeders and seed technologists; regulatory and government agencies; seed industry professionals, and research policy makers. It consolidates fundamental concepts as well as the latest advances in seed biology, production and quality assurance research with a wide coverage of the subject of seed science and technology. This open access book has 17 chapters contributed by the eminent experts not only from India, but also from Germany, the Netherlands and the USA blending the diverse experiences of Indian and global seed systems. The chapters are developed around the themes of the lectures delivered by the experts during capacity-building programmes under the 'Indo-German Cooperation on Seed Sector Development'- a component of the bilateral cooperation between the Governments of India and Germany.

The book has covered every aspect of seed technology starting from basics of seed development, maturation, dormancy, germination, vigour and invigoration, and seed deterioration; variety maintenance and production of genetically pure seed of open-pollinated and hybrid varieties of field and vegetable crops, seed processing, packaging and storage; and seed quality assurance systems, including seed health testing being followed globally. New and emerging areas like use of molecular tools and technologies for variety identification, genetic purity and seed health testing, and seed quality enhancement technologies, and future trends in seed technology have also been covered adequately.

I congratulate the editors Drs. Malavika Dadlani and Devendra Kumar Yadava along with the contributors of different chapters for bringing out this book, which I hope will bridge the knowledge gaps in various aspects of seed research.

(T. MOHAPATRA)

Dated the 22ⁿᵈ July, 2022
New Delhi

Preface

Seed security is a prerequisite to food security, which in turn depends much on the success of crop improvement programmes. Strong plant breeding programmes lead to the continued development of improved crop varieties that perform well under optimum as well as erratic climatic conditions, meet the market demands fulfilling changing consumer preferences, and fit into newer farming system models. However, development of improved crop varieties is only of limited consequence without a sound and resilient seed supply system to achieve success in agricultural production. It is seen that countries that laid emphasis on continued investment in capacity building, and the creation of necessary infrastructure in crop improvement and seed system development, achieved success in meeting the growing food demands and challenges of climate change in a sustainable manner.

Concurrent with the introduction of semi-dwarf, high-yielding varieties of wheat and rice in the 1960s to 1970s, ushering an era of "Green Revolution", and a well-planned programme of hybrid development in maize, sorghum, and pearl millet in the 1960s, India created a seed supply system by investing in the multidisciplinary research and education on seed technology, introducing appropriate legislation, policy support, and building necessary infrastructures for seed production, processing, and quality control. These were supported by various programmes, projects, and international collaborations, all of which contributed to capacity building and infrastructure development.

In the fast advancing field of Seed Science and Technology, continued efforts are needed for the upgradation of the capacity of seed professionals. Viewing this, in the frame of a Bilateral Cooperation between the governments of India and Germany, and the project "Indo-German Cooperation on Seed Sector Development", an extensive capacity-building programme was undertaken, mostly online, during the years 2020 and 2021, for the benefit of scientists, faculty members, and seed sector professionals working in agricultural research institutions, universities, certification and seed-production agencies in India. Nearly 100 lectures were delivered by 70 eminent experts from India and EU countries during these training programmes on various aspects of seed science and technology including Seed Production; Post-harvest Technologies and Storage; Seed Quality Enhancement; Seed Testing, and Seed Certification.

Considering the huge response to these trainings, it was decided to bring out a book by using, collating, updating, and restructuring the training lectures and elaborating the topics further in a manner that the book will not only be useful to the seed professionals in India but will also serve as a base reference to those in other parts of the world, particularly in South and SE Asia. The chapters in this book have been contributed by the experts in their respective fields, most of whom were involved in these capacity-building programmes. Many of the contributors are from the seed industry having practical knowledge of the gaps and the process of technology development. Some chapters have also been contributed by the experts who were not associated with these training programmes, but have long practical experience in their respective fields.

The introductory chapter presents a brief overview of the multidisciplinary structure of seed science and technology. The next 15 chapters have been organised into three major sections, Seed Biology; Seed Production; and Seed Quality Assessment and Enhancement. The concluding chapter discusses the most significant recent advances, potential areas of seed research, and the developments of promising technologies. Keeping the multidisciplinarity of the subject, there are some overlapping between the chapters. As the chapters were prepared around the lectures delivered during the training programmes, some degree of overlapping provides interlinking, which will help provide a comprehensive understanding of seed technology.

In bringing out this publication we are indebted to the Governments of India and Germany, and the project on "Indo-German Cooperation on Seed Sector Development", and its key functionaries, Sh. Ashwani Kumar, Joint Secretary (Seed), Department of Agriculture and Farmers Welfare, GOI, India, and Dr. Ulrike Mueller, Deputy Programme Director, GFA Consulting Group/General Agent of the Bilateral Cooperation Programme of the BMEL, Germany, without whose constant guidance and support this project would not have materialised. We are profoundly grateful to Mr. Ekkehard Schroeder, Managing Director, ADT Project Consulting GmbH, Adenauerallee 174, 53113 Bonn, Germany, for making this book happen. Concerted efforts made by the Nodal Officers of Indo-German Training Programme from Northern Region (Dr. D.K. Yadava and Dr. Sandeep K Lal), Eastern Region (Dr. Sanjay Kumar), and Southern Region (Dr. Vilas A. Tonapi) are also duly acknowledged. Our sincere thanks are due to Sh. M. Gunasekaran, Asst. Commissioner (Seed), Dr. Sowmini S. and Dr. Raghvendra Kavali, for their constant support in conducing the trainings and providing relevant materials for the preparation of this book.

Though the book primarily addresses the situation prevailing in India, it also compares, wherever relevant, the Indian seed system with that of OECD member countries and the USA. It is hoped that this book will serve the purpose of a reference book for graduate students, postgraduate researchers, scientists, officials involved in seed production and certification, and all other seed professionals working in the public and private sectors not only in India but elsewhere too. We are grateful to all the experts who reviewed the chapters meticulously and provided critical comments and valuable suggestions which helped improve the contents.

We express our profound gratitude to Dr. Trilochan Mohapatra, Secretary DARE, and Director General, ICAR, Min. of Agriculture & Farmers' Welfare, Govt. of India, a plant geneticist, and breeder *par excellence*, for a Foreword to the book.

We are indebted to our family members for their patience and unconditional support throughout the preparation of this book. Malavika specially thanks Dr. N.K. Dadlani and Ms. Deepika for providing many valuable inputs and technical help.

New Delhi, India   Malavika Dadlani
June, 2022   Devendra K. Yadava

Contents

Editors and Contributors

About the Editors

Malavika Dadlani is a renowned seed scientist with an active career of four decades in the field of Seed Science and Technology. Educated at some of the most prestigious academic institutions in India, including the Indian Agricultural Research Institute (IARI), New Delhi, Indian Institute of Technology (IIT), Kharagpur, and Calcutta University, Dr. Dadlani dedicated her entire professional career to research and postgraduate teaching in seed science and technology at IARI, where she held the positions of the Head of the Division of Seed Science and Technology, and the Joint Director (Research). A seed physiologist by training, Dr. Dadlani has also made significant contributions to hybrid seed production, plant variety identification, and purity testing and supervised the postgraduate and doctoral research of 21 students in the disciplines of seed science and technology, plant physiology, and plant biotechnology. She has more than 200 publications, including papers in peer-reviewed journals, chapters, and books. She is a Fellow of the National Academy of Agricultural Sciences (NAAS), India, and a recipient of many other recognitions and awards. Dr. Dadlani served as the President of the Indian Society of Seed Technology and a member of the Vigour Committee, ISTA, and presently is the Editor of NAAS. She served as a Consultant at Bioversity International, India Office, and a short-term FAO consultant in the seed programs of Cambodia and Myanmar.

Devendra K. Yadava is a noted plant breeder and seed technologist with 26 years of research experience, which includes more than 6 years as the Head, Division of Seed Science & Technology at Indian Agricultural Research Institute (IARI), New Delhi. Presently he is the Assistant Director General (Seed), Indian Council of Agricultural Research (ICAR), New Delhi. He has contributed to the development and release of 21 plant varieties (mustard 18; pulses 3) which include early maturing, low erucic acid and canola oil quality mustard varieties that provided better choice to both farmers and consumers. Dr. Yadava contributed to the development of state seed rolling plans leading to enhanced seed and varietal replacement rates and significant improvements in the national breeder seed production program. As a member of many regulatory bodies of the Govt. of India, he contributed to policy formulation/amendments in the Seed Bill, Genome editing guidelines, proposed amendments of Biodiversity Act, etc. He is a Fellow of the National Academy of Agricultural Sciences, India, and a recipient of many prestigious awards, e.g., Rafi Ahmad Kidwai Award of ICAR; Dr. B.P. Pal Memorial and Dr. A.B. Joshi Memorial Awards of IARI, New Delhi; NAAS Recognition Award; and Dr. P.R. Kumar Brassica Outstanding Scientist Award, Bharatpur. He has published 90 research papers in high impact factor peer-reviewed journals, authored eight books along with 170 other publications.

Contributors

Theerthagiri Anand Plant Pathology, TNAU, Coimbatore, India

Sudipta Basu Division of Seed Science and Technology, ICAR-Indian Agricultural Research Institute, New Delhi, India

Alexander Buehler Operational Excellence – Seed Applied Solutions EME, Bayer Crop Science, Langenfeld, Germany

Shyamal K. Chakrabarty Division of Seed Science and Technology, ICAR-Indian Agricultural Research Institute, New Delhi, India

Bhavya Chidambara Division of Basic Sciences, ICAR Indian Institute of Horticultural Research, Bengaluru, India

P. Ray Choudhury Crop Sci. Division, Indian Council of Agricultural Research, New Delhi, India

Malavika Dadlani Formerly at ICAR-Indian Agricultural Research Institute, New Delhi, India

Steven P. C. Groot Wageningen University & Research, Wageningen Seed Centre, Wageningen, Netherlands

Anuja Gupta Formerly at ICAR-Indian Agricultural Research Institute, Regional Station, Karnal, India

Arnab Gupta Wageningen University & Research, Wageningen, The Netherlands

Nakul Gupta ICAR-Indian Institute of Vegetable Research, Varanasi, India

Adelaide Harries Seed Net Inter-American Institute for Cooperation on Agriculture (IICA), Miami, FL, USA

Raghavendra Kavali Indo–German Cooperation on Seed Sector Development, ADT Consulting, Hyderabad, India

Ashwani Kumar ICAR-Indian Agricultural Research Institute, Regional Station, Karnal, India

Sanjay Kumar ICAR - Indian Institute of Seed Science, Maunath Bhanjan, Uttar Pradesh, India

Manjit K. Misra Seed Science Center, BIGMAP, GFSC, IOWA State University, Ames, IA, USA

P. C. Nautiyal Formerly at Seed Science & Technology, ICAR-Indian Agricultural Research Institute, New Delhi, India

Vinod K. Pandita Formerly at ICAR-Indian Agricultural Research Institute, Regional Station, Karnal, India

Manish Patel Incotec India Pvt Ltd, Ahmedabad, India

Kuppusami Prabakar Centre for Plant Protection Studies, TNAU, Coimbatore, India

K. Raja Department Seed Science & Technology, TNAU, Coimbatore, India

S. Rajendra Prasad Formerly at University of Agricultural Sciences, Bengaluru, India

Umarani Ranganathan Seed Centre, TNAU, Coimbatore, India

Kundapura V. Ravishankar ICAR - Indian Institute of Horticultural Research, Bengaluru, India

Nagamani Sandra Division of Seed Science and Technology, ICAR-Indian Agricultural Research Institute, New Delhi, India

Santhy V. Seed Technology, ICAR-Central Institute for Cotton Research, Nagpur, India

W. Schipprach Maize Breeding, University of Hohenheim, Stuttgart, Germany

Rakesh Seth ICAR-Indian Agricultural Research Institute, Regional Station, Karnal, India

P. M. Singh Div. Vegetable Improvement, ICAR-Indian Institute of Vegetable Research, Varanasi, India

J. P. Sinha Water Technology Centre, ICAR-Indian Agricultural Research Institute, New Delhi, India

S. N. Sinha Formerly at ICAR-Indian Agricultural Research Institute, Regional Station, Karnal, India

K. Sivasubramaniam Formerly at Seed Science & Technology, TNAU, Madurai, India

K. V. Sripathy ICAR-Indian Institute of Seed Science, Maunath Bhanjan, Uttar Pradesh, India

S. Sundareswaran Department of Seed Science & Technology, TNAU, Coimbatore, India

K. Udaya Bhaskar ICAR-Indian Institute of Seed Science, Regional Centre, Bengaluru, India

C. Vanitha Seed Centre, TNAU, Coimbatore, India

Banoth Vinesh ICAR-Indian Institute of Seed Science, Maunath Bhanjan, Uttar Pradesh, India

Karuna Vishunavat Department of Plant Pathology, College of Agriculture, GBPUAT, Pantnagar, India

Elmar A. Weissmann HegeSaat GmbH & CoKG, Singen, Germany

Devendra K. Yadava Indian Council of Agricultural Research, New Delhi, India

R. N. Yadav ICAR-Indian Agricultural Research Institute, Regional Station, Karnal, India

Seed Quality: Variety Development to Planting—An Overview

S. Sundareswaran, P. Ray Choudhury, C. Vanitha, and Devendra K. Yadava

Abstract

The importance of availability of quality seed of high-yielding varieties in achieving food security has been recognized globally. The chapter presents an overview of the activities and requirements of seed production system globally, with an emphasis on quality, and highlights the linkages between variety development and seed production programmes. The seed development in angiosperm through the process of fertilization has been briefly touched and understanding the processes underlying pollination, fertilization, seed development, and maturation, which are vital for production of quality seed, has been highlighted. System of variety development and release, maintenance of variety purity during seed multiplication, and their importance have been enumerated to benefit those associated with any seed programme. Seed quality parameters including physical and genetic purity, physiological quality, seed vigour, and health, along with factors determining seed quality, have been presented in a holistic manner. Regulatory mechanism for seed quality assurance including steps in seed certification, seed testing and various field and seed standards has been outlined comparing the Indian system with other major international systems working globally. Procedures for seed health testing and application of advanced molecular marker technologies for varietal identity, genetic purity of seed and detection

S. Sundareswaran (✉)
Department of Seed Science & Technology, TNAU, Coimbatore, India

P. Ray Choudhury
Crop Sci. Division, Indian Council of Agricultural Research, New Delhi, India

C. Vanitha
Seed Centre, TNAU, Coimbatore, India

D. K. Yadava
Indian Council of Agricultural Research, New Delhi, India

M. Dadlani, D. K. Yadava (eds.), *Seed Science and Technology*,
https://doi.org/10.1007/978-981-19-5888-5_1

of seed-borne pathogens, which are becoming increasingly relevant in the present seed scenario, have been discussed. Fundamentals of seed processing for quality upgradation, and improvement of seed quality through enhancement technologies, have been explained. The chapter presents an overview of the importance of seed quality, its indicators, regulations, systems of development of varieties and their maintenance and use of modern tools and techniques for assurance and enhancement of seed quality.

Keywords

Seed quality · Variety · Standards · International organizations · Maintenance · Regulation · Enhancement · Seed testing

1   Introduction

Quality seed is the basic and most critical input for sustainable agriculture. The intensive efforts in plant breeding by the breeders in the public research institutions, international organizations (such as the Consultative Group institutions), private seed industry and farmers/farming community resulted in the development of a large number of new and improved crop varieties with high productivity, better resource use efficiency, varying levels of tolerance to biotic and abiotic stresses and specific quality traits which have played a vital role in enhancing the global food production ensuring food and nutritional security. The growth in food grain production and productivity in some of the developing economies, like India, has been remarkable with the adoption of modern plant breeding, development of high-yielding varieties (HYVs), and improved production and protection technologies, bringing its status from food-dependent to food secure in what is termed as 'Green Revolution'. The success of Green Revolution in India to a great extent is credited to linking crop improvement to the seed production system which led to 6.19 times enhancement in production of food grains, 3.30 times in pulses, 7.46 times in oilseeds, 10.31 times in cotton and 7.55 times in sugarcane since 1950. A clear-cut direct correlation between availability of good quality seed and food grain production can be witnessed. With the availability of 350,000 tons of quality seed during 1980–1981, the food grain production was 129.29 million tons, oilseeds production was 9.37 million tons and pulse production was 10.63 million tons whereas, with the availability of 4,836,600 quintals quality seeds, the food grains (310.74 million tons), oilseeds (35.95 million tons) and pulses (25.46 million tons) production have attained all time highest level (Agricultural Statistics at a Glance, DAC & FW, MA & FW, GOI, 2021, www.agricoop.nic.in, http://eands.dacnet.nic.in).

A synchronized development of the crop improvement programmes through All India Coordinated Research Programmes (AICRP) and public seed system, along with enabling policies and regulatory framework, coupled with the contribution of the private sector R&D in breeding and seed production, helped bring such a change. Similar approaches were also adopted successfully in many countries in the SAARC region, African continent and elsewhere. As a result, the demand for quality seeds of improved varieties is growing fast around the world (Tony et al. 2002). However, the production and distribution of seeds is a complex process involving breeders, seed technologists, farmers, growers, government agencies, research institutions, the private seed industry, the farmers' cooperatives and other stakeholders. Mostly the public organizations play a dominant role in the production and distribution of high volume-low value seeds of food security crops such as cereals, pulses and oilseeds, whereas, the private sector seed industry focuses more on high-value segments comprising vegetables, hybrids in field crops and other horticultural crops (Hanchinal 2017). Hence, the production and supply of high-quality seed of improved crop varieties to the growers have become a high priority in agricultural growth and development.

However, the lack of timely availability of quality seed of improved varieties, and also varietal mismatches at times (meaning that despite sufficient availability of seeds of a crop, requirements of the seeds of certain varieties in demands are not fulfilled), particularly in the developing world, is one of the greatest impediments in achieving the potential yields and production targets. Even with ideal conditions of soil, water, and climate, low-quality seeds compromise the proper plant stand, which would directly influence the productivity of the crop. In situations where plant population is drastically below the recommended stand, re-sowing increases the overall cost of production, besides the loss of the best sowing period, problems of herbicide efficiency, and risks of product overlapping (Asif 2016). All these factors contribute to low productivity. Hence, to approach the potentially realizable yield of a cultivar, production and distribution of required quantity of high-quality seed of improved crop varieties are critical. This depends on a thorough understanding of the processes underlying the seed biology from its formation to death; pollination behaviour and its management for production of genetically pure seeds; maintenance of variety purity; a sound system of quality assurance and upgradation, following seed regulations, appropriate quality evaluation procedures and enhancement technologies. This chapter intends to present a brief overview of the aspects that are vital for a sustainable seed system, while these are dealt in more detail in the succeeding chapters.

2 Seed Development

Seeds of the angiosperms develop as the result of a unique phenomenon called 'double fertilization' in which after the pollination pollen grains germinate on the stigma and the pollen tube carrying the two male gametes penetrate the embryo sac, wherein one male gamete fuses with the egg cell to produce the diploid embryo (zygote), while the other fuses with the polar nuclei to form the triploid endosperm (Maheshwari 1950; Raghavan 2005). A series of programmed cell divisions after fertilization, and differentiation results in the development of the miniature seed that undergoes several molecular, cellular and metabolic changes to mature into a fully functional and viable seed. The success of seed setting and development of healthy and vigorous seed is influenced by the genetic constitution, physiological status of the mother plant, and its interaction with the environment. Hence, understanding the processes underlying pollination, fertilization, seed development and maturation is vital in the production of quality seed (see chapter "Seed Development and Maturation" for details).

3   System of Variety Development, Release, and Notification

It is well accepted that a continuous flow of improved varieties is vital for the agricultural development of a country. The new improved varieties are bred by the breeders in public institutions, private seed companies, as well as those working independently following four major steps:

1. Selection of breeding lines, making a large number of crosses, and evaluation of the progenies.
2. Identification of promising genotypes.
3. Development of new strains and evaluation of their performance. Various breeding methods can be used for the development of new strains in self and cross-pollinated species as given below:

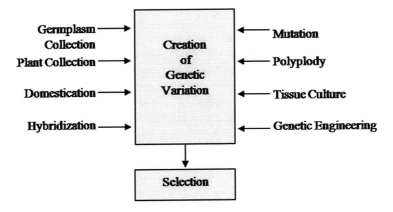

The value in cultivation and use (VCU) of new strains needs to be evaluated by crop specialists independently at several locations, following a set of performance criteria for a minimum period of 3 years, and simultaneously developing DUS characters, before considering their release for cultivation. Though the system of variety release/registration/notification recommended by the concerned official departments may differ in different countries, the common principles are as follows:

- The new varieties with desired attributes developed by the breeders are expected to be pure, homozygous (except for hybrids) and homogeneous (uniform in its expression), and having clearly distinguishing features.

- Varieties are tested for VCU, indicating its agronomic performance in different agro-climatic regions, including tolerance/resistance against major biotic and abiotic stresses specific to the crop species, and their performances are compared with the existing best checks and qualifying varieties in the test trials.
- Varieties showing improvement and advantage in performance over the existing ones are identified for release at the state (or regional)/national level. The identified varieties are considered for release and notification and also for registration as per extant guidelines of the country.
- In the EU and some other countries, VCU and DUS testing are to be performed simultaneously before release. However, in some countries, like India, the official release is not compulsory for the commercialization of a variety, though registration for VCU in the intended region for 1 or 2 years may be necessary.
- Whether officially registered/released or not, the breeder/breeding organization is responsible for maintaining a small quantity of pure seed stock (breeders' stock/nucleus seed) for further multiplication.

Thus, the first criterion that determines seed quality is the homogeneity of the genotype (variety) and genetic purity of the source/stock seed (the Indian varietal evaluation and release system is described in chapter "Principles of Quality Seed Production"). The All India Coordinated Research Projects of different field and horticulture crops play a very important role in conducting VCUs at multiple locations in different agro-climatic zones specified for various crops/crop groups. This provides an inbuilt mechanism of early screening of the material for various biotic and abiotic stresses. Considering the diverse agro-ecologies in different parts of the country, plant varieties are released at central or state levels, while both are notified in the Gazette of India (Tandon et al. 2015; Virk 1998). Once a variety is registered or released for commercial use, it is the joint responsibility of the breeder/breeding organization and the seed-producing agencies to take necessary measures to maintain varietal purity, which conforms to the genetic constitution of the variety, by applying a system of variety maintenance through the generations of seed multiplication. Seed Technology is the link between variety improvement and realization of such improvement by the growers/farmers following scientific methods of seed multiplication, processing and storage, and maintaining a high quality of seed.

4 Seed Production

Profitable production of high-quality and genetically pure seed requires adherence to a set of practices which are not only crop and climate specific, but also the genetic constitution of the variety in use. Thus, knowledge of the flowering and pollination behaviour of the species in question, i.e., self-pollinated, cross-pollinated or often-cross-pollinated; mode of pollination, requirement of supplementary aids to pollination by maintaining bee hives in the seed fields, etc.; and genetic constitution, i.e., whether it is a pure line, hybrid, clonally propagated or genetically modified variety, are vital for the success in seed production. Seed production of vegetable crops is fundamentally different from the field crops in many ways, which must be known well. Similarly, seed production of hybrids takes into account the mode of hybridization, whether based on the male sterility or self-incompatibility systems, or requires manual emasculation and pollination operations; the type of male sterility system operating, i.e., CMS, GMS, CGMS; the mechanism of restoration, etc., all of which are critical for undertaking seed production without compromising the genetic purity of seed (see chapters "Principles of Quality Seed Production", "Vegetable Seed Production", and "Principles of Variety Maintenance for Quality Seed Production" for more).

5   Variety Maintenance/Maintenance Breeding

The life of a released/registered crop variety may extend from 10 years (cross-pollinated species) to 20 years (self-pollinated species) or even more. As long as a variety is in active seed multiplication chain, its genetic constitution must be maintained unaltered. Every care is to be taken during maintenance of a variety to ensure that only 'true-to-type' plants are selected for seed multiplication and the progenies from such plants/ear heads/pods/panicle are further examined for their trueness before the next step of multiplication. Variety maintenance must be undertaken in its area of adaptation to minimize the influence of environmental stress. Repeated seed multiplication in unfavourable conditions of growth exerts abiotic and biotic pressure resulting in genetic drift causing deviations in the plant type. Sometimes some minute levels of residual heterozygosity may exist in the genotype at the time of release, which remains unexpressed in its phenotype. However, in successive generations, such plant progenies express deviation from the characteristics of the original variety. Similarly, even in self-pollinated species, there are chances of some cross-pollination resulting in genetic variation. Such impurities and off-types must be cleaned by following the procedure of maintenance breeding. Zeven (2002) defined maintenance breeding as 'all breeding measures taken to conserve the genetic composition of a variety'. In case of varieties that are bred for special quality traits, such as erucic acid content in brassicas or gliadin profile in bread wheat, precise laboratory analysis is followed after table examination of the harvested pods/ears/seed/panicle. Molecular markers can be an effective tool, especially for maintaining the purity of varieties bred with genes for specific trait introgressed. Hence, maintenance breeding followed by the generation system of seed multiplication is the key to ensure varietal purity.

6   Seed Quality

Quality seed plays a vital role in the success of crop production (Singh 2011) as it

- ensures (physical) purity of the seed
- gives desired plant stand and ensures uniform growth and maturity

- enables the crop to withstand adverse conditions and responds well to fertilizer and other inputs
- produces vigorous, fast-growing seedlings that can withstand pest and disease incidence to a certain extent
- establishes well-developed root system that will be more efficient in the uptake of nutrients.

6.1   Physical Quality

The seed with good physical quality should have uniform size, weight and colour, and should be free from undesired materials like stones, debris, dust, leaves, twigs, stems, flowers, other crop seeds, weed seeds and inert material. It also should be devoid of shrivelled, diseased, mottled, moulded, discoloured, damaged and empty seeds (Du and Sun 2004). The quality seed should also be free from seed of other distinguishable varieties (ODVs) of the same crop (see chapters "Seed Processing for Quality Upgradation" and "Testing Seed for Quality" for more). Purity is maintained by field and seed inspections for maintaining standards, and adhering to the processing specifications. Hence, knowledge of seed morphology and identification of weed seeds are important. Purity testing and labelling are therefore, mandatory for quality seed.

6.2   Genetic Purity

Varietal purity is an essential quality requirement, which refers to the true-to-type seed, and is important in obtaining pure plant population of a specific variety. Varietal mixtures can cause uneven maturity, lower yield potential, increased susceptibility to disease and insect pests and poor adaptability to specific environmental conditions (Sendekie 2020). For all quality parameters, specific standards are set by the national and international authorities, which may vary to some extent from one country to the other. This is ensured by complying with the requirements and rules with respect to the source of seed for the specified class, conducting field inspections at specified stages, meeting the field and seed standards at production and processing, post-control plot test and laboratory tests (see chapters "Seed Development and Maturation" and "Role of Seed Certification in Seed Quality Assurance" for more).

6.3   Physiological Quality

Physiological quality determines the actual planting value of seed for cultivation. Physiological quality of seed comprises of two important attributes viz., seed germination and vigour, that determine the potential of the seed to germinate and establish well in the field under a wide range of growing conditions, with optimum population ensuring desired production. The ability of a dry viable seed to remain viable and germinate under favourable conditions is determined by its genetic constitution and modulated by external factors. The highest level of germination and vigour of mature seed undergoes gradual decline subsequently, till it becomes completely non-viable. In oilseeds like groundnut and soybean viability loss is fast if not stored under optimum conditions. Seeds in a lot vary with respect to their physiological status and hence, upon testing some seeds are capable of producing robust and normal seedlings, some produce abnormal or weak seedlings, and others may not produce seedlings at all. In some species, seeds are unable to germinate due to the phenomenon of dormancy. Hence, it is important to evaluate the ability of a seed lot to produce healthy and normal seedlings under favourable conditions by performing germination tests. Germination tests are conducted for a prescribed time period under laboratory conditions that ensure optimum moisture, temperature and light following the ISTA Rules for Seed Testing or as recommended by the other official procedures (see chapters "Seed Dormancy and Regulation of Germination", "Seed Longevity and Deterioration", and "Testing Seed for Quality" for more). Germination testing is one of the mandatory requirements for quality assurance in seed.

Since the field conditions at sowing are not always optimum for germination of seeds, assessment of seed vigour is advisable wherever a standard and reliable test procedure is available. Seed vigour not only influences the final plant stand, but also the rate of emergence, uniformity of emergence and post-harvest storability. Seeds low in vigour are more susceptible to biotic and abiotic stresses and would often produce weak seedlings, which have poor chance of survival under sub-optimal or fluctuating growth conditions (Sundareswaran et al. 2021). Vigour testing is applicable only to seed lots meeting the minimum prescribed germination, as all vigorous seeds will be germinable but all germinable seeds need not be vigorous (Finch-Savage and Bassel 2016). Hence, testing seed vigour is important for raising good crop, specially under less favourable or fluctuating conditions of growing, though this is not a mandatory test (see chapter "Seed Vigour and Invigoration" for more). There is need to standardize at least one reliable vigour test for all important food crops and vegetables.

6.4   Seed Health

Health status of seed determines the ability to raise a healthy crop and is determined by the absence of insect pests and pathogens borne on, or transmitted with the seed. The health status of seed also influences its storability and vigour. The causal organisms of designated diseases carried over from one generation to the other affect the growth habit, reproductive efficiency and cause drastic reduction in the seed yield. Hence, it is mandatory to perform appropriate and precise tests for detecting the presence and load of designated pathogens to meet the permissible limits. Several fungi such as *Phomopsis* spp., *Colletotrichum truncatum*, *Cercospora kikuchii*, *Fusarium* spp. and *Aspergillus* spp. are among the most frequently associated with seed quality. Besides detection of seed insect pests and pathogens during seed testing, equally important is to follow appropriate measures of seed health management during production and storage (see chapters "Seed Storage and Packaging" and "Seed Health: Testing and Management" for more).

7 Factors Influencing Seed Quality

The seed quality can be influenced by several factors occurring during the production phase in the field before harvesting, during harvesting, drying, processing, storage, transport and sowing. These factors include extreme temperatures during maturation, fluctuation of moisture conditions including drought and excess rains at maturation, weathering, plant nutrition deficiencies, occurrence of pests and diseases; improper handling, drying and storage (Cavatassi et al. 2010). Seed ageing or deterioration is a natural process over time that results from the interaction of cytological, physiological, biochemical and physical changes in seed causing reduction in vigour, germination and eventually leading to the loss of viability. The extent and rate of seed deterioration depends on species, storage environment, length of storage period and the initial quality of the stored seeds (see chapters "Seed Longevity and Deterioration" and "Seed Storage and Packaging"). Therefore, testing the quality of seeds stored for different lengths of time is important to determine the effect of ageing on seed quality (Finch-Savage and Bassel 2016).

8 Seed Quality Assurance

Quality of the seed can be assured and improved in the following ways:

1. Seed quality regulation
2. Seed quality maintenance
3. Seed quality upgradation and enhancement
4. Genetic improvement

Regulatory mechanism to control seed quality is an important component of the seed programme both for the domestic and international seed trade. Notwithstanding the differences in the seed legislations of different countries, most legislations follow the basic processes of

- Registration for multiplication
- Field inspection
- Seed sampling
- Seed testing
- Certification of the seed lot

The primary objective of seed legislation is to make sure that seeds should conform to the minimum prescribed standards of physical and genetic purity, seed health and germination either by compulsory labelling or certification. Further, an

Act provides a system for seed quality control through independent Certification agencies authorized or designated by the government department of agriculture. It also prescribes certain requirements to be complied by the persons/organization carrying out the business of selling seed. Labelling of each class of seed is distinguished by a specific tag/label of a certain colour and dimension. In countries such as India, the USA and many others, where certification is not compulsory, labelling of any notified kind of variety of seed is made compulsory. This category is known as Truthfully labelled/Labelled seed (TL seed), which should maintain the same quality standards as that of CS (see chapter "Role of Seed Certification in Seed Quality Assurance" for more).

9   Seed Certification and Quality Testing

During seed certification, which is the key to seed quality control, the following steps for verification of field standards and laboratory analysis for seed standards are followed in India before the grant of the certificate.

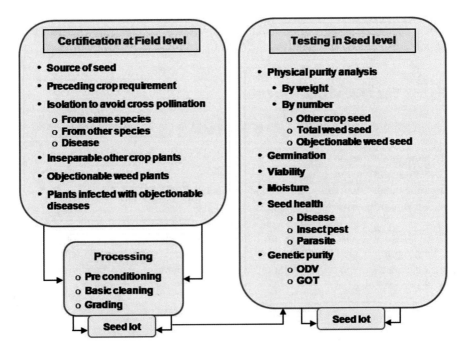

The principles are broadly similar in other countries, though there could be some procedural differences. The Indian Minimum Seed Certification Standards (IMSCS) updated in 2013 cater to the needs of domestic seed certification system and also satisfy many requirements of OECD rules and directions for field inspection to ensure varietal identity and purity. These include both field standards and seed standards (see chapters "Principles of Quality Seed Production" and "Role of Seed Certification in Seed Quality Assurance" for more).

10 Role of International Organizations

The following international organizations are involved in the promotion of quality seed production, testing, certification and trade at the international level (see chapter "Role of Seed Certification in Seed Quality Assurance" for more).

1. International Crop Improvement Association (ICIA): It is an organization of seed certification agencies in the USA and Canada which was changed to the Association of Official Seed Certifying Agencies (AOSCA) in 1969 and is responsible to establish minimum standards for certification in the USA, promote production, identification and distribution of seed and educate seed growers.
2. International Seed Testing Association (ISTA): ISTA is an international network of member laboratories which promotes 'uniformity in seed quality evaluation'. Through the participation of member laboratories and Working Groups under the technical guidance and approval of the Technical Committees, it develops standard procedures for sampling and testing of seed after multi-laboratory testing and validation. The ISTA rules describe the principles and standard procedures for sampling, testing and reporting of results. It also accredits member laboratories after verifying their technical competence and proficiency in conducting various tests. It promotes research, organizes training for capacity building and disseminates knowledge in seed science and technology. This ensures seed quality and facilitates national and international seed trade. ISTA-accredited laboratories are authorized to issue orange and blue international seed analysis certificate which supports global seed trade (Masilamani and Murugesan 2012).
3. Organization for Economic Cooperation and Development (OECD): Develops certification system for international trading acceptable for member countries. Despite the existence of a number of regulatory bodies, whose objective is to harmonize regulations and encourage regional or international seed trade, there are still considerable differences in the seed laws and regulations governing the national and international seed system (Cortes 2009). These differences restrict the seed trade between countries. Hence, there is a need for defining, endorsing and enforcing minimum criteria for the International seed trade. The OECD Seed Schemes prescribe a set of procedures, methods and techniques for certifying the variety identity, purity and quality standards, which are applicable to the varieties from the member countries included in the OECD list under its seven schemes. It provides legal framework for the certification of the movement of seeds in the international trade.

   The schemes ensure the varietal identity and purity of the seed through appropriate requirements and controls through the production, processing and labelling operations. The OECD certification provides for official recognition of "quality-guaranteed" seed, thus facilitating international trade and contributing to the removal of technical trade barriers (Rajendra Prasad et al. 2017). The National Designated Authority (NDA) may issue certificates for each lot of Pre-Basic seed (White label with diagonal violet stripe), Basic seed (White label), Certified seed (C1—Blue label; C2—Red label), Not Finally Certified seed (Grey label) and Standard seed (Dark yellow label) approved under the Scheme.

4. International Plant Protection Convention (IPPC) is a multilateral treaty that promotes effective actions to prevent and control the introduction and spread of pests of plants and plant products. It allows countries to evaluate the risks of their plant resources and promote use of science-based measures for their safety. Implementation of this convention is based on the exchange of technical and official phytosanitary information among member states and adoption of the International Standard for Phytosanitary Measures (ISPM) that is crucial in the international seed movement.

11   Seed Quality Maintenance

Seed quality can be maintained by the following measures:

1. Seed production under suitable environment
2. Adoption of proper harvesting and post-harvest management techniques
3. Efficient and effective quality control procedures
4. Use of advanced storage techniques

Among the various environmental factors, moisture and temperature stress have direct influence on reproduction. Early reproductive processes like pollen and stigma viability, anthesis, pollination, fertilization, and early embryo development are all highly prone to moisture stress and/or temperature stress. Failure of any of these processes increases early embryo abortion, leading to lower seed setting, thus limiting the seed yield. Harvesting seed crop at the right stage of maturity, using proper methods of drying and post-harvest handling, and safe storage are important to maintain vigour and viability of seed.

12   Seed Quality Upgradation

A seed lot can be upgraded by the removal of defective and poor-performing seeds to the extent that is practical and economical. In the commercial seed system, good quality seeds are referred as accept fraction, while low-quality seeds are considered as reject fraction of the seed lot. Application of too stringent norms may reduce the economic seed yield, whereas laxity in these will lead to poor quality seed. This makes seed processing a highly specialized activity, which can be effectively performed by taking into consideration parameters which are specific to the seed morphology, structure and composition of the species.

Seed processing narrows down the level of heterogeneity of the harvested seed lot and improves the physical purity of the seed lot by eliminating the undersized, shrivelled, immature, ill-filled seeds using appropriate methods of separation and fine cleaning. The germination and vigour of the seed lot can be improved by grading the seeds based on size, specific gravity, length and density of the seeds (see chapter "Seed Processing for Quality Upgradation" for more). Copeland and McDonald (1995) proposed the following basic steps in seed processing for upgrading the quality of seed lots, which is followed with some modifications as per the specific needs:

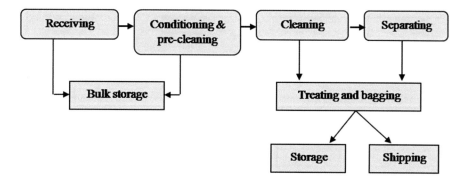

13   Genetic Improvement for Seed Quality

Varietal improvement through plant breeding harbours desirable traits by broadening the genetic base through combination of desired genes/alleles, which increases the farm productivity, improve quality, genetic diversity in agro-ecosystems, and thus ensures sustainable food production systems under the climate change scenario. Emphasis is given to breed crops for improvement in yield, quality, adaptability, abiotic and biotic stress resistance, synchronized flowering and maturity, and amenability to mechanical operations. Genetic improvement can also be targeted to increase seed quality traits in many ways by

- Facilitating the maintenance of physiological quality of seeds by increasing their inherent resistance or tolerance to factors which are responsible for seed deterioration, such as seed coat characteristics
- Selecting for inherent physiological or physical properties which contribute to better seedling vigour
- Selecting for longer viability in breeding lines
- Selecting strains performing well under biotic and abiotic stress conditions and exhibiting tolerance to climate change.

14   Seed Quality Enhancement

Seed enhancement is defined as post-harvest treatments that improve germination or seedling growth, or facilitate the delivery of seeds and other materials required at the time of sowing (Taylor et al. 1998). Seed quality enhancement through advanced technologies of coating, pelleting, time and target-oriented seed additives, electron treatment, magnetic treatment, and plasma coating are used for improved performance and better adaptability to biotic and abiotic stresses. Use of third-generation seed quality augmentation strategies viz., nanotechnology for external as well as internal designing of seeds and bio-priming technology, offers enhanced seed performance for eco-friendly and safer agriculture (see chapter "Seed Quality Enhancement" for more).

15   Scope of Molecular Technologies

With the advancement in breeding technologies and expansion of global seed trade, more precise, rapid and reliable diagnostic tools and techniques are needed for variety identification and genetic purity testing, establishing distinctiveness among closely related varieties, seed health testing, assessing the trait purity in GM varieties and detecting adventitious presence of GM seeds, and maintenance of breeding lines. Hence, cost-effective molecular technologies with greater accuracy need to be standardized for assessment of various parameters of seed quality (see chapter "Molecular Techniques for Testing Genetic Purity and Seed Health" for more).

Characterization of released varieties for seed traits like seed viability, longevity and seed dormancy is poorly documented. This information is much required for better seed multiplication, timely harvesting and safe storage. With the advancement of new molecular technologies, different seed quality traits need to be identified to undertake breeding programmes to improve the cultivar performance under variable conditions, uniform maturity; no/low seed shattering; prolonged seed longevity; intermediate or controlled seed dormancy; improved seed coat permeability, etc. Seed quality testing and detection of GM seeds is an area which needs more attention and can be better addressed through molecular tools. Likewise, the genome editing products are also likely to be available for general cultivation in the times to come. Issues of seed quality of gene-edited varieties will be addressed through the existing seed system. The use of molecular tools for genetic purity and implementation of seed traceability will be the two landmark developments during the current decade which will revolutionize the seed systems by ensuring the seed quality.

The role of seed biology on seed production and performance, importance of production and post-production management to achieve highest quality standards, its maintenance through storage and improvement of performance by various seed enhancement technology, testing seed quality using appropriate standard protocols including seed health testing, and application of molecular technologies for precision in quality assessment have been discussed at length in different chapters, along with the regulatory mechanisms for seed quality assurance in different systems.

References

Asif AA (2016) Role of seed and its technological innovations in Indian agricultural sector. Biosci Biotechnol Res Commun 9:621–624

Cavatassi R, Lipper L, Narloch U (2010) Modern variety adoption and risk management in drought prone areas: insights from the sorghum farmers of eastern Ethiopia. Agric Econ 42:279–292

Copeland LO, McDonald MB (1995) Principles of seed science and technology, 3rd edn. Chapman and Hall, New York, NY, p 409

Cortes J (2009) Overview of the regulatory framework in seed trade. In: Proceedings of the second world seed conference. Responding to the challenges of a changing world: the role of new plant varieties and high quality seed in agriculture. FAO, Rome, p 201

Du CJ, Sun DW (2004) Recent developments in the applications of image processing techniques for food quality evaluation. Trends Food Sci Technol 15:230–249. https://doi.org/10.1016/j.tifs.2003.10.006

Finch-Savage WE, Bassel GW (2016) Seed vigour and crop establishment: extending performance beyond adaptation. J Exp Bot 67:567–591. https://doi.org/10.1093/jxb/erv490

Hanchinal RR (2017) The Indian Seed Industry: achievements and way forward. In: Souvenir-6th M S Swaminathan Award 2015-16: Environment, energy and sustainable production. Retired ICAR Employees' Association, Hyderabad, pp 12–22

Maheshwari P (1950) An introduction to the embryology of angiosperms. McGraw-Hill, New York, NY, p 453

Masilamani P, Murugesan P (2012) ISTA accreditation and strength and weakness of Indian seed testing system. In: National Seed Congress on Welfare and Economic Prosperity of the Indian Farmers through Seeds, Raipur, Chhattisgarh, pp 45–59

Raghavan V (2005) Double fertilization-embryo and endosperm development in flowering plants. Springer, Berlin, p 234

Rajendra Prasad S, Chauhan JS, Sripathy KV (2017) An overview of national and international seed quality assurance systems and strategies for energizing seed production chain of field crops. Indian J Agric Sci 87:287–300

Sendekie Y (2020) Seed genetic purity for quality seed production. Int J Sci Eng Sci 4:1–7

Singh BD (2011) Plant breeding: principles and methods, 4th edn. Kalyani Publisher, Ludhiana, p 916

Sundareswaran S, Vanitha C, Raja K (2021) Seed quality - an overview. In: Sundareswaran S, Raja K, Jerlin R (eds) Seed quality enhancement, 1st edn. TNAU Press, Coimbatore, pp 1–16

Tandon JP, Sharma SP, Sandhu JS, Yadava DK, Prabhu KV, Yadav OP (2015) Guidelines for testing crop varieties under the All-India Coordinated Crop Improvement Projects. Indian Council of Agricultural Research, New Delhi

Taylor AG, Allen PS, Bennett MA, Bradford JK, Burris JS, Mishra MK (1998) Seed enhancements. Seed Sci Res 8:245–256

Tony JG, Gastel V, Bill R, Gregg, Asiedu EA (2002) Seed quality control in developing countries. J New Seeds 4(1–2):117–130. https://doi.org/10.1300/J153v04n01_09

Virk DS (1998) The regulatory framework for varietal testing and release in India. In: Seeds of choice: making the most of new varieties for small farmers. Oxford, IBH Pub, New Delhi, pp 69–84

Zeven AC (2002) Traditional maintenance breeding in landraces: practical and theoretical considerations on maintenance of varieties and landraces by farmers and gardeners. Euphytica 123:147–158

Seed Development and Maturation

K. V. Sripathy and Steven P. C. Groot

Abstract

In plants, a fascinating set of post-fertilization events result in the development of a dispersal unit known as a seed. During the maturation phase, seeds accumulate storage reserves and acquire desiccation tolerance, followed by an increase in seed vigour during maturation drying. Physiological (or mass) maturity may be attributed to the stage of seed maturation when maximum seed dry matter accumulation has occurred, marking the end of the seed-filling phase. The stage of maturity at harvest is one of the most important factors that can influence the quality of seeds. Recent studies established that seed vigour and longevity continue to increase even after physiological maturity, signifying the importance of the late maturation phase for maximizing seed quality. Among the plant hormones, abscisic acid (ABA) has been studied extensively for its role during seed development and maturation. Apart from ABA, gibberellic acid (GA), cytokinin and auxin also play a critical role during the development of seeds. Desiccation tolerance in seeds begins much before the attainment of physiological maturity. Acquisition of desiccation tolerance is associated with embryo accumulation of oligosaccharides of the raffinose family, low molecular weight antioxidants, late embryogenesis abundant proteins and heat shock proteins coupled with structural changes at the cellular level. To obtain seeds of maximum quality (in terms of germination, vigour and longevity), harvesting needs to be performed at or slightly after harvest maturity a period at which seed moisture content stabilizes with environmental factors. In this chapter, an attempt has been

K. V. Sripathy (✉)
ICAR-Indian Institute of Seed Science, Maunath Bhanjan, Uttar Pradesh, India
e-mail: sripathy.v@icar.gov.in

S. P. C. Groot
Wageningen University & Research, Wageningen Seed Centre, Wageningen, The Netherlands
e-mail: steven.groot@wur.nl

© The Author(s) 2023
M. Dadlani, D. K. Yadava (eds.), *Seed Science and Technology*,
https://doi.org/10.1007/978-981-19-5888-5_2

17

made to present the current understanding of seed development and maturation concentrating on various aspects viz. phases of seed development, the role of plant hormones, other factors affecting seed development, concepts of seed maturity, and its relevance to seed quality, maturity indices in crop plants and acquisition of desiccation tolerance in seeds.

Keywords

Seed development · Maturation drying · ABA · Germination · Seed moisture · CF sorting · Desiccation tolerance

1   Introduction

In plants, seed development commences with the initiation of flowering followed by formation of floral structures and effective pollination. In angiosperms, after fertilization, as a result of cell division, expansion and histo-differentiation, there is a stage in embryogenesis, in which seed structure primordia are formed and future embryo parts can be envisaged. Delouche (1971) defined seed development and maturation as a process comprising of a series of morphological, physical, physiological and biochemical changes that occur from ovule fertilization to the time when seeds become physiologically independent of the parent plant. Seed moisture content increases during the initial part of development after fertilization and later begins to decline until equilibrium is established with environmental factors. Harrington (1972) defined that a seed attains physiological maturity when the dry weight reaches its maximum and that after this stage, the flow of nutrients to seeds from the mother plant generally ceases. However, significant seed quality processes occur in seeds even after the end of seed filling. In this regard, seed physiologists consider the late maturation phase as an additional phase, to demarcate seed development into three main phases viz. embryogenesis, seed filling and late seed maturation. The late maturation phase is also often called the maturation drying phase, where seeds prepare themselves for survival after shedding by the acquisition of several protection mechanisms. The relative lengths of seed developmental phases differ between species (Leprince et al. 2017).

A significant decline in seed moisture content occurs at the end of maturation, whereas the acquisition of desiccation tolerance begins during mid and late maturation. At the same time, massive structural and physiological changes occur within the seed, with a strong reduction in metabolic activity and a transition to a quiescent and frequently dormant state at the end of late maturation. Recalcitrant seeds usually do not show this quiescent state. Seed maturation is one of the main factors of seed quality and a prerequisite for successful germination and emergence. The plant hormones play a crucial role in histo-differentiation, pattern formation and embryo maturation. Extensive studies have led to the understanding of the role of ABA during seed development and maturation as compared to other plant hormones. Environmental factors viz. soil fertility, soil water content, photoperiod, temperature and position of the seed in the inflorescence or on the mother plant also affect the process of seed development and maturation. Most seeds degrade chlorophyll during maturation drying, hence chlorophyll fluorescence (CF) sorting is employed in some crops for upgrading seed lots exhibiting heterogeneity in seed maturity. Physiological maturity is marked as the time when seeds attain maximum dry weight and thereby, maximum yield when it concerns crop production. Seed quality for propagation purposes increases during the late maturation phase and reaches maximum vigour at a stage called harvest maturity, which in natural systems most often coincides with seed shedding. In a seed crop, to secure maximum quality, harvesting needs to be performed at the end of the late maturation, although care has to be taken for preventing the onset of pre-harvest sprouting that can be induced after maturation under moist conditions with non-dormant seeds. The maturity indicators presented on the plant or seed can often serve as a sign to determine the harvest maturity. The morphological, physiological and biochemical changes associated with seed development and maturation right from the stage of fertilization have been contemplated in this chapter.

2 Double Fertilization

Following the events of mega and microsporogenesis, the functional megaspore (n) undergoes three mitotic cell divisions and develops into an embryo sac also known as ovule (one egg cell and two synergids constitute the egg apparatus at the micropylar end, two polar nuclei at the centre and three antipodal cells at the chalazal end). Similarly, the haploid nucleus of functional microspore (n) (pollen) undergoes mitotic cell division to form bi-cellular pollen grain. The mature anthers dehisce and release pollen grains. After successful contact of pollen grains on the receptive stigma, it germinates and the pollen tube traverses along the length of the style. The haploid generative cell divides again to form two haploid male cells, also known as sperm cells or male gametes. In angiosperms, both male gametes participate in fertilization. One male gamete fuses with egg cell to produce diploid zygote (2n) likewise, the other male gamete fuses with two polar nuclei to form triploid nucleus, also known as primary endosperm nucleus (3n). Together, these two fertilization events in angiosperms are known as double fertilization (Fig. 1). The testa (seed coat) develops from the outer and inner integuments, which is an ovular tissue. The fertilized ovule forms the seed, whereas the tissues of the ovary become the fruit or pericarp, usually enveloping one or more seeds. Whereas the embryo and endosperm are a combination of maternal and paternal genetics, the testa and pericarp originate from maternal tissues and these provide the same genetic constitution for all seeds originating from the same mother plant.

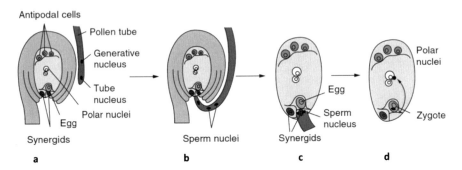

Fig. 1 Schematic representation of events during double fertilization in an angiosperm ovule. (**a**) Organization of cells in the functional megaspore and pollen tube prior to fertilization. (**b**) The pollen tube grows into the style and the generative nucleus divides to form two sperm cells inside the pollen tube. (**c**) Sperm cells are released into the embryo sac near the egg apparatus. (**d**) One sperm cell fuses with the egg cell to form a zygote (2n) and the other sperm cell migrates and fuses with the central cell to form the primary endosperm nucleus (3n). (Source: Sliwinska and Bewley 2014)

3   Embryogenesis

After fertilization, embryonic development begins. In the first stage of embryonic development, the zygote divides to form two cells, the upper cell (apical cell) and the lower cell (basal cell). The division of the basal cell gives rise to the suspensor, which finally makes a connection with the maternal tissue, providing a path for nutrition to be transported from the mother plant to the growing embryo. The apical cell undergoes multiple mitotic divisions, giving rise to a globular-shaped proembryo.

3.1   Embryogenesis in Monocot

Embryogenesis in monocot occurs through four distinctive stages in succession. Proembryo is the first stage in the embryogenesis followed by globular, scutellar and coleoptilar stage.

Firstly, the fertilized egg cell (2n) undergoes one cycle of mitotic cell division to produce two diploid celled structure known as the proembryo. The lower basal cell of the pro-embryo undergoes further division to form a structure known as suspensor (not well developed in monocots) (de Vries and Weijers 2017). The upper apical cell undergoes mitotic cell divisions to form a 16 dipliod celled globular structure, this stage is referred as globular stage. At this stage, cells at one side of globular structure divide faster to form the embryonic axis. Whereas, mitotically divided cells at the other end results in a single cotyledon (scutellum), the stage is referred as scutellar stage. In monocot seeds, the scutellum is the interface between embryonic axis and endosperm. As the scutellum is derived from the apical cell of the proembryo, its

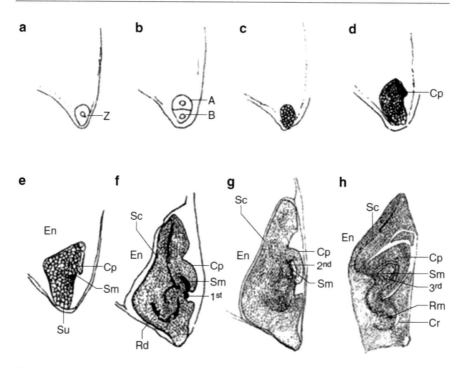

Fig. 2 Embryo development in a monocot (rice). (**a**) Formation of zygote (Z) post fertilization. (**b**) First mitotic division of zygote to form an apical (A) and a basal (B) cell, both undergo series of mitotic divisions to form multicellular (**c**) proembryo. (**d, e**) Differentiation of proembryo into a suspensor (Su), shoot meristem (Sm) and coleoptile (Cp). (**f–h**) The mature embryo is formed with distinct scutellum (Sc), radicle (Rd), root meristem (Rm), coleorhiza (Cr) and third leaf (third). Proembryo stage (**b, c**); Globular stage (**d, e**); Scutellar stage (**f, g**); Coleoptilar stage (**h**). (Source: Bewley et al. 2013)

genetic constitution is same as embryo. In the later stage of embryogenesis, the embryonic axis differentiates and plumule and radicle can be distinguished (Fig. 2).

Meanwhile, the triploid endosperm nuclei undergoes repeated nuclear division, later free nuclei migrate to the periphery of the cell and cellularization occurs upon cell wall formation. The inner layers of cells develop into endosperm and peripheral cells form an epidermis-like layer called aleurone. The aleurone layer is the outermost layer of the endosperm but is very distinct from starchy endosperm cells in terms of morphology and biochemical composition (Becraft and Yi 2011). The aleurone cells remain metabolically active after maturity and its role during seed germination is vastly studied. During seed germination, the scutellum produces gibberellin, which triggers aleurone cells to produce enzymes for hydrolysis of starch (α amylase) and storage protein (proteases). In monocot seeds, the funicle is not well developed and functional. During the seed-filling stage, storage reserves (photosynthates) are transferred from source to triploid endosperm cells via transfer cells and stored predominantly as starch. The endosperm cells expand with the accumulation of food material.

3.2   Embryogenesis in Dicot

The embryogenesis in monocot and dicot seeds is mostly similar up to the globular stage. In dicots, fully functional suspensor is developed from basal cell of the proembryo which pushes the globular structure into the embryo sac and also aids in transfer of food reserves to globular cells.

A depression is formed at the tip of globular structure (heart-shaped stage), which is the initiation of cotyledon differentiation. Relative deepening of the depression at the tip and elongation of cotyledons gives the torpedo stage. The radical and hypocotyls are well developed at the cotyledonary stage (Fig. 3). The rudimentary suspensor absorbs food material from surrounding tissue and transfers it to the cotyledons. The food reserves are stored in the cotyledons predominantly as proteins or lipids. To aid in the transfer of photosynthates to cotyledons, a vascular strand runs through the funicle and connects at one part of the seed coat. From seed coat, nutrients are diffused to nucellus tissues and later absorbed by the suspensor. Removal of one cotyledon may show between both the cotyledons the presence of the early stages of the first true leaves, the plumule, as can be seen in bean seeds. However, with many other seeds, as in tomato and cabbage, the shoot apical meristem does not develop beyond a dome-shaped structure and the formation of leaf primordia is seen only after the commencement of germination.

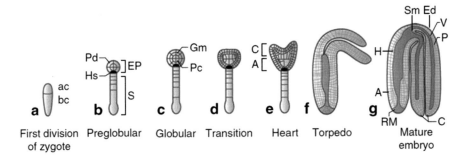

Fig. 3 (a–g) Embryo development in a dicot (Arabidopsis). (**a**) First mitotic division of zygote to form apical cell (ac) and basal cell (bc). (**b**) Further divisions lead to the formation of suspensor (S) and embryo proper (EP) at pre-globular stage, the protoderm (Pd) develops into epidermal layer and hypophysis (Hs) develops into root meristem (RM). Shoot meristem (Sm) differentiates from the apical-central region of the embryo. The ground meristem (Gm) of the globular stage develops into storage parenchyma cells (P) of cotyledons (C). The pre-cambium (Pc) forms vascular tissue (V). The axis (A) develops into radicle, plumule and hypocotyl (H). (Source: Bewley et al. 2013)

4 Acquisition of Desiccation Tolerance During Seed Maturation

After completion of series of cell division and cell differentiation, seed development shifts to the maturation phase that can be divided into early (reserve accumulation) and late maturation (maturation drying) (Fig. 4). During early maturation, the seed can acquire desiccation tolerance. Desiccation tolerance is defined as the ability of a living entity to deal with extreme moisture loss to levels below 0.1 g water per gram dry weight, or drying to relative humidity below 50%, and subsequent re-hydration without accumulation of lethal damage (Leprince and Buitink 2010). Based on the desiccation tolerance, seeds can be classified into orthodox and recalcitrant types. The maturation in the orthodox seeds is accompanied with a water loss up to 5–10% w/w, which allows them to sustain unfavourable environmental conditions, such as extremely high and low temperatures and drought. In contrast, recalcitrant seeds are sensitive to dehydration and desiccation leads to damage and loss of viability (Azarkovich 2020). In orthodox seeds, the mechanisms behind the onset of desiccation tolerance are activated at the early stages of maturation (Leprince et al. 2017). Later on, desiccation tolerance is lost during germination, at the moment of radicle emergence.

Desiccation tolerance is acquired by seeds through accumulation of an array of small molecules and proteins that enables them to maintain the structural integrity of critical cellular organelles, membranes and proteins so that they can persist during the dry state and resume their biological functions upon hydration (Bewley et al. 2013). The embryo accumulates specific molecules that are associated with the cells' ability to tolerate extreme water stress viz. low molecular weight antioxidants, oligosaccharides such as raffinose, stachyose, late embryogenesis abundant proteins (LEAs) and heat shock proteins (HSPs). Further, structural changes occur at the

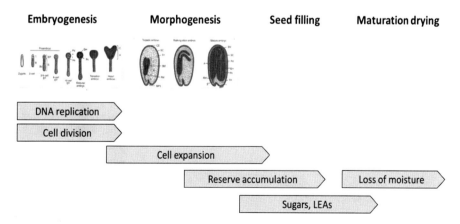

Fig. 4 Seed developmental stages signifying the series of events towards reserve accumulation, acquisition of desiccation tolerance and maturation drying. (Source: SPC Groot, Wageningen University & Research, The Netherlands)

cellular level such as folding of cell walls, condensation of chromatin and dismantling of thylakoids in chloroplasts (Ballesteros et al. 2020). These physiological and structural changes reduce metabolic activity while mitigating the mechanical stress of cell shrinkage during dehydration (maturation drying process). Changes at this stage correspond with a gradual increase in seed longevity (Verdier et al. 2013).

LEA proteins have relative high content of glycine, alanine, glutamate, lysine, arginine and threonine, while low amounts of cysteine and tryptophan residues (Battaglia et al. 2008). Due to this primary nature, LEA proteins are stable in a broad temperature range. During cell dehydration, LEA proteins act as chaperons, i.e., involved in structural stabilization of denatured proteins and promote their refolding through intensive hydrogen bond formation (Smolikova et al. 2021). LEA proteins are also responsible for sequestration of ionic compounds, accumulating during cell dehydration, and protection of membrane proteins and enzymes from the deleterious effects of increased salt concentrations. Non-reducing sugars fill the free volume between large molecules, created during dehydration and the dehydrated cytoplasm forms a glassy matrix with very low molecular mobility (Ballesteros et al. 2020). Other structural adaptations that occur during this stage are chromatin compaction and nuclear size reduction, which are reversed during germination (van Zanten et al. 2011). Furthermore, metabolic activity is reduced and chlorophyll is degraded towards the end of seed maturation thereby minimizing the production of reactive oxygen species (ROS) (Pammenter and Berjak 1999). To protect the seed against oxidative damage, which cannot be repaired by enzyme activity under dry conditions, seeds accumulate during seed maturation many antioxidants, such as ascorbate, glutathione, polyols, tocopherols, quinones, flavonoids and phenolics (Kranner and Birtić 2005).

5   Seed Development and Maturation in Relevance to Seed Quality

Since protection mechanisms are mainly built during the late seed maturation phase, the stage of harvest becomes the most critical factor for seed quality and storability (Jalink et al. 1998; Demir et al. 2008). Harvesting seeds too early when there is inadequate development of essential structures and protection mechanisms may result in poor quality (Ekpong and Sukprakarn 2008). Similarly, harvesting too late may increase the risk of shattering and may decrease the quality of seed due to ageing. If harvesting is delayed, incidence of adverse environmental conditions such as rain and humidity may result in precocious germination (Elias and Copeland 2001). Additionally, under high humid conditions, delayed harvest of seeds can also result in infection by saprophytic fungi resulting in discolouration of seeds. Low quality of seeds can potentially decrease the rate and percentage of germination and seedling emergence, leading to poor stand establishment in the field and consequently yield loss, as has been found with many crops such as rice, corn, wheat, cotton, barley and garden pea. Therefore, it is necessary to examine and identify the

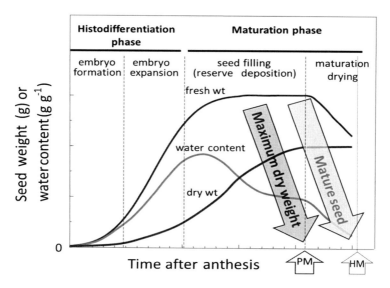

Fig. 5 Progression of seed fresh weight, dry weight and moisture content during seed development and maturation. *PM* physiological maturity, an index used for yield, *HM* harvest maturity, an index used for seed quality. (Source: SPC Groot, Wageningen University & Research, The Netherlands)

suitable stage of harvest (maturity indices) in crops for the production of high-quality seeds (Fig. 5).

Even though seeds attain maximum dry weight at physiological maturity, the maximum vigour, which is responsible for performance of seed under stress conditions, is not attained until the end of maturation drying. As seed production is dealing with a population of seeds, the relationship between germination, desiccation tolerance, vigour and longevity during seed development and maturation is expressed through a sigmoid curve (Fig. 6). Hence, bulk harvested seeds are always heterogeneous pertinent to seed maturity.

In crops viz. rapeseed and mustard, cole crops, pigeon pea, onion, carrot, etc. indeterminate flowering results in wide variation in seed developmental stages within the inflorescence (Singh and Malhotra 2007). During a bulk harvest, few over-mature, mature and immature seeds shall always persist in crops having indeterminate flowering habit (Fig. 7). In such cases, seed harvesting needs to be done at a stage when most of seed-bearing structures are mature or nearing maturity and flower initiation stops in the inflorescence. Prolonging the harvest beyond certain stage may result in seed shattering. With some crops, it is possible to cut the mother plant from the roots and let the plant, with developing fruits, slowly dry (in field or a threshing yard) before harvesting the seeds. This practice provides ample time with enough moisture for the less mature seeds to complete the process in their late maturation phase.

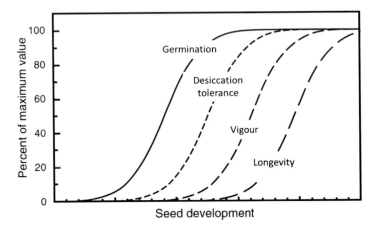

Fig. 6 Pattern of development of seed quality attributes during seed development and maturation. Seed vigour and longevity progressively developed during the late maturation phase and reaches maximum during harvest maturity. (Source: Bewley et al. 2013)

Fig. 7 Varied seed developmental stages observed in plants as a result of indeterminate flowering behaviour. (Source: H. Jalink, Wageningen University & Research, The Netherlands)

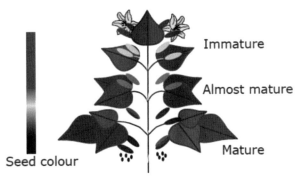

5.1   Hormonal Regulation of Seed Development and Maturation

Plant hormones are signal molecules that are produced in the plant and are active at very low concentrations. The hormones abscisic acid (ABA), gibberellins (GAs), auxin (IAA), cytokinins, ethylene and brassinosteroids regulate cellular processes in targeted cells, which may or may not be the cells in which they are synthesized. The most important role of plant hormones is to control and coordinate cell division, growth and differentiation (Hooley 1994).

Measurements of endogenous hormone concentration have suggested the high transient expression of cytokinins, GAs and IAA during the early phase of seed development (Bewley et al. 2013; Davies 2013). In studies with tomato, GAs were found essential to produce fertile pollen, but pollination of a GA-deficient (female) mutant with mutant pollen, obtained upon spraying the male plant with GAs, resulted in the development of normal-looking healthy seeds that only needed GAs for germination (Groot and Karssen 1987; Groot et al. 1987). Fruits could develop on GA-deficient mutant mother plants, but they remained small without fertilization and seed development. The transport cells (that aid transfer of nutrition from source to sink) like those of the suspensors are important for nutrition of early embryos. Studies with *Phaseolus* have shown that exogenous GAs can substitute for a detached suspensor in promoting embryonic growth, suggesting that the suspensor may normally provide GAs as well as nutrients to the developing embryo. Cytokinins have also been implicated in promoting suspensor function, but may be even more significant in promoting endosperm growth and grain filling via promotion of cell division (Bewley et al. 2013). In contrast, during early embryogenesis, auxins play a major role in establishing the embryonic body plan via effects on apical-basal polarity or pattern formation (transition of embryo from globular to heart shape and cotyledon separation at later stages) and vascular development (Vogler and Kuhlemeier 2003).

During maturation, seeds of most species acquire the capability to endure desiccation. The maturation phase begins when the embryo and endosperm have accomplished the morphogenesis and patterning stages (Wobus and Weber 1999). This phase is categorized by a growth arrest, followed by the synthesis and accumulation of reserves, whose degradation upon germination will provide nutrients to the growing seedling before the photosynthetic capacity is fully acquired (Baud et al. 2002). Early and mid-phases of maturation are controlled by the action of ABA, initially synthesized in the maternal tissues and later on, although to a lower extent, in the embryo and endosperm (Nambara and Marion-Poll 2003). Seed maturation coincides with an increase in seed ABA content; consistent with the fact that ABA induces expression of a cyclin-dependent kinase inhibitor (ICK1) that could lead to cell cycle arrest (Finkelstein et al. 2002).

In the later stage, a decline in ABA level occurs and synthesis of LEA proteins follows, which is characteristic to the late maturation phase. Maturation is not always an obligatory process, if ABA effects are eliminated by removing the embryo from the seed would lead to development of seedlings (Berger 2003). But due to their low vigour, planting these immature seeds in the field will not result in the development of a healthy seedling.

The ethylene pathway studies in relation to seed development and maturation are extremely limited. In plant tissues, ethylene affects chlorophyll metabolism (Matilla 2000). Because chlorophyll loss is triggered during the final stages of embryogenesis (during acquisition of seed vigour), this process may be affected by ethylene. Mustard and canola seeds produce significant amounts of ethylene during embryogenesis, specifically in the early pre-desiccation stages (Child et al. 1998). Hence the role of ethylene can be attributed as minor during seed development and maturation and may be associated with the embryo de-greening process.

6   Physiological Maturity, Mass Maturity and Harvest Maturity

Seed development and maturation passes through a series of distinct (or overlapping) events such as histo-differentiation, embryogenesis, morphogenesis, reserve accumulation and maturation. Developmental stages could be monitored through relative changes in traits such as seed moisture content, dry weight accumulation, acquisition of desiccation tolerance, development of germinability, attainment of maximum vigour and longevity at harvest maturity. Two stages of maturity have been defined viz. physiological (or mass) maturity and harvest maturity. Physiological maturity is the end of the seed-filling period (Harrington 1972), whereas harvest maturity is the point of time that coincides with the end of maturation drying.

Physiological maturity was defined as the seed developmental stage at the end of seed filling (Shaw and Loomis 1950), when seed dry weight is at its maximum. Physiological maturity is more relevant for agronomic purpose to highlight the seed developmental stage beyond which none of the agronomic intervention could increase seed yield, because the funicular functionality is lost and nutrients cannot be transferred from mother plant to seed (Ellis 2019). Harrington (1972) hypothesized that at physiological maturity stage seed quality is greatest since seed quality improves during the seed-filling phase reaching a maximum but seeds do deteriorate thereafter. This concept of physiological maturity was supported by investigators for more than two decades in many crop species. Still and Bradford (1998) identified that physiological maturity was said to occur some days after maximum seed dry weight was attained in two Brassica seed crops, retracting the definition of Shaw and Loomis (1950). Black et al. (2006) abridged the definition of physiological maturity from that of Harrington's, to the stage of development at which a seed, or the majority of a seed population, has reached its maximum viability and vigour. Finch-Savage and Bassel (2016) also detached the definition of physiological maturity from the original to define it solely as the point of maximum seed quality. This led to some distortion or inconsistency in the use of term physiological maturity per se. Moreover, by earlier researchers the final stage of seed maturation, i.e., maturation drying has not been taken into consideration to be important to development of seed quality.

Lately, many researchers have confirmed that maximum seed quality does not occur until sometime after physiological maturity in cereals (Rao et al. 1991; Ellis and Pieta Filho 1992; Ellis et al. 1993; Sanhewe and Ellis 1996; Ellis 2019), Brassica (Still and Bradford 1998) and vegetable crops (Demir and Ellis 1993; Jalink et al. 1998). Ferguson (1993) established that in different cultivars of peas, maximum seed quality was attained 14–19 days after the stage of maximum seed dry weight and seed quality did not decline before harvest maturity. In these studies, seed quality continued to improve for considerable time and maximum seed quality was achieved, after attainment of physiological maturity, just prior to (or at) harvest maturity. Assuming the fact that, the definition of physiological maturity (stage of development marked with maximal seed dry weight, germinability, vigour and viability) had become compromised and misleading, the term mass maturity was proposed to designate the end of the seed-filling phase alone (Ellis and Pieta Filho 1992). Mass maturity represents the stage of seed development that Shaw and Loomis (1950) termed as physiological maturity. The physiological maturation is represented for individual seed and this maturation will not be the same for all seeds in the population, due to differential flowering habit.

The term harvest maturity represents the stage of maturity at which a seed crop is ready for harvest. The maturation drying phase ends at harvest maturity. The moisture content of the seed at this stage is significantly lower than at mass maturity. For obtaining seeds of maximum quality (germination, vigour and longevity), crop should be harvested at or slightly after harvest maturity; a period at which seed moisture content equilibrates with environment humidity. Black et al. (2006) defined harvest maturity as stage of development at which a seed, or majority of the seed population is best suited to harvesting in high quality and yield, considering its storage, its handling characteristics to minimize mechanical injury, and potential field losses due to inefficient collection by harvesting equipment. In practice, harvest maturity dates vary among the crops.

6.1 Seed Maturity Indices in Relation to Harvest Maturity

Indicators of maturity that could predict the right stage of harvest in a given crop are termed as maturity indices, although, for a bulk harvesting, a quick estimation of maturity at field level is quite challenging. Harvest maturity can be determined in a variety of crops by visual indicators, such as apparent visual changes in seed, fruit, panicle or through testing seed brittleness. Below is the list of common visual indicators associated with harvest maturity in few field and horticultural crops (Table 1).

6.2 Trackable Parameters During Seed Development and Maturation

Significant efforts were made during 1960s and 1970s by seed technologists across the globe to study the maturation process and primary changes associated with seed development. The research was oriented towards determination of morphological characteristics presented in either plant or seed during maturation process. This approach allowed an identification of harvest maturity on a plant population basis. Chlorophyll degradation resulting in a less green colour of the seeds was noticed. In the late 1990s, chlorophyll fluorescence was discovered as a very sensitive marker for seed maturity that can also be used on individual seed basis (Jalink et al. 1998). Given below are the plant characteristics that could be monitored during seed development process.

Table 1 Harvest maturity indicators for select field and horticultural crops

Crops	Seed maturity indicators as a criterion for harvesting
Field crops	
Rice	Nearly 85% of panicle turn straw-coloured and seeds are in hard dough stage in the lower portion of panicle
Wheat	Seeds in hard dough stage and yellowing of spikelets
Sorghum	Yellow-coloured ears with hard seeds
Pearl millet	Ears turn compact and hard seed comes out upon pressing
Finger millet	Hard seeds are embedded in brown-coloured earhead
Maize	Husk colour turns pale brown
Red gram	Nearly 80% of pods exhibit brown colour
Black gram/green gram	Seeds become hard and pods turns to brown or black colour
Cowpea	Brown-coloured seeds are observed in light straw-coloured pods
Groundnut	Pods become dark, patchy appearance inside the shell and upon pressing the kernel, oil is observed on fingers
Cotton	Black-coloured seeds exposed in fully opened boll
Horticultural crops	
Onion	Seeds become black on ripening in silver-coloured capsules
Carrot	Second and third order umbel turn brown
Radish	Pods become brown and parchment like
Turnip	Plants turn to brown and parchment colour
Peas	Pods become parchment like
Beans	Earliest pods dry and parchment like and remaining have turned yellow
Eggplant	Fruit turn to straw yellow colour
Tomato	Skin colour turn to red and the fruits are softened
Cucumber	Fruit becomes yellowish brown in colour, and stalk adjacent to the fruit withers for confirming actual seed maturity
Squash, Pumpkin	Rind becomes hard and its colour changes from green to yellow/orange or golden yellow to straw colour
True potato seed	Berries of potato become green to straw coloured and soft
Anise	Tips of fruits greyish green in colour
Celery	Pick the mature umbels periodically when they turn brown

Source: Malarkodi and Srimathi (2007), Singh and Malhotra (2007)

6.2.1   Seed Moisture Content

In both monocots and dicots, ovule moisture content at the time of fertilization is very high. The moisture content decreases during maturation process, but still it remains relatively high throughout most of the maturation period because water is the vehicle for transferring nutrients from the parent plant to the developing seeds. Moreover, enzymes, including those needed to produce the storage compounds, can be more active at high water contents. Dehydration was observed to be slow during the initial phase and gets accelerated after seed attains mass maturity. This decrease in moisture content proceeds during the maturation drying phase until hygroscopic equilibrium is attained with the environment. From that point onwards, changes in seed moisture content are the function of environmental factors. However, developing recalcitrant seeds do not show significant changes in desiccation during maturation phase and possess moisture levels usually over 60% on fresh weight basis.

6.2.2 Seed Size

As a result of intense cellular division and expansion during initial phases of embryogenesis, seed size increases gradually. The ovule during fertilization is a smaller unit in comparison to final seed size. The seed size reaches maximum at the end of reserve accumulation phase. For instance, in soybean, the maximum-sized seeds were observed at the full seed stage (R6) (Du et al. 2017). Thereafter, reduction in seed size was observed depending on cultivar and intensity of drying process coinciding with late maturation phase. In case of legumes, seed size reduction is more obvious than in cereals.

6.2.3 Seed Dry Weight

Post fertilization, as a result of accumulation of food reserves and water uptake, progressive increase in weight is observed in developing seeds. During the early phase of seed development, the accumulation of food reserves occurs at slower pace due to intense cell division and elongation. Thereafter, dry matter accumulation increases until seed attains maximum dry weight at mass maturity stage.

6.2.4 Germination

Seeds of various cultivated species are able to germinate a few days after ovule fertilization. Here, germination refers to radicle emergence, not the formation of a normal seedling, because histo-differentiation has not been completed and reserve accumulation is still incipient at this phase. Therefore, this germination does not lead to the production of vigorous seedlings. In the absence of seed dormancy, generally, the proportion of germinable seeds in the plant increases during maturation and reaches maximum when seeds attain harvest maturity.

6.2.5 Vigour

During maturation drying, the seed prepares itself for survival in the dry state, when moisture levels are too low for enzymatic repair. For an optimal survival in the dry state several protection mechanisms are imposed, as mentioned in the above section. The percentage of vigorous seeds significantly increases during the later maturation phase, reaching a maximum around the time when seeds attain harvest maturity (end of maturation drying phase).

6.3 Chlorophyll Fluorescence (CF) Sorting vis-à-vis Seed Maturation

During photosynthesis reactive oxygen species are produced, which are directly scavenged enzymatically. But in a dry seed, enzymatic scavenging is not possible and formation of reactive oxygen species would result in oxidation of organic molecules, including DNA and membrane lipids. In the late maturation phase, chlorophyll present in the seed is degraded (Ward et al. 1992). The amount of chlorophyll in the seed or seed coat will therefore serve as a marker for assessing the level of maturity. Chlorophyll in white-seeded *Phaseolus vulgaris* seeds is visible and may be detected and sorted using conventional colour-sorting equipment (Lee et al. 1998), but with most seeds, colour sorters are not sensitive enough to discriminate subtle differences in chlorophyll. At the end of the last century, it was discovered that chlorophyll levels could very sensitively be measured for individual seeds using its fluorescent properties (Jalink et al. 1998). That technique makes use of laser technology, narrow optical bandwidth filters, detection of chlorophyll-a in the seed coat, measuring the resulting chlorophyll fluorescence (CF), and linking it with the quality of the seeds. Chlorophyll-a in the seed coat is excited by laser radiation (at 650 nm) and the resulting fluorescence is measured instantaneously and non-destructively (at 730 nm). An exponential decrease in CF during maturation of seeds was found. This decline in CF signal was directly related to the germination performance under laboratory and greenhouse conditions. Equipment has been developed for analysing and sorting seeds individually based on their CF signal. Cabbage seeds with high CF signal are of lower quality and seeds with the lowest CF signal of better quality (Fig. 8). This analysis has been performed with seeds of many different species, including cabbage (*Brassica oleracea*) (Jalink et al. 1998; Dell'Aquila et al. 2002), tomato (*Solanum lycopersicum*) (Jalink et al. 1999), barley (*Hordeum vulgare*) (Konstantinova et al. 2002), carrot (*Daucus carota*) (Groot et al. 2006), and pepper (*Capsicum annuum*) (Kenanoglu et al. 2013).

6.4   External Factors Affecting Seed Development and Maturation

Environmental stresses can occur at any point of time during seed development. Responses to these stress factors are diverse and complex and largely dependent on the intensity and duration of stress, stage of incidence and the position of seed on the mother plant. The environment factors that influence seed development and maturation include soil fertility, water, temperature and light.

6.4.1   Soil Fertility

In general, plants that have been well nourished with the three major elements (N, P and K) produce larger seeds than those which have not been well nourished. The increase in seed size is due to enhanced seed development rate during the seed-filling period as a consequence of increased nutrient availability. Soils deficient in minor elements may cause seed quality issues. Calcium and boron deficiencies are known to cause cotyledonary discolouration in field beans. According to Copeland and McDonald (2001) when the effects of individual elements on seed development are considered, nitrogen has the greatest influence on seed size, seed germination and vigour.

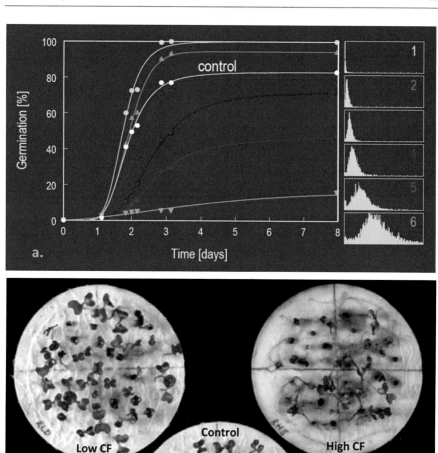

Fig. 8 CF sorting in relation to germination in cabbage. (**a**) Germination studies in six CF-sorted cabbage seed lots, peaks represented in the box on right-hand side indicate level of chlorophyll in seeds. (**b**) Differences in germinability among CF sorted seeds in Brassica. (Source: H Jalink, Wageningen University & Research, The Netherlands)

6.4.2 Water

Water deficits largely affect plant metabolic processes including seed development. The manifestations of water deficit are reduced leaf area, photosynthesis, excessive flower drop and abortion of embryos coupled with poor photosynthate production and translocation to developing seeds. Prolonged droughts leading to poor water availability result in reduced seed size more particularly when these stress factors occur during the seed-filling phase. Similarly, if water deficit coincides with flowering, it results in excessive flower drop and poor seed set (Copeland and McDonald 2001).

6.4.3 Temperature

High temperatures during seed development produce smaller seeds, while suboptimal temperatures retard seed growth. Seed germination and vigour are also adversely affected by exposure to low temperatures during development. High temperatures are considered the principal reason for the forced maturation resulting in wrinkled and deformed seeds in some plants. This phenomenon is also caused by water deficits during maturation. The occurrence of greenish seeds due to forced maturation is undesirable because this abnormality translates into decline in seed germination and vigour (Copeland and McDonald 2001).

6.4.4 Light

The intensity and duration of solar radiation and its seasonal distribution is crucial for plant development. It is well established fact that diffused or reduced light to mother plant results in smaller seeds due to decreased photosynthesis (Copeland and McDonald 2001).

6.4.5 Seed Position on the Plant

The development rate with seeds is largely affected by its relative positioning in the inflorescence. For instance, wheat seeds located at the distal end of spike exhibit slower growth rate and shorter filling periods as compared to the seeds positioned at proximal end. Similarly, in maize, developing seeds positioned at the tip of ear are smaller in comparison to seeds located at the base which is attributed to poor supply of photosynthates. Further, in soybean, pods in the lower and upper branches are produced in different time frames and they experience different environmental conditions during development, and this results in differences in seed performance. The poorly filled seeds (smaller seeds) produced during fag-end of crop growth period exhibit decreased germination and vigour.

7   Conclusion

Induction of flowering and differentiation of flower parts are considered the starting points of seed development. The course of seed development and maturation is controlled genetically and involves an organized sequence of events starting from ovule fertilization to the point in which the seed becomes independent from the parent plant. Earlier studies on seed maturation were mainly focused on recognizing phenotypical differences among species and cultivars in order to identify reliable parameters to determine the best time for seed harvest. When flowers in the same inflorescence are not pollinated at the same time, uniformity of seed maturation is never achieved, especially when a plant population is considered. Seed moisture content, seed size, germination, dry matter accumulation and seed vigour are considered to be the best parameters for evaluation of maturity status of developing seed. During the process of seed development and maturation, seeds of most crops acquire desiccation tolerance at mid or late maturation stages, which allow maintaining seed viability even after the loss of water up to 95%. The mechanisms behind the desiccation tolerance mostly rely on LEA proteins, small heat shock proteins, non-reducing oligosaccharides, antioxidants and structural changes at the cellular level. Orthodox seeds are tolerant to desiccation, which can be dried without loss of viability and the metabolic processes can be resumed upon subsequent rehydration. During late maturation, protection mechanisms needed for survival after shedding are strengthened to obtain maximum vigour at harvest maturity.

Seed is the basic unit of multiplication, but it should possess quality characteristics in terms of physical, physiological soundness, genetic purity and seed health. Hence, production of quality seed depends on genetic, environmental, edaphic and biotic factors prevailing in the production site. One of the environmentally influenced genetic factors deciding the quality of seed is the period and pattern of development and maturation of seed in any particular crop. Seed quality is a complex trait and high viability and vigour attributes of a seed enable the emergence and establishment of normal seedlings under a wide range of environments. In general, the seeds harvested at harvest maturity will have the greater seed yield and quality. In crops, the maturation will not be always uniform but there will be mingling of immature, matured and over-matured seeds based on the time of anthesis and fertilization. Hence, optimum timing of harvest for a given seed crop is necessary as beyond this point losses will be greater than the potential seed yield. The mechanism behind the intriguing events of development of seeds from point of fertilization till acquisition of desiccation tolerance, role of hormones and secondary metabolites and underlying genes controlling the whole set of events needs to be studied in a holistic manner. This might provide a new insight into the entire set of sequence or pathways associated with the various events during seed development stages.

References

Azarkovich MI (2020) Dehydrins in orthodox and recalcitrant seeds. Russ J Plant Physiol 67(2): 221–230

Ballesteros D, Pritchard HW, Walters C (2020) Dry architecture: towards the understanding of the variation of longevity in desiccation-tolerant germplasm. Seed Sci Res 30(2):142–155

Battaglia M, Olvera-Carrillo Y, Garciarrubio A, Campos F, Covarrubias AA (2008) The enigmatic LEA proteins and other hydrophilins. Plant Physiol 148(1):6–24

Baud S, Boutin JP, Miquel M, Lepiniec L, Rochat C (2002) An integrated overview of seed development in *Arabidopsis thaliana* ecotype WS. Plant Physiol Biochem 40(2):151–160

Becraft PW, Yi G (2011) Regulation of aleurone development in cereal grains. J Exp Bot 62(5): 1669–1675

Berger F (2003) Endosperm: the crossroad of seed development. Curr Opin Plant Biol 6(1):42–50

Bewley JD, Bradford KJ, Hilhorst HWM, Nonogaki H (2013) Seeds: physiology of development, germination and dormancy, 3rd edn. Springer Science & Business Media, New York, NY, p 381

Black M, Bewley JD, Halmer P (2006) The encyclopedia of seeds. Science, technology and uses. CABI, Wallingford

Child RD, Chauvaux N, John K, Onckelen HV, Ulvskov P (1998) Ethylene biosynthesis in oilseed rape pods in relation to pod shatter. J Exp Bot 49(322):829–838

Copeland LO, McDonald MB (2001) Principles of seed science and technology, 4th edn. Springer, Boston, MA, p 390

Davies PJ (ed) (2013) Plant hormones: physiology, biochemistry and molecular biology. Springer Science & Business Media, New York, NY

Dell'Aquila A, Van der Schoor R, Jalink H (2002) Application of chlorophyll fluorescence in sorting controlled deteriorated white cabbage (Brassica oleracea L.) seeds. Seed Sci Technol 30(3):689–695

Delouche JC (1971) Determinants of seed quality. In Proceedings of the Short Course for Seedsmen. https://scholarsjunction.msstate.edu/seedsmen-short-course/255

Demir I, Ellis RH (1993) Changes in potential seed longevity and seedling growth during seed development and maturation in marrow. Seed Sci Res 3(4):247–257

Demir I, Ashirov AM, Mavi K (2008) Effect of seed production environment and time of harvest on tomato (Lycopersicon esculentum) seedling growth. Res J Seed Sci 1:1–10

Du J, Wang S, He C, Zhou B, Ruan YL, Shou H (2017) Identification of regulatory networks and hub genes controlling soybean seed set and size using RNA sequencing analysis. J Exp Bot 68(8):1955–1972

Ekpong B, Sukprakarn S (2008) Seed physiological maturity in dill (Anethum graveolens L.). Agric Nat Res 42(5):1–6

Elias SG, Copeland LO (2001) Physiological and harvest maturity of canola in relation to seed quality. Agron J 93(5):1054–1058

Ellis RH (2019) Temporal patterns of seed quality development, decline, and timing of maximum quality during seed development and maturation. Seed Sci Res 29:135–142

Ellis RH, Pieta Filho C (1992) The development of seed quality in spring and winter cultivars of barley and wheat. Seed Sci Res 2(1):9–15

Ellis RH, Hong TD, Jackson MT (1993) Seed production environment, time of harvest, and the potential longevity of seeds of three cultivars of rice (Oryza sativa L.). Ann Bot 72(6):583–590

Ferguson AJ (1993) The agronomic significance of seed quality in combining peas (Pisum sativum L.). Doctoral Thesis submitted to University of Aberdeen (United Kingdom)

Finch-Savage WE, Bassel GW (2016) Seed vigour and crop establishment: extending performance beyond adaptation. J Exp Bot 67(3):567–591

Finkelstein RR, Gampala SS, Rock CD (2002) Abscisic acid signaling in seeds and seedlings. Plant Cell 14(Suppl 1):S15–S45

Groot SPC, Karssen CM (1987) Gibberellins regulate seed germination in tomato by endosperm weakening: a study with gibberellin-deficient mutants. Planta 171(4):525–531

Groot SP, Bruinsma J, Karssen CM (1987) The role of endogenous gibberellin in seed and fruit development of tomato: studies with a gibberellin-deficient mutant. Physiol Plant 71(2): 184–190

Groot SPC, Birnbaum Y, Rop N, Jalink H, Forsberg G, Kromphardt C, Werner S, Koch E (2006) Effect of seed maturity on sensitivity of seeds towards physical sanitation treatments. Seed Sci Technol 34(2):403–413

Harrington JF (1972) Seed storage and longevity. In: Kozlowski TT (ed) Seed Biology, vol III. Academic Press, New York, NY, pp 145–245

Hooley R (1994) Gibberellins: perception, transduction and responses. In: Palme K (ed) Signals and signal transduction pathways in plants. Springer, Dordrecht, pp 293–319

Jalink H, Frandas A, Schoor RVDA, Bino JB (1998) Chlorophyll fluorescence of the testa of Brassica oleracea seeds as an indicator of seed maturity and seed quality. Sci Agric 55:88–93

Jalink H, Van der Schoor R, Birnbaum YE, Bino RJ (1999) Seed chlorophyll content as an indicator for seed maturity and seed quality. Acta Hortic 504:219–228

Kenanoglu BB, Demir I, Jalink H (2013) Chlorophyll fluorescence sorting method to improve quality of Capsicum pepper seed lots produced from different maturity fruits. Hortic Sci 48(8): 965–968

Konstantinova P, Van der Schoor R, Van den Bulk R, Jalink H (2002) Chlorophyll fluorescence sorting as a method for improvement of barley (*Hordeum vulgare* L.) seed health and germination. Seed Sci Technol 30(2):411–421

Kranner I, Birtić S (2005) A modulating role for antioxidants in desiccation tolerance. Integr Comp Biol 45(5):734–740

Lee PC, Paine DH, Taylor AG (1998) Detection and removal of off-colored bean seeds by color sorting. Seed Technol 20(1):43–55

Leprince O, Buitink J (2010) Desiccation tolerance: from genomics to the field. Plant Sci 179(6): 554–564

Leprince O, Pellizzaro A, Berriri S, Buitink J (2017) Late seed maturation: drying without dying. J Exp Bot 68(4):827–841

Malarkodi K, Srimathi P (2007) Seed physiological maturity. Int J Plant Sci 2(1):222–230

Matilla AJ (2000) Ethylene in seed formation and germination. Seed Sci Res 10(2):111–126

Nambara E, Marion-Poll A (2003) ABA action and interactions in seeds. Trends Plant Sci 8(5): 213–217

Pammenter NW, Berjak P (1999) A review of recalcitrant seed physiology in relation to desiccation-tolerance mechanisms. Seed Sci Res 9(1):13–37

Rao NK, Rao SA, Mengesha MH, Ellis RH (1991) Longevity of pearl millet (*Pennisetum glaucum*) seeds harvested at different stages of maturity. Ann Appl Biol 119(1):97–103

Sanhewe AJ, Ellis RH (1996) Seed development and maturation in *Phaseolus vulgaris* II. Post-harvest longevity in air-dry storage. J Exp Bot 47(7):959–965

Shaw RH, Loomis WE (1950) Bases for the prediction of corn yields. Plant Physiol 25(2):225

Singh R, Malhotra SK (2007) Harvesting and maturity indices in seed spices crops. In: Malhotra SK, Vashishtha BB (eds) Production, development, quality and export of seed spices. NRCSS, Ajmer, pp 195–200

Sliwinska E, Bewley JD (2014) Overview of seed development, anatomy and morphology. In: Gallagher RS (ed) Seeds: the ecology of regeneration in plant communities, 3rd edn. CAB International, Wallingford, pp 1–17

Smolikova G, Leonova T, Vashurina N, Frolov A, Medvedev S (2021) Desiccation tolerance as the basis of long-term seed viability. Int J Mol Sci 22(1):101

Still DW, Bradford KJ (1998) Using hydrotime and ABA-time models to quantify seed quality of Brassicas during development. J Am Soc Hortic Sci 123(4):692–699

Verdier J, Lalanne D, Pelletier S, Torres-Jerez I, Righetti K, Bandyopadhyay K, Buitink J (2013) A regulatory network-based approach dissects late maturation processes related to the acquisition of desiccation tolerance and longevity of *Medicago truncatula* seeds. Plant Physiol 163(2): 757–774

Vogler H, Kuhlemeier C (2003) Simple hormones but complex signalling. Curr Opin Plant Biol 6(1):51–56

de Vries SC, Weijers D (2017) Plant embryogenesis. Curr Biol 27(17):R870–R873

Ward K, Scarth R, McVetty PBE, Daun J (1992) Effects of genotype and environment on seed chlorophyll degradation during ripening in four cultivars of oilseed rape (*Brassica napus*). Can J Plant Sci 72(3):643–649

Wobus U, Weber H (1999) Sugars as signal molecules in plant seed development. Biol Chem 380(78):937–944. https://doi.org/10.1515/BC.1999.116

van Zanten M, Koini MA, Geyer R, Liu Y, Brambilla V, Bartels, Soppe WJ (2011) Seed maturation in *Arabidopsis thaliana* is characterized by nuclear size reduction and increased chromatin condensation. Proc Natl Acad Sci 108(50):20219–20224

Seed Dormancy and Regulation of Germination

P. C. Nautiyal, K. Sivasubramaniam, and Malavika Dadlani

Abstract

Seed germination and dormancy are vital components of seed quality; hence, understanding these processes is essential for a sound seed production system. The two processes are closely interrelated and regulated, both by genetic as well as environmental factors. While dormancy provides an inherent mechanism aimed at the survival of the plant species to withstand adverse external conditions by restricting the mature seed from germinating, the ability of the dehydrated seed to remain viable and produce a vigorous seedling upon hydration under favourable conditions is the key to the survival and perpetuation of the plant species. In addition, quality seed is expected to result in timely and uniform germination under favourable field conditions after sowing to establish a healthy crop stand. Therefore, in seed technology, dormancy is not considered a desirable trait in the seed lots used for sowing. Thus, to achieve the highest germination percentage, understanding the factors controlling these two interlinked and contrasting processes is vital. In seed testing and seed trade, knowledge of seed germination and dormancy is needed for a reliable assessment of seed quality and its planting value, and to make right decisions. Though much is yet to be understood, the present status of knowledge on these aspects has made significant advances, especially in genetic control, molecular mechanism, and physiological and environmental factors influencing germination and dormancy. The

P. C. Nautiyal (✉)
Formerly at Seed Science & Technology, ICAR-Indian Agricultural Research Institute, New Delhi, India

K. Sivasubramaniam
Formerly at Seed Science & Technology, Tamil Nadu Agricultural University, Madurai Campus, Madurai, India

M. Dadlani
Formerly at ICAR-Indian Agricultural Research Institute, New Delhi, India

© The Author(s) 2023
M. Dadlani, D. K. Yadava (eds.), *Seed Science and Technology*,
https://doi.org/10.1007/978-981-19-5888-5_3

39

information compiled in this chapter may help the seed technologists in developing new methods for breaking dormancy and testing germination,

Keywords

Types of dormancy · Regulation of dormancy · Release of dormancy · Types of germination · Hormonal regulation · GA and germination · ABA and dormancy · Induction of dormancy

1   Introduction

Both inductions of seed dormancy and onset of germination are crucial physiological states of seed which determine the success of field establishment after sowing in several crop plants. Seed dormancy and germination are mainly regulated through several physiological processes and environmental factors. Though contrasting in their expression, both processes are equally important for the management and planning of crop cultivation. A low degree of dormancy is seen to be of vital advantage in preventing in situ germination in various crop species, for example, Spanish-type groundnut (Nautiyal et al. 1993, 2001), wheat (Mares 1983), maize (Neill et al. 1987), and rice (Sohn et al. 2021). On the other hand, a deep dormancy would prevent normal germination even under favourable conditions, resulting in poor crop stand. There are reports that interaction between environmental factors (i.e., light, temperature, water status) and growth hormones (i.e., abscisic acid, gibberellic acid and ethylene) play an essential role in dormancy vs. germination. Germination may be seen as a chain of processes transforming a quiescent embryo into a metabolically active one and developing well-differentiated tissues (viz., apical meristems, root and shoot), which are essential for establishing a healthy seedling. The germination process starts with the imbibition of water, followed by a metabolically active phase, during which a series of biochemical reactions take place providing energy and supporting the cellular processes leading to radicle emergence and seedling growth. The metabolically active phase begins with enzyme activation, hydrolysis and mobilisation of stored food reserves. After mobilisation of reserve food material, embryo growth is initiated followed by the weakening and rupture of the seed coat to make way for radicle emergence. These are crucial steps directly related to the seedling establishment and, thus, crop productivity. This chapter discusses fundamental processes underlying regulation of seed dormancy and germination, including the external environmental and internal hormonal factors.

2   Seed Dormancy

2.1   Definition

Seed dormancy has been defined differently by different researchers. Several workers have broadly described it as a state of temporary suspension of the ability of a viable seed to initiate germination, even under favourable environmental conditions (Bewley 1997; Baskin and Baskin 2004; Hiroyuki et al. 2018). On the other hand, a non-dormant and viable seed would germinate when favourable environmental conditions are given (Baskin and Baskin 1998). Thus it is presumed that a dormant seed is unable to germinate due to some inadequacies or possessing some inherent blocks that must be overcome and released to initiate germination (Bewley and Black 1994). The state of dormancy may be differentiated from quiescence—a term commonly used for the phenomenon, where a non-hydrated seed remains non-germinated because of not being provided with adequate conditions like water, temperature or air that are essential for germination. This characteristic is the basis of prolonged storability (also termed longevity) of desiccation-tolerant 'orthodox' seeds. Further, it is clear that dormancy is a mechanism that organises the distribution of germination in time or space and having immense ecological significance. It is evident from the fact that seed of several plant species shows variability in the degree of dormancy, which exhibits sporadic release from dormancy resulting in irregular germination. Hence, the significance of temporal dispersal is clearly to help in enhancing spread and survival of the species. The temporal dispersal thus enhances the spread and survival of the species and is more evident in monocarpic than polycarpic species. In nature, seasonal environmental changes (particularly light and temperature), depth of burial in the soil or light penetration through the canopy above ground, and interventions of birds and animals are some of the factors determining seed dormancy and germination processes. Induction of dormancy during the late stage of maturation also renders the seed protection against pre-harvest sprouting which is a desirable feature in many cereal species. It is a well-established fact that environmental temperature regulates both dormancy and germination, while light germination alone (Vleeshouwers et al. 1995; Batlla et al. 2004). Moreover, the role of light in the induction of dormancy is a debatable issue (Bewley and Black 1994; Baskin and Baskin 2004).

2.2   Classification of Seed Dormancy

Different classification systems have been proposed by seed ecologists for different forms of dormancy, whereas seed technologists need an internationally acceptable hierarchical system of and the underlying mechanisms to develop suitable treatments for breaking seed dormancy. Baskin and Baskin (2004) have suggested a modified version of the scheme of the Russian seed physiologist Marianna G. Nikolaeva for classifying seed dormancy. The modified system includes three hierarchical layers, i.e., class, level and type and includes five classes of dormancy: physiological dormancy (PD), morphological dormancy (MD), morphophysiological dormancy (MPD), physical dormancy (PY) and combinational dormancy (PY + PD). The most extensive classification schemes are for PD, which includes three levels and five types of dormancy in the nondeep level, whereas MPD includes eight levels but no difference in their types. PD (non-deep level) is the most common kind of dormancy that occurs in gymnosperms (Coniferales, Gnetales) and in all major clades of angiosperms. Recently, the diversity in kind of seed dormancy and its classification has been reviewed by Ordonez-Parra King (2022) and Baskin and Baskin (2021).

2.2.1 Morphological Dormancy

Seeds with morphological dormancy (MD) are often characterised by small and underdeveloped embryos, but well differentiated into cotyledon(s) and hypocotyl-radicle (Baskin and Baskin 1998). Embryos in seeds with MD are not physiologically dormant and do not require a dormancy-breaking pre-treatment in order to germinate, these simply need time to complete the development of embryo to grow to full size and then germinate. The period of dormancy in such seeds (that is the time between incubation of fresh seeds and radicle emergence) may extend from a period of a few days to 1–2 weeks, and germination would complete in about 30 days. In addition, the following features are also commonly seen in seeds with morphological dormancy: (a) physical factors such as impermeability of seed coat or pericarp to water and oxygen; (b) chemical factors such as chemical inhibitors (phenols) in pericarp or in seed coat; and (c) mechanical factors (i.e., mechanical resistance of pericarp, seed coat or endosperm to embryo growth). Such seeds also show light sensitivity, at times.

2.2.2 Physiological Dormancy

Most of the species with physiological dormancy (PD) exhibit non-deep dormancy. Further, based on the responses to temperature in breaking dormancy, five types of non-deep PD are recognised. Physiological dormancy is best explained in terms of hormonal balance, in which ABA acts as an inhibitor and GA as a promoter. These hormones simultaneously and antagonistically regulate the onset, maintenance and termination of dormancy (Amen 1968; Wareing and Saunders 1971). Hilhorst et al. (2010) proposed a model for onset of germination and release of dormancy following ABA and GA interaction in response to environment. In this model GA and ABA do not interact directly. While ABA produced by the embryo induces dormancy during seed development, GA promotes germination of non-dormant seeds. Further, the amount of GA required for germination of ripe seeds is determined by ABA concentrations during seed development. Thus, seeds with a low level of ABA produced during their development exhibit low dormancy and require low concentrations of GA to promote germination, whereas those with a high concentration of ABA exhibit deep dormancy, and require higher concentrations of GA. Experimental results on tomato (Groot and Karssen 1992), sunflower (Le Page-Degivry and Garello 1992) and wild oat (Fennimore and Foley 1998) support this model. Evidences have been presented for the involvement of both ABA and GA in dormancy break in seeds of potato (Alvarado et al. 2000), groundnut

(Nautiyal et al. 2001) and other crops. In addition to ABA and GA, ethylene, a third plant hormone, is also involved in the regulation of seed dormancy and germination in some species (Nautiyal et al. 1993; Kepczynski and Kepczynska 1997; Matilla 2000), possibly by decreasing the responsiveness of the seed to endogenous ABA. Thus, ethylene may promote germination by interfering with the action of ABA (Beaudoin et al. 2000). Studies on wild oats have shown that ABA-responsive genes specific to mRNAs and heat-stable proteins are upregulated, synthesised, and maintained in embryos of imbibed dormant seeds, and decline in non-dormant or after-ripened seeds as a result of GA synthesis and signalling and disappear during seed germination (Li and Foley 1994; Holdsworth et al. 1999; Hilhorst et al. 2010).

2.2.3 Physical Dormancy

Physical dormancy (PY) is mainly referred to the state in which seeds are dormant because of some physical barriers that do not allow imbibition of water into the seed which is the primary requirement of germination. It is mainly due to the water-impermeability of the seed or fruit coat (Jayasuriya et al. 2009). Physical dormancy is reported to be caused by one or more water-impermeable layers of palisade cells in the seed or fruit coat (Baskin Jerry et al. 2000). Typically, PY dormancy breaks both under natural environmental conditions (ambient), or by subjecting to some physical or chemical treatments that helps in the formation of an opening (water gap) or a specialised anatomical structure on the seed (or fruit) coat such as strophiole, through which water moves to the embryo (Baskin Jerry et al. 2000). Similarly in some taxa of Fabaceae, dormancy is reported to be broken by heating which weakens the seed coat in region(s) other than the strophiole (lens), making it water permeable (Morrison et al. 1998). Thus, mechanical or chemical scarification are often effective and simple methods for releasing non-deep physiological dormancy and promote germination.

2.2.4 Combinational Dormancy

In this type of dormancy, two factors are involved to impart dormancy such as seeds with physiological and physical dormancy which can be represented as PY + PD; or morpho-physiological dormancy. This type of dormancy is a combination of imper-meability of the seed coat for water uptake, and physiologically dormant embryo. Morpho-physiological dormancy is mostly the result of underdeveloped embryos, combined with physiological inadequacy during embryo development (Baskin and Baskin 2004). Such seeds, therefore, require a combination of dormancy-breaking treatments, for example, a combination of warm and/or cold stratification, followed by GA application for dormancy release.

The physiological component of such dormancy appears to be of non-deep level, for example, freshly matured seeds of some winter annuals, *Geranium* (Geraniaceae) and *Trifolium* (Fabaceae) have some conditional dormancy, which can be broken by after-ripening in dry storage or in the field within a few weeks after maturity, even while the seed coat remains impermeable to water (Baskin and Baskin 1998). Embryos in *Cercis* (Fabaceae) and *Ceanothus* (Rhamnaceae) are more dormant (but non-deep) and require a few weeks of cold stratification, after the PY is broken and seeds imbibe water before they can germinate. Seshu and Dadlani (1991) observed that dormancy in rice (*Oryza sativa* L.) seed is a result of certain physical and chemical factors associated with both the hull and the pericarp, a type of combinational dormancy. They also reported the role of nonanoic acid, a short chain (C 9) fatty acid, in imposing dormancy and its release by dry heat treatment, a common practice in releasing low dormancy in rice. The relative significance of these substances in cultivars of tropical and temperate origins and their implications in terms of ecogeographic adaptability were suggested. Embryos in *Cercis* (Fabaceae) and *Ceanothus* (Rhamnaceae) are more dormant (but non-deep) and require a few weeks of cold stratification, after PY is broken and seeds imbibe water before they germinate.

2.3   Types of Seed Dormancy in Legumes and Cereals

The contrasting germination and seed dormancy patterns have extensively been reported in chickpea (*Cicer arietinum*) by Sedlakova et al. (2021). In groundnut seed dormancy, which causes huge economic losses due to in situ sprouting at times in Spanish-type varieties (Fig. 1a), is directly associated with the duration of crop maturity, i.e., longer the duration, prolonged is the seed dormancy and vice versa (Nautiyal et al. 2001; Bandyopadhyay et al. 1999). The long-duration varieties of Virginia type showed dormancy for longer duration (63 days), than the Spanish and Valencia types, which showed no dormancy or exhibit it for a shorter period (maximum of 30 days). It was also seen that while in Spanish groundnut varieties,

A   B

Fig. 1 (**a**) Pre-harvest sprouting (in situ seed/pod germination) in groundnut due to unseasonal rain before or during harvest. (**b**) Pre-harvest sprouting (in situ germination) in sorghum millet due to rain before harvest or during harvest

dormancy is mainly controlled by testa (or seed coat), in Virginia types the seed coat, cotyledons and embryonic axis all contribute to it. The term fresh seed dormancy (FSD) was used for seed germination at high moisture content (>22%) seeds before or during harvest under field condition (Nautiyal et al. 2001).

Pre-harvest sprouting in wheat, barley, maize, rice, sorghum (Fig. 1b), rye and triticale poses a serious problem in crop cultivation (Gualano et al. 2014) due to environmental factors such as temperature, rains and light quality. It seems that a certain degree of dormancy is essential to prevent yield losses in several crop species (Rodriguez Marıa et al. 2015). As in seeds of many other species, the antagonism between the plant hormones abscisic acid and gibberellins is instrumental for the inception, expression, release and re-induction of dormancy. Thus, resistance to pre-harvest sprouting could be associated with a number of physiological, developmental, and morphological features of the grains on the spike, including pericarp colour, transparency, hairiness, waxiness, permeability of water, α-amylase activity, and levels of growth substances such as ethylene, ABA, and GA in the embryo (Sohn et al. 2021).

The seed dormancy and pre-harvest sprouting in cereals are important issues in crop production. Pre-harvest sprouting is a matter of serious concern as it results in total loss of seed viability as the seed loses its desiccation tolerance in several cereals such as wheat, barley, maise, rice, sorghum, rye and triticale (Gualano et al. 2014) (Fig. 1b). These cereals mainly originated from both temperate and tropical regions and developed diverse responses to environmental factors such as temperature, rains and light quality. Hence, a short period of dormancy is a desirable agronomic trait to prevent pre-harvest sprouting in cereals (Rodriguez Marıa et al. 2015). As in seeds of many other species, the antagonism between the plant hormones abscisic acid and gibberellins is instrumental in cereal grains for the inception, expression, release and re-induction of dormancy, though its induction and regulatory mechanism varies in each species. In oat (*Avena sativa* L.) freshly harvested seeds are dormant at relatively high temperatures (>20–25 °C), which results partly from the structures surrounding or adjacent to the embryo (pericarp, testa and endosperm) and partly due to the embryo itself.

In rice, the pre-harvest sprouting is regulated both by genetic and environmental factors, as well as interactions between these. Further, pre-harvest sprouting resistance could be associated with a number of physiological, developmental and morphological features of the grains on the spike, including pericarp colour transparency, hairiness, waxiness, permeability of water, α-amylase activity and levels of growth substances such as ethylene, ABA and GA in the embryo, all of which play a role in the resistance against pre-harvest sprouting (Sohn et al. 2021).

2.4   Induction of Dormancy

Though dormancy is genetically predisposed, its induction is controlled by various environmental factors, such as temperature, humidity and light during seed development, which significantly influence the onset and release of seed dormancy. This underlines the need to understand the effect of changing environmental conditions on seed germination under field conditions. Integrating research techniques from different disciplines of biology, i.e., transcriptomics, proteomics and epigenetics, could help in understanding the mechanisms of the processes controlling seed germination, and induction of dormancy in developing seed (Klupczyńska and Pawłowski 2021). Seed dormancy is influenced by both short- and long-term effects of climate. Long-term effects may result in inheritable dormancy differences through species, ecotype and clonal variation (Montague et al. 2008), whereas short-term effects are specific to weather conditions during seed maturation (Huang et al. 2015), or germination. Dormancy, that is observed early in the seed development process or induced by the mother plant, is known as primary dormancy, while secondary dormancy is imposed by the harsh environmental conditions unfavourable for seed germination (Nadella and Foley 2003).

2.4.1   Primary Dormancy

Several workers have shown that ABA synthesis in the embryonic axis during seed maturation and its continued synthesis de novo is an essential requirement for the induction and persistence of dormancy (Ketring and Morgan 1969). It is also reported that ABA controls the expression of specific ABA-responsive genes, such as a set of late embryo abundant proteins (LEAs), which control dormancy in the developing seed, as seen in many cereal crops with some degree of primary dormancy that do not germinate to escape unfavourable conditions (Hilhorst 2007; Hilhorst et al. 2010).

2.4.2   Secondary Dormancy

There are several reports mentioning that secondary dormancy may be induced by environmental conditions, such as high temperature and hypoxia (Grahl 1965; Leymarie et al. 2008; Corbineau and Come 2003). The experiments conducted with barley and oat found that induction of thermodormancy was apparent after 3–8 h of incubation at 30 °C, and was maximum after 1–3 days (Leymarie et al. 2008). In addition, induction of thermodormancy requires a critical moisture content in the embryo (approx. 40–50% dry weight basis) (Hoang et al. 2012) and develops concomitant with activation of the cell cycle (Gendreau et al. 2008). Secondary dormancy can also be imposed during imbibition at low temperatures (10–15 °C), combined with low oxygen tensions in seeds having primary dormancy (Hoang et al. 2014).

2.5   Phytochrome and Seed Dormancy vs. Germination

Photoperiod sensitivity in plants, a quantitatively inherited trait which has many biologically important consequences, is regulated through the phytochromes—receptor of the light stimulus. Hence phytochromes, members of a duplicated-gene family of photoreceptors, are the most important environmental sensors in plants. Phytochromes exist in two forms. Synthesised in dark, phytochrome remains in a biologically inactive (Pr) form, which upon exposure to the light get converted into biologically active form (Pfr). This process of photo conversion to the far-red light-absorbing (Pfr) form is optimised at red wavelengths. Photo conversion of Pfr, back to the biologically inactive Pr form is optimised at far red wavelengths, resulting in a dynamic photo equilibrium of Pr and Pfr in natural light conditions. The basic mechanism of phytochrome signalling involves a physical interaction of the photo-receptor with the PHYTOCHROME INTERACTING FACTORS (PIFs)—a sub-family of bHLH (basic Helix Loop Helix) transcription factors (Ngoc et al. 2018). This system is reported to enable plants to alter gene expression rapidly in response to fluctuations in the light environment. In addition, five phytochrome genes (PHY, A-E) are considered to be mainly associated with seed dormancy (Goosey et al. 1997; Tomoko 1997).

The enabling role of gibberellins (GAs) in seed germination is well documented (Yamauchi et al. 2007), and the effect of light on GA signalling proteins is also suggested. The GA-deficient mutants were unable to germinate without exogenous GA application, whereas GA signalling mutants produced defective phenotypes (Steber et al. 1998; Bassel et al. 2004). Yamaguchi et al. (1998) reported that one of the two 3-β-hydroxylase enzymes (encoded by the GA4H gene) is induced by the phytochrome in germinating seeds. Crucial in this process are the DELLA proteins which repress GA action. Among various DELLA proteins, RGL2 (repressor of GA1-3 like 2) plays the major role in regulating seed germination. During dormancy release and seed germination, the change in the GA content and its expression is gene-regulated (Yamauchi et al. 2007). An important signal transduction component of light-induced germination is the bHLH transcription factor PIF1 (Phytochrome Interacting Factor 1), also known as PIL5 (Phytochrome-Interacting factor 3-Like 5) (Castillon et al. 2007), which binds to Pfr and causes proteosome degradation. Similarly, alteration in GA level is also brought about by the repression of GA biosynthesis genes, activation of GA2ox gene, and genes encoding DELLA proteins. Somnus (SOM) which encodes a nucleus-localised CCCH-type zinc finger protein, is another gene acting downstream of PIL5 (Kim et al. 2008), suggesting that PIL5 regulates genes regulating ABA and GA metabolism partly through SOM.

Molecular control of seed dormancy and germination has been reviewed extensively at the gene action level by Garello et al. (2000), Finch-Savage William and Leubner-Metzger (2006), Matilla Angel (2020) and Faiza et al. (2021), which indicate a major role of the DELAY OF GERMINATION (DOG) genes in the regulation of germination and dormancy. Available information also suggests that binding of DOG1 to Protein Phosphatase 2C ABSCISIC ACID (PP2C ABA) Hypersensitive Germination (AHG1) and heme are independent processes, but both are essential for DOG1's function in vivo. AHG1 and DOG1 constitute a regulatory system for dormancy and germination.

2.6   Methods to Release Dormancy

The above discussion elaborated the significance of seed dormancy, its regulation and release in nature. However, from the seed technological perspectives, germinability remains the most crucial factor in crop establishment ensuring uniform emergence and for maintaining optimum plant population, for which knowledge of dormancy behaviour in different species, and methods to release these are vital. Various physical, chemical or physiological treatments may be applied to release different levels of dormancy.

2.6.1   Scarification

Scarification is the most common method of breaking seed dormancy in hard seed coat type. This literally mean creating a scar on the seed coat so that water gets its entry into the seed and hydrates the embryo. Seeds can be scarified following any of the means such as physical and chemical treatment. In chemical treatment, seeds can be treated with commercial-grade (95%) H_2SO_4 @ 100 mL/kg or a strong alkali, i.e., NaOH (20%) for 2–3 min followed by washing in running water to remove the traces of acid or alkali. The duration of treatment may vary depending on the degree and proportion of hardness of the seed coat. Acid scarification is employed in legumes (blackgram, green gram), rose, tamarind, and *Acacia* spp., etc., whereas alkali treatment is effective in scarifying cashew drupes. Physical methods using heat or mechanical abrasion are also employed to release hard seededness. Also, exposing seeds to high temperatures ruptures the seed coat, whereas puncturing or clipping the abaxial end of the seed coat helps imbibition. Seeds of chickpea, *Acacia* spp. hedge lucerne respond well to soaking in hot water for 10–15 min to soften the seed coat. Exposure to high temperature (~50 °C) also enables dry seeds to overcome dormancy in many cereals. Dry heat treatment for 7–14 days removes dormancy of rice seeds. In some tree species (i.e., Teak) the heat generated by natural or created fire scorches the pericarp of the seed making the seed coat permeable, and resulting in germination. Besides, compounds produced by the charring of plant materials are also reported to have a stimulating effect on dormancy release in many woody species (Keeley and Fotheringham 2000; Baskin et al. 2002).

2.6.2   Stratification

The seeds having morpho-physiological dormancy may be treated following stratification. In this process, seeds are treated with GA_3 or thiourea to counter the influence of ABA. Cold stratification is another way to remove dormancy and it involves placing the seed in stratified layers of wet sand/soil/sawdust/absorbent cotton wool in 1:4 ratio, and exposing them to temperatures of 3–5 °C for 2–3 days to several months, depending on the type of seed. This treatment is common to overcome seed dormancy in seeds requiring after-ripening period. In this case seed is shed with underdeveloped embryos. In some species, seeds are fully soaked in cold water, i.e., 3–5 °C for 48 h. After draining the water, seeds are mixed with moist sand (1:4 ratios at 20–25 °C) or alternating temperatures of 20–30 °C for a period of 2–6 weeks, depending on the species to overcome the morphological dormancy (i.e., oil palm).

2.6.3 Leaching of Metabolites (Inhibitors)

Leaching simply is a washing off the inhibitors present in seed or fruit coat by soaking or placing under running water. It mainly, involves soaking seeds in water for some time (i.e., 2–4 h to 3 days). In prolonged soaking change of water once in 12 h is a must to avoid fermentation or decay. Seeds soaked in running water for a day to leach out the inhibitors such as cyanides and phenolic compounds as reported in sugar beet (Kumar and Goel 2019).

2.6.4 Treating Seed with Chemical Activators and Growth Hormones

Seed dormancy when present mainly due to the presence of inhibitors (mainly ABA) may be subjected to prolonged washing. Also, such dormancy may be released by application of GA and kinetin (100–1000 ppm). In addition, GA_3 and KNO_3 are used as light substituting chemicals and exhibited germination in tomato seed that require light for germination. Thiourea can be used for seeds that require both light and chilling treatment in lettuce (see chapter "Testing Seed for Quality").

3 Seed Germination

Germination could be defined in the physiological terms as the metabolic activation of seed upon hydration, culmination with 'chitting' or protrusion of the radicle through the seed coat. However, a seed technologist may define it as 'the emergence and development of seedling from embryo, having the essential functional structures indicative of the ability to produce a normal seedling under favourable conditions' (Lawrence and Miller 2001; Parihar et al. 2014). Under field conditions a seed is considered as germinated when it emerges completely out of the soil which is referred as 'field emergence'. Thus, seed germination is not only the first crucial step in the life cycle of plants, it also determines the crop stand for obtaining an optimum plant population, and hence is crucial in agricultural production.

However, even for agronomic success, a basic understanding of the process of germination could be useful. Nature has evolved an intrinsic mechanism of regulation of germination and dormancy that has many ecological advantages, a knowledge of which could be applied by the seed technologists for practical use. This includes a number of mechanisms evolved by the mother plant, such as the hard outer tissues of seeds; the seed coat which is derived from the ovule integuments, or the pericarp, which is the maternal fruit tissue; which regulates the physiological behaviour of the progeny seeds. In addition, physiological and biochemical changes, responses of seeds to environmental cues that can trigger germination, and morphological changes during germination, all of which can impart a direct or an indirect influence on the survival and growth of seeds and seedlings and vegetative growth which consequently affect yield and quality. There are several reviews on physiological aspects of seed germination including dormancy, plant-water-relations, environmental factors and hormonal control (Finch-Savage William and Leubner-Metzger 2006; Holdsworth et al. 2008; Karin et al. 2011; Gerardo et al. 2020). Similarly, the speed of germination is an important parameter to measure seed vigour and models have been developed for calculating this (Jardim et al. 2021; Chao et al. 2021), which could provide a more accurate assessment of the planting value of the seed.

Several intrinsic factors including genetic control, regulation by hormones and chemical stimulants, and cellular repair processes (Loïc et al. 2012) have been examined to fully understand the process of germination. The current advances through the post-genomics approaches have drawn special attention to three main aspects: (a) the translational control of germination and the role of stored components, (b) the importance of metabolic transitions in germination, and (c) the search for biomarkers of seed vigour.

3.1   Morphology of Seed Germination

Irrespective of the number and structure of the cotyledons, two types of seed germination are seen in nature based on their fate upon germination. It is basically of two types as could be seen in bean and pea seeds (Fig. 2). Although these seeds are similar in structure and are from the same taxonomic family. These two forms of seed germination and seedling emergence are commonly known as epigeal and hypogeal. As literary meaning of epigeal is defined as germination above ground, is a characteristic feature of bean seed and is considered more primitive than the hypogeal germination, which is below into the soil (King Keith 2003). Further, it could be elaborated that in epigeal germination the cotyledons are raised above the ground where they continue to provide nutritive support to the growing points. On the other hand, in hypogeal germination, as in pea seeds, cotyledons or comparable storage organs remain under the soil while the plumule pushes upward and emerges

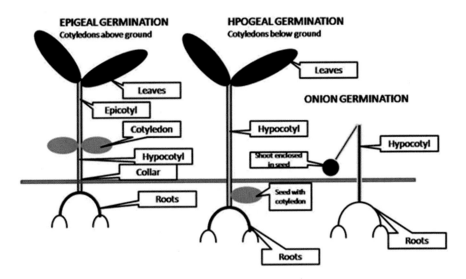

Fig. 2 Types of germination

above the ground. Understanding the type of germination is useful during preparation of beds, sowing, planning, as well as evaluating the field emergence (i.e., counted based on germinated and healthy seedlings).

3.2 Metabolic Processes During Seed Germination

Seed germination is the result of a series of physiological, biochemical and morphological changes which are responsible for germination, seedling survival and vegetative growth of various plant parts that influence both field emergence and stand establishment. Dry seeds are usually at very low water potential, in the range of − 350 to −50 MPa. The metabolic activation of a quiescent seed starts with the absorption of water, which is majorly determined by the tissues surrounding the radicle. In soft-coated seeds (meaning water-permeable seed coat) water imbibition initially is almost entirely a physical phenomenon up to the start of a steady phase II.

The weakening of these tissues by enzymatic action is a key event regulating the timing of radicle emergence. And it is suggested that endo-β-mannanase is involved in the weakening of tissues in this process. It is still speculated that it may not be the only determinantal factor for radicle emergence (Loıc et al. 2012) as several other factors are also influencing the radicle emergence under natural conditions.

The process of seed germination, thus, may be categorised into three major phases: phase I, characterised by rapid water imbibition; phase II, characterised by reactivation of metabolism, cell elongation and chitting; and phase III, characterised by rapid cell division and coinciding with radicle growth. The phase II is most crucial during which physiological and biochemical processes such as hydrolysis of food reserves, macromolecular biosynthesis, respiration, reorganisation of subcellular structures, and cell elongation are reactivated that play a key role in the onset of germination (Bonsager et al. 2010). In addition, phase II is also critical in invigoration of seed after a period of drying (see chapter "Seed Vigour and Invigoration" for more). After the onset of germination, stored and de novo synthesised messenger RNAs (mRNAs) play important roles, as germination relies both on the quality and specificity of the transcription and translational apparatus and on posttranslational modifications such as carbonylation, phosphorylation and ubiquitination (Loıc et al. 2012).

3.2.1 Hydrolytic Enzymes and Seed Germination

Various metabolic processes, including activation or synthesis of hydrolytic enzymes for breaking down of stored starch, protein, lipid, hemicellulose, polyphosphates and other reserve food materials into simple carbohydrates, get activated in a fully imbibed seed. There is an elevation of oxygen requirement, with the rise in the activities of mitochondrial enzymes involved in the Krebs cycle and electron transport chain (Mayer and Poljakoff-Mayber 1989). The hydrolysed food material is the source of energy that provides the carbon skeleton for growth and development of juvenile seedling.

Such metabolic processes during seed germination in several crop species have been reviewed from time to time by several workers (Loıc et al. 2012; Renu 2018).

3.2.2 Hydrolysis of Starch

In monocot seed carbohydrates represent the primary storage reserve. It is well documented that α-amylase in the aleurone layers of the monocot seeds plays an important role in hydrolysing the endosperm starch into sugars, which provide the energy for the growth of roots and shoots. Starch, stored in the endosperm is reported to be hydrolysed rapidly during germination, in *Oryza sativa* (Palmiano and Juliano 1972), *Sorghum bicolour* (Elmaki et al. 1999) and *Avena sativa* (Xu et al. 2011) seeds. In these seeds, most hydrolytic enzymes are produced in the aleurone or scutellum in response to germination signals. Kaneko et al. (2002) using mutants defective in shoot formation, and epithelium cell development reported that synthesis of active GA in the epithelium is important for α-amylase expression in the endosperm. In addition, the role of calcium might be expected to involve amylase stability, and to have a much more complex involvement in regulating enzyme activities. The amylase activity is regulated by the concentration of reducing sugars in vivo in both cotyledons and axes which increases gradually, while the starch decreases (Wang et al. 2011). Though starch, protein and fat reserves in dry seeds were not found significantly correlated with germination percentage or speed of germination, soluble sugars and soluble protein contents at different germination stages are reported to be positively correlated with germination rate in some grassland species (Zhao et al. 2018).

3.2.3 Hydrolysis of Proteins

During seed germination and seedling development, hydrolysis of proteins from reserve food material generates free amino acids, these are used as the building blocks for protein synthesis in endosperm and embryo (Tully and Beevers 1978). In addition, carboxypeptidase in combination with cysteine proteinase facilitates the flow of protein hydrolysis. The other key enzymes catalysing protein hydrolysis and biosynthesis in germinating seed are proteinases, proteases, legumin-like proteinase (LLP), metalloproteinase and aminopeptidases.

3.2.4 Hydrolysis of Lipids

During the process of germination in the oil-rich seeds, simple sugars are the end products of the lipid breakdown, which in growing tissues acts as an important regulatory agent. In developing legume seeds a strong negative association is commonly seen between accumulation of storage proteins and storage lipids. Lipids are stored in the form of triacylglycerols in oleosomes and may comprise 35–40% of seed dry weight in oil-rich seeds (Graham 2008). As the germination progresses, triacylglycerols are hydrolysed to free fatty acids and glycerol molecules. The glycerol, thus released, is further metabolised to the glycolytic intermediate dihydroxyacetone phosphate (DHAP) for the synthesis of sugars, amino acids (mainly asparagine, aspartate, glutamine and glutamate) and carbon chains required for other macromolecules for sustaining the embryonic growth (Quettier and Eastmond

2009). The flow of lipid-derived carbon skeletons to amino acids was illustrated in lupin seeds, and at least four alternative or mutually complementary pathways of carbon flow from the breakdown of storage lipids to newly synthesised amino acids have been suggested (Sławomir et al. 2015). Thus, lipases are the most important group of hydrolytic enzymes associated with the lipid metabolism during germination which catalyse the hydrolysis of ester carboxylate bonds releasing fatty acids and organic alcohols (Perreira et al. 2003). There are reports of a sugar-dependent lipase in Arabidopsis (Quettier and Eastmond 2009), activity of which may be regulated by the available sugars. However, the activities of the key enzymes in glycolysis, pentose phosphate pathway (PPP), the tricarboxylic acid cycle (TCA cycle), and amino acid metabolism investigated during germination in several species, reveal that glyoxylate cycle predominated over the TCA-cycle pathway in seed mitochondria.

The glyoxylate cycle (GC) plays a crucial role in breaking down the lipids in germinating seeds. The beta-oxidation of fatty acids releases acetyl CoA, which is metabolised to produce 4- and 6-C compounds. Thus, besides providing NADPH for biosynthetic reactions in germinating seeds (Perino and Come 1991), in the process of GC, the stored lipids are converted to glucose (gluconeogenesis), the main respiratory substrate utilised during germination and seedling establishment (Muscolo et al. 2007). The activity of the key enzymes of glyoxylate-cycle, i.e., isocitrate lyase and malate synthetase found to be remarkably high during seed germination after the emergence of radicle and reaching a maximum at 8 days in a sesame an oil-rich seed (Hyun-Jae et al. 1964). The action of the two glyoxylate cycle enzymes isocitratelyase (ICL) and malate synthase (MS) is essential in the process of germination oilseeds in bypassing the decarboxylation steps of the TCA cycle. In this reaction, 2 mol of acetyl-CoA are introduced with each turn of the cycle, resulting in the synthesis of 1 mol of the four-carbon compound succinate that is transported from the glyoxysome into the mitochondrion and converted into malate via TCA cycle. The malate, thus formed, is exported to cytosol in exchange for succinate and is converted to oxaloacetate. In addition, phosphoenol pyruvate carboxykinase (PEP-CK) catalyses the conversion of oxaloacetate to phosphoenol-pyruvate which fuels the synthesis of soluble carbohydrates necessary for germination (Muscolo et al. 2007).

3.2.5 Hydrolysis of Phytic Acid

The phytic acid (inositol hexaphosphate or IP6) is the principal form of storage of total phosphorus (P) in legumes and cereal seeds (Jacela et al. 2010). Phytin is mainly stored in protein bodies in the aleurone layer and scutellum cells of most seeds. In germinating seeds phytin is hydrolysed by phytase, an acid phosphatase enzyme (Raboy et al. 1991) releasing phosphate, cations and inositol which are utilised by the developing seedling. In addition, IP6-related compounds such as pyrophosphate containing inositol phosphates (PP-IP) play a vital role in providing Pi for ATP synthesis during the early stages of germination. Phosphate metabolism is one of the negatively affected processes under various stressful environments. Under stressed conditions, the restriction of growth and phosphorus availability result in enhancement of the activity of phosphatases to produce Pi by hydrolysis that modulate mechanism of free phosphate uptake.

The hydrolysis of major seed reserves during germination can be broadly classified as:

Stored seed reserve	Key enzymes	Primary products
Starch	Amylases, glucosidases	Glucose
Proteins	Proteinases	Amino acids
Fats	Lipases	Fatty acids and glycerol
Phytin	Phytase	Inositol, PO_4^{2-}, Ca^{2+}, Mg^{2+}

In addition, complex carbohydrates such as hemicelluloses, mannans and galactomannans, which are primarily structural components of the cell wall, and play an important role in seed germination, are also known to undergo hydrolysis and provide respiratory substrate to the germinating seeds in some species (Fincher 1989; Pandey et al. 2009).

3.3 ROS Function

Oxygen, the basic life support system for respiration during germination, also generates reactive oxygen species (ROS), such as singlet oxygen ($^1O_2$), superoxide ($O_2^{\bullet-}$) or hydroxyl ($\cdot OH$) radicals, and hydrogen peroxide (H_2O_2). These are considered as the main ROS involved in cellular signalling (Petrov and Van Breusegem 2012). The ROS is accumulated in seed during dry storage also, but much of these are quenched during the initial water imbibition phase. The controlled generation of ROS plays an important role in the perception and transduction of environmental conditions that control germination. When these conditions are permissive for germination, ROS levels are maintained at a level which triggers cellular events associated with germination, such as hormone signalling. It has also been reported that the spatiotemporal regulation of ROS production, in combination with hormone signalling, influences cellular events involved in cell expansion during germination (Christophe 2019). Accumulation of H_2O_2 on the other hand, in association with the oxidative damages to the antioxidant machinery, is regarded as the source of stress that may suppress germination (Chmielowska-Bkak et al. 2015). Matilla (2020) suggested a link between the homeostasis of the reactive oxygen species (ROS) and seed after-ripening (AR) process, wherein the oxidation of a portion of seed long-lived (SLL) mRNAs appears to be related to dormancy release.

3.4 Nitrogenous Compounds and Seed Germination

Promotion of seed germination by the exogenous application of nitrates is known in a wide range of plant species. The effect of nitrate compounds on promotion of seed germination is best realised in combination with other optimising factors such as temperature or light (Eremrena and Mensah 2016). Among all such compounds potassium nitrate (KNO_3) is the most widely used chemical for releasing seed dormancy and promoting germination. For this purpose, the Association of Official Seed Analysts (AOSA) and the International Seed Testing Association (ISTA) recommend solutions of 0.1–0.2% KNO_3 for optimising germination tests of many species, which show low/shallow dormancy patterns (see chapter "Testing Seed for Quality"). The detrimental effect of salts on seed germination, seedling growth, mitotic activity and chromosomal aberrations were found significantly reduced by the application of KNO_3 (Kursat et al. 2017). Similarly, KNO_3 solutions are among the most commonly applied priming chemicals for enhancing seed germination by improving uniformity and speed of germination (Shin et al. 2009; Thongtip et al. 2022).

3.5   Mobilisation of Reserve Food Material

Once the seed storage reserves are hydrolysed into usable metabolites, mobilisation of these from storage tissues to growing points is a critical step ensuring supply of nutrients for the growth of the developing seedling until it becomes autotrophic. This process determines early vigour in developing seedling for crop establishment under field conditions. The first step in the mobilisation of food reserve is hydrolysis of sugar and to provide it to the germinating seed embryo. In addition to the hydrolysis of carbohydrates in the endosperm or cotyledons, it is also released from the catabolism of lipids. Similarly, a part of the complex carbohydrates are mobilised and utilised by the seed to support early seedling development (Bewley et al. 2013). Similarly, sucrose constitutes a significant reserve in the scutellum which is efficiently consumed by the growing embryonic axis, as seen in maize seed, where a net flow of sucrose takes place from the scutellum to the growing embryo axis during germination. Sucrose and hexose transporters, as well as H^+-ATPase, become fundamental in the transport of nutrients during radicle elongation, required for successful germination (Sanchez-Linares et al. 2012). In cereals, where starch is the major food reserve, starch—the primary source of nutrients stored in the endosperm—are broken down by enzymes synthesised and secreted by the aleurone and scutellar tissues, and transported to the endosperm. The sugars are transported back to nourish the growing embryo. However, often the increase in soluble sugars in the embryo is not concomitant with a decrease in starch in the endosperm, suggesting that sugars originated mostly from the catabolism of lipids.

In oil-rich seeds such as sunflower in which soluble sugars and starch constitute only about 2.2% of the cotyledon dry mass, lipids are the main seed reserves mobilised during germination, while proteins are the second most utilised reserves (Alencar et al. 2012). In groundnut seeds lower seed-weight was found associated with insufficient supply of reserve food materials resulting into poor vigour of seedlings (Nautiyal 2009; Nautiyal and Yadav 2019). In addition, the hormonal metabolism and signalling leads the seed to germinate. Germination essentially depends on the resumption of cell cycle after a period of quiescence. Hence, entry into G1 and progression from G1 to S states may represent an important control of the cellular events in early seedling development (Maria et al. 2005). Based on the expression analysis of cell cycle genes with mRNA in situ localisation, β-glucuronidase assays, and semi-quantitative reverse transcription-polymerase chain reaction (RT-PCR), Maria et al. (2005) showed that transcription of most cell cycle genes was detected only after completion of germination. They, however, suggested spatial and temporal expression profiles of cell cycle control genes of importance in germinating seeds of *Arabidopsis* and white cabbage (*Brassica oleracea*).

3.6   Hormone Metabolism and Signalling

To understand the molecular mechanisms underlying the process of after-ripening (seed dry storage) in triggering hormone-related changes and dormancy decay in wheat, temporal expression patterns of genes related to abscisic acid (ABA), gibberellin (GA), jasmonate and indole acetic acid (IAA), their role in signalling and levels of the respective hormones were established in dormant and after-ripened seeds in both dry and imbibed states (Liu et al. 2013; Wang et al. 2020).

To unravel the complex interplay of plant hormones and novel signalling molecules that regulate germination, and to determine the mechanisms of signal transduction are the major aims of seed biology studies. Studies on the involvement of plant hormones have revealed the novel aspects of seed physiology and the complex metabolic balance between multiple hormones during germination and development (Preston et al. 2009; Kanno et al. 2010). Of various plant hormones, ABA and GA unquestionably play the most critical roles in regulating the processes of dormancy and germination. ABA induces the expression of LEA proteins, which become abundant during the late stage of seed maturation, which is characterised by rapid dehydration. Hence, LEA proteins are suggested to act as chaperones to protect macromolecular structures against desiccation injury (Nautiyal and Shono 2010). In addition, it also exerts an inhibitory effect on the metabolic processes triggering precocious germination of developing seeds on the mother plant (pre-harvest sprouting) thereby allowing the maturation process to be completed and maintaining the economic value as well as quality of seeds at harvest maturity. However, to germinate successfully a non-dormant seed must undergo certain changes counteracting the inhibitory effects of ABA and synthesise tetracyclic diterpenes, the gibberellins (GAs), which are essential to activate germination (Loïc et al. 2012), decrease the sensitivity and biosynthesis inhibition to reduce the active level of ABA (Faiza et al. 2021). In addition, synthesise tetracyclic diterpenes, the gibberellins (GAs), which are essential germination activators (Loïc et al. 2012). High levels of bioactive ABA found in imbibed dormant seeds of *Arabidopsis thaliana* (Ali-Rachedi et al. 2004) further confirmed the idea that the ABA/GA ratio regulates the metabolic transition required for germination.

Production of GA in the embryos of the germinating wheat seed, on the other hand, was found to diffuse to aleurone layer initiating a signalling cascade in the cellular system. This result in the synthesis of α-amylases and other hydrolytic enzymes which are then transported to the endosperm to hydrolyse the food reserves. A variety of cellular processes in plants are under the control of phytohormones which play key roles and coordinate various signal transduction pathways under abiotic stress, as well as regulating germination. During the early phase of seed germination in *Arabidopsis*, a decrease in the contents of germination inhibitors jasmonic acid (JA) and salicylic acid (SA) and an increase in the levels of auxins were recorded (Ali-Rachedi et al. 2004), suggesting that both JA and SA act as negative regulators of germination. Auxins are considered to be regulating the process of germination via a crosstalk with GAs, ABA and ethylene (ET). Similarly, the brassinosteroid signals could stimulate germination by decreasing the sensitivity to ABA. Molecular and genetic bases of hormonal interactions or crosstalk regulating seed germination have recently been reviewed by Gerardo et al. (2020).

Further, there are reports suggesting epigenomic modulation of the expression of genes regulating or influencing dormancy, maturation and germination. For example, specific chromatin modifiers and re-modellers have been reported to promote seed dormancy or germination by enhancing or repressing the expressions of specific gene subsets, respectively (Rajjou et al. 2004; Loïc et al. 2012; Wang et al. 2013; Matilla 2020). Also, dry seeds are shown to accumulate RNAs whose abundance change towards a "germination-friendly" transcriptome (Nakabayashi et al. 2005). Similarly, Karrikins, a group of plant growth regulators found in the smoke of burning plant material is reported to stimulate seed germination in *Arabidopsis* spp. by enhancing light response (Nelson et al. 2010), mimicking a signalling hormone.

3.7 Ethylene and Other Growth Regulators

Ethylene (ET), the smallest gaseous hormone, plays an important role via crosstalk with other hormones in various activities of plants such as seed germination, seed dormancy, plant growth and developmental processes. It can stimulate seed germination and overcome dormancy in many species. For instance, the inhibitory effects of high temperature on seed germination of lettuce can be overcome by exogenous ethylene. Significant progress has been made in unveiling ET crosstalk with other hormones and environmental signals, such as light (Matilla 2007; Jalal et al. 2020; Loïc et al. 2012). Similarly, several other plant hormones, i.e., brassinosteroids, salicylic acid, cytokinin, auxin, jasmonic acid and oxylipins are also known to influence seed germination process in different ways. These hormones may form an interlocked signalling network interacting with one another and influencing seed germination directly or indirectly. Their response is particularly significant in response to environmental stresses.

3.8 Environmental Factors Influencing Seed Germination

Regulation of germination (and dormancy) by environmental factors is considered to be an adaptive mechanism of evolutionary significance, and has assumed a greater significance in the present scenario of climate change. The response to any type of stress is related to the application of an external factor that exercises a detrimental influence on the metabolism of the plant tissue, causing injury, disease or physiological abnormalities (Jaleel et al. 2009). Hence, abiotic and biotic stresses can significantly influence seed germination, seedling growth and crop stand leading to a decline in biological and economic yields.

Under field conditions, optimum soil moisture is a prerequisite for satisfactory seed germination. Upon sowing in the field seeds start absorbing water to prepare for germination. The amount of water absorbed by the seed will depend on the water-energy status of the seed and the soil water potential (Bradford 1995). The condition of deficit water during seed germination may delay primary root protrusion, reduce the percentage of seeds that complete the process, or completely inhibit germination (Lei et al. 2017). At low soil moisture levels, thus, the initial stages of germination may initiate, but may not be sufficient to complete the process of germination. On the other hand, excessive moisture level inhibits germination by creating anoxia (i.e., depletion of oxygen in the soil). Similarly, the reduced germination caused by soil salinity results from the combined action of two types of stress: the water deficit created by the osmotic effect of the salt in the soil, also known as 'osmotic drought', and the toxicity as a result of the excessive influx of ions, such as Cl^- and Na^+ into the tissues (Munns et al. 1995; Zhu 2003). In addition, under saline conditions, osmotic and ionic stress leads to the excessive production of reactive oxygen species (ROS) in chloroplasts, mitochondria and the apoplastic space (Nazar et al. 2011). As the sensitivity to salinity is more pronounced at the seedling stage, it can be used in some crops as a reliable criterion for the selection of genotypes with better tolerance to salinity stress.

Large thermal variations that occur during the day, when the air or the soil surface temperatures reach 40 °C or more at mid-day, can considerably reduce the germination of hydrated seeds (Boero et al. 2000). Every crop species exhibit specific requirement of the optimum, maximum, minimum and base temperatures for germination. Hence, the base temperature for crops cultivated during winter season is lower than that for summer/rainy season. The effects of temperature on the rate of germination and the total germination have been explained exhaustively by Mayer and Poljakoff-Mayber (1989), which are also applied in standardisation of germination test procedures. The combined effects of water stress and temperature may vary among species (Benech-Arnold Roberto and Augusto 2017). Besides, germination is greatly affected by the interactions between temperature, water potential, and water flow within the soil and by variations in the Q_{10} factors (i.e., temperature coefficient—a method of comparing effective seed biological activity rates).

Further, the aeration regimes (i.e., rates of gaseous exchange) in the soil affect the biological activity of the microflora and enhance the competition for oxygen with germinating seeds. As oxygen is required in germination as a terminal electron receptor in respiration and other oxidative processes of a regulatory nature, low oxygen availability can reduce or even prevent germination in many crop species (Bradford et al. 2007). Mayer and Poljakoff-Mayber (1989) reported that there is a sharp rise in oxygen requirement for metabolic activities at an early stage of germination. It is followed by a second peak that marks the beginning of the growth stage and radicle emergence. Also, oxygen supply is greatly influenced by the thickness of the water film covering the germinating seed, and of the hydrated seed coat, especially in seeds that have a swollen mucilaginous cover with very low diffusivity to oxygen. In seeds rich in fatty or starchy storage substances, germination is inhibited if the oxygen level falls below 2%. Oxygen requirements also increase with the rise in soil temperature and under light and/or water stress. Therefore, good aeration and gaseous exchange attained in well-structured, aggregated soil beds support healthy growth of germinating seeds, while soil crusting and compaction may have deleterious effects on gaseous exchange and, in turn, on seed germination, besides exerting mechanical pressure by the germinating seeds.

Germination testing protocols, a vital component of seed quality assurance, are therefore, standardised taking into account the type of germination (epigeal or hypogeal) and root morphology (adventitious/fibrous and tap root systems) to identify the most suitable substratum; optimum temperature regimes; requirement of light or other triggers; dormancy behaviour (deep or shallow; coat imposed or embryo dormancy; hard seededness; requirement of after-ripening or stratification based on morpho-physiological behaviour, etc.); speed of germination, and time to complete seedling development to derive at the days of first and final counting days (see chapter "Testing Seed for Quality" for more details).

4   Conclusions and Future Thrust Areas

Dormancy is the mechanism that supports seed to escape unfavourable environmental conditions for germination and seedling establishment, and favours survival and distribution over time and space. However, for successful crop cultivation, it is important to understand various processes of seed dormancy that are regulating the germination. Moreover, to mitigate the serious threats of climate change, a low degree of seed dormancy with subsequent rapid germination could be an important strategy for crops grown in arid and semi-arid regions. On the other hand, by adopting ways and means to manipulate dormancy through external treatments, it could be possible to maintain seeds in quiescent state for as long as the growing conditions are not favourable. It is well established that during and after germination, early seedling growth is supported by the catabolism of stored reserves (proteins, lipids, or starch) accumulated during seed maturation, which requires necessary attention of seed technologists and agronomists to ensure satisfactory conditions during grain filling stage, especially in seed multiplication plots. The molecular pathways, recognised by omics and molecular biology analyses, may elucidate more about the effects of plant hormones on seed germination and dormancy. In addition, the role of soil bacteria interacting with plant hormones, and hence promoting seed germination, also needs to be explored for better use as an effective tool to enhance seed germination under field conditions.

References

Alencar NL, Innecco R, Gomes-Filho E, Gallao MI, Alvarez-pizarro JC, Prisco JT et al (2012) Seed reserve composition and mobilisation during germination and early seedling establishment of *Cereus jamacaru* D.C. ssp. *jamacaru* (Cactaceae). An Acad Bras Cienc 84:823–832. https://doi.org/10.1590/S0001-37652012000300024

Ali-Rachedi S, Bouinot D, Wagner MH, Bonnet M, Sotta B et al (2004) Changes in endogenous abscisic acid levels during dormancy release and maintenance of mature seeds: studies with the Cape Verde Islands ecotype, the dormant model of *Arabidopsis thaliana*. Planta 219:479–488

Alvarado V, Hiroyaki H, Bradford KJ (2000) Expression of endo--mannanase and SNF-related protein kinase genes in true potato seeds in relation to dormancy, gibberellin and abscisic acid. In: Viemont JD, Crabbe J (eds) Dormancy in plants: from whole plant behavior to cellular control, vol 26. CABI Publishing, Wallingford, pp 347–364

Amen RD (1968) A model of seed dormancy. Bot Rev 34:1–31

Bandyopadhyay A, Nautiyal PC, Radhakrishnan T, Gor HK (1999) Role of testa, cotyledons and embryonic axis in seed dormancy of groundnut (*Arachis hypogaea* L.). J AgroCrop Sci 182:37–41

Baskin CC, Baskin JM (1998) Seeds–ecology, biogeography, and evolution of dormancy and germination. Academic Press, San Diego, CA

Baskin JM, Baskin CC (2004) A classification system for seed dormancy. Seed Sci Res 14:1–16

Baskin JM, Baskin CC (2021) The great diversity in kinds of seed dormancy: a revision of the Nikolaeva–Baskin classification system for primary seed dormancy. Seed Sci Res 34(4): 249–277. https://doi.org/10.1017/S096025852100026X

Baskin Jerry M, Baskin C, Li X (2000) Taxonomy, anatomy and evolution of physical dormancy in seeds. Plant Spec Biol 15(2):139–152. https://doi.org/10.1046/j.1442-1984.2000.00034.x

Baskin CC, Zackrisson O, Baskin JM (2002) Role of warm stratification in promoting germination of seeds of Empetrumhermaphroditum (Empetraceae), a circumboreal species with a stony endocarp. Am J Bot 89:486–493

Bassel GW, Zielinska E, Mullen RT, Bewley JD (2004) Down-regulation of DELLA genes is not essential for germination of tomato, soybean, and *Arabidopsis* seeds. Plant Physiol 1366(1): 2782–2789

Batlla D, Kruk BC, Benech-Arnold RL (2004) Modelling changes in dormancy in weed soil seed banks: implications for the prediction of weed emergence. In: Benech-Arnold RL, Sanchez RA (eds) Handbook of seed physiology: applications to agriculture. Food Product Press and the Haworth Reference Press, New York, NY, pp 245–270

Beaudoin N, Serizet C, Gosti F, Giraudat J (2000) Interactions between abscisic acid and ethylene signaling cascades. Plant Cell 12:1103–1115

Benech-Arnold Roberto L, Augusto SR (2017) Modeling weed seed germination. In: Kigel J (ed) Seed development and germination. Routledge, London, pp 545–566. https://doi.org/10.1201/9780203740071-21

Bewley JD (1997) Breaking down the walls – a role for endo-β-mannanase in release from seed dormancy. Trends Plant Sci 2:464–469

Bewley JD, Black M (1994) Seeds physiology of development and germination, 2nd edn. Plenum Press, New York, NY

Bewley JD, Bradford Kent J, Hilhorst HWM, Nonogaki H (2013) Seeds: physiology of development, germination and dormancy, 3rd edn. Springer, New York, NY. https://doi.org/10.1007/978-1-4614-4693-4

Boero C, González J, Prado F (2000) Efecto de la temperatura sobre la germinación de diferentes variedades de quinoa (*Chenopodium quinoa* Willd.). Lilloa 40:103–108

Bonsager BC, Shahpiri A, Finnie C, Svensson B (2010) Proteomic and activity profiles of ascorbate–glutathione cycle enzymes in germinating barley embryo. Phytochemistry 71: 1650–1656

Bradford KJ (1995) Water relations in seed germination. In: Kigel J, Galili G (eds) Seed development and germination. Marcel Dekker, Inc, New York, NY, pp 351–396

Bradford KJ, Come D, Corbineau F, Kent J (2007) Quantifying the oxygen sensitivity of seed germination using a population-based threshold model. Seed Sci Res 17:33–43. https://doi.org/10.1017/S0960258507657389

Castillon A, Shen H, Huq E (2007) Phytochrome interacting factors: central players in phytochrome-mediated light signaling networks. Trends Plant Sci 126(1):514–521

Chao S, Mitchell J, Fukai S (2021) Factors determining genotypic variation in the speed of rice germination. Agronomy 11:1614. https://doi.org/10.3390/agronomy11081614

Chmielowska-Bkak J, Izbiańska K, Deckert J (2015) Products of lipid, protein and mRNA oxidation as signals and regulators of gene expression. Front Plant Sci 6:405. https://doi.org/10.3389/fpls.2015.00405

Christophe B (2019) The signalling role of ROS in the regulation of seed germination and dormancy. Biochem J 476:3019–3032. https://doi.org/10.1042/BCJ20190159

Corbineau F, Come D (2003) Involvement of energy metabolism and ABA in primary and secondary dormancies in oat (*Avena sativa* L.) seeds—a physiological approach. In: Nicolas G, Bradford KJ, Come D, Pritchard H (eds) The biology of seeds: recent research advances. CAB International, Oxon, pp 113–120

Elmaki HB, Babiker EE, Tinay AHE (1999) Changes in chemical composition, grain malting, starch and tannin contents and protein digestibility during germination of sorghum cultivars. Food Chem 64:331–336. https://doi.org/10.1016/S0308-8146(98)00118-6

Eremrena PO, Mensah SIJ (2016) Effect of plant growth regulators and nitrogenous compounds on seed germination of pepper (Capsicum frutescens L) Appl Sci Environ Manage 20(2):242–250. www.ajol.info, www.bioline.org.br/ja

Faiza A, Qanmber G, Li F, Wang Z (2021) Updated role of ABA in seed maturation, dormancy, and germination. J Adv Res 35:199. https://doi.org/10.1016/j.jare.03.011. www.elsevier.com/locate/jare

Fennimore SA, Foley ME (1998) Genetic and physiological evidence for the role of gibberellic acid in the germination of dormant *Avena fatua* seeds. J Exp Bot 49:89–94

Fincher GB (1989) Molecular and cellular biology associated with endosperm mobilisation in germinating cereal grains. Annu Rev Plant Physiol Plant Mol Biol 40:305–346

Finch-Savage William E, Leubner-Metzger G (2006) Seed dormancy and the control of germination. New Phytol 171:501–523. https://doi.org/10.1111/j.1469-8137.2006.01787.x

Garello G, Barthe P, Bonelli M, Bianco-Trinchant J, Bianco J, Le Page-Degivry MT (2000) Abscisic acid-regulated responses of dormant and non-dormant embryos of *Helianthus annuus*: role of ABA-inducible proteins. Plant Physiol Biochem 38:473–482

Gendreau E, Romaniello S, Barad S, Leymarie J, Benech-Arnold R, Corbineau F (2008) Regulation of cell cycle activity in the embryo of barley seeds during germination as related to grain hydration. J Exp Bot 59:203–212

Gerardo C-C, Calleja-Cabrera J, Pernas M, Gomez L, Oñate-Sánchez L (2020) An updated overview on the regulation of seed germination. Plants 9:703. https://doi.org/10.3390/plants9060703. www.mdpi.com/journal/plants

Goosey L, Palecanda L, Sharrock RA (1997) Differential patterns of expression of the Arabidopsis PHYB, PHYD and PHYE phytochrome genes. Plant Physiol 115:959–969

Graham IA (2008) Seed storage oil mobilization. Annu Rev Plant Biol 59:115–142

Grahl A (1965) Lichteinfluss auf die keimung des getreides in abhangigkeit von der keimruhe. Landbauforschung Volkenrode 2:97–106. (in Russian)

Groot SPC, Karssen CM (1992) Dormancy and germination of abscisic acid-deficient tomato seeds. Plant Physiol 99:952–958

Gualano NA, Del Fueyo PA, Benech-Arnold RL (2014) Potential longevity (Ki) of malting barley (*Hordeum vulgare* L.) grain lots relates to their degree of pre-germination assessed through different industrial quality parameters. J Cereal Sci 60:222–228

Hilhorst HWM (2007) Definitions and hypotheses of seed dormancy. In: Bradford K, Nonogaki H (eds) Seed development, dormancy and germination. Annual plant reviews, vol 27. Blackwell Publishing Ltd, Oxford, pp 50–71. https://doi.org/10.1002/9780470988848.ch3

Hilhorst HWM, Finch-Savage WE, Buitink J, Bolingue W, Leubner-Metzger G (2010) Dormancy in plant seeds. In: Lubzens E, Cerdà J, Clarck M (eds) Dormancy and resistance in harsh environment. Springer, Berlin, pp 43–67. https://doi.org/10.1007/978-3-642-12422-8_4

Hiroyuki N, Barrero Jose M, Li C (2018) Editorial: Seed dormancy, germination, and pre-harvest sprouting. Front Plant Sci 9:1783

Hoang HH, Sotta B, Gendreau E, Bailly C, Leymarie J, Corbineau F (2012) Water content: a key factor of the induction of secondary dormancy in barley grains as related to ABA metabolism. Physiol Plant 148:284. https://doi.org/10.1111/j.1399-3054.2012.01710.x

Hoang HH, Sechet J, Bailly C, Leymarie J, Corbineau F (2014) Inhibition of germination of dormant barley (*Hordeum vulgare* L.) grains by blue light as related to oxygen and hormonal regulation. Plant Cell Environ 37:1393–1403. https://doi.org/10.1111/pce.12239

Holdsworth M, Kurup S, McKibbin R (1999) Molecular and genetic mechanisms regulating the transition from embryo development to germination. Trends Plant Sci 4:275–280

Holdsworth MJ, Bentsink L, Soppe WJJ (2008) Molecular networks regulating *Arabidopsis* seed maturation, after-ripening, dormancy and germination. New Phytol 179:33–54

Huang Z, Olçer-Footitt H, Footitt S, Finch-Savage WE (2015) Seed dormancy is a dynamic state: variable responses to pre- and post-shedding environmental signals in seeds of contrasting Arabidopsis ecotypes. Seed Sci Res 25:159–169. https://doi.org/10.1017/S096025851500001X

Hyun-Jae L, Kim SJ, Lee KB (1964) Study on the glyoxylate cycle in germinating sesame seed embryos. Arch Biochem Biophys 107(3):479–484. https://doi.org/10.1016/0003-9861(64)90304-2

Jacela JY, De Rouehey Tokach MD, Goodband RD, Nelssen JL, Renter D, Dritz SS (2010) Feed additives for swine: fact sheets-prebiotics and probiotis and phytogenics. J Swine Health Prod 18:87–91

Jalal AG, Saikat G, Mitra M, Youxin Y, Xin L (2020) Role of ethylene crosstalk in seed germination and early seedling development: a review. Plant Physiol Biochem 151(Suppl l):124. https://doi.org/10.1016/j.plaphy.2020.03.016

Jaleel CA et al (2009) Drought stress in plants: a review on morphological characteristics and pigments composition. Int J Agric Biol 11(1):100–105

Jardim A, Pereira dos Santos D, Nunes da Piedade AR, Quelvia de Faria G, Amaral da Silva R, Pereira EA, Sartori MM (2021) The use of the generalised linear model to assess the speed and uniformity of germination of corn and soybean seeds. Agronomy 11:588. https://doi.org/10.3390/agronomy11030588

Jayasuriya KM, Baskin GG, Geneve RL, Baskin CC (2009) Sensitivity cycling and mechanism of physical dormancy break in seeds of *Ipomoea hederacea* (Convolvulaceae). Int J Plant Sci 170(4):429–443. https://doi.org/10.1086/597270

Kaneko M, Itoh H, Ueguchi-Tanaka M, Ashikari M, Matsuoka M (2002) The alpha-amylase induction in endosperm during rice seed germination is caused by gibberellin synthesised in epithelium. Plant Physiol 128(4):1264–1270. https://doi.org/10.1104/pp.010785

Kanno Y, Jikumaru Y, Hanada A, Nambara E, Abrams SR, Kamiya Y et al (2010) Comprehensive hormone profiling in developing *Arabidopsis* seeds: examination of the site of ABA biosynthesis, ABA transport and hormone interactions. Plant Cell Physiol 51:1988–2001

Karin W, Muller K, Leubner-Metzger G (2011) Darwin review first off the mark: early seed germination. J Exp Bot 62(10):3289–3309. https://doi.org/10.1093/jxb/err030

Keeley JE, Fotheringham CJ (2000) Role of fire in regeneration from seeds. In: Plant Communities, Fenner M (eds) Seeds: the ecology of regeneration. CAB International, Wallingford, pp 311–330

Kepczynski J, Kepczynska E (1997) Role of enzymes in seed germination. Ethylene in seed dormancy and germination. Physiol Plant 101:720–726

Ketring D, Morgan PW (1969) Ethylene as a component of emanations from germinating peanut seeds and it's effect on dormant Virginia type sees. Plant Physiol 44:326–330

Kim DH, Yamaguchi S, Lim S, Oh E, Park J, Hanada A, Kamiya Y, Choi G (2008) SOMNUS, a CCCH-type zinc finger protein in *Arabidopsis*, negatively regulates lightdependent seed germination downstream of PIL5. Plant Cell 206(1):12

King Keith E (2003) Analysis of the effects of hypogeal and epigeal emergence on seedling competition in legumes. McCabe Thesis Collection. Paper 20. http://dclu.langston.edu/mccabe_theses

Klupczyńska EA, Pawłowski TA (2021) Regulation of seed dormancy and germination mechanisms in a changing environment. Int J Mol Sci 22(3):1357. https://doi.org/10.3390/ijms22031357

Kumar N, Goel N (2019) Phenolic acids: natural versatile molecules with promising therapeutic applications. Biotechnol Rep 24:e00370. https://doi.org/10.1016/j.btre.2019.e00370

Kursat C, Cadıl S, Cavusoglu D (2017) Role of potassium nitrate (KNO_3) in alleviation of detrimental effects of salt stress on some physiological and cytogenetical parameters in *Allium cepa* L. Cytologia 82(3):279–286

Lawrence C, Miller MD (2001) Principles of seed science and technology. Springer, New York, NY, pp 17–38. https://doi.org/10.1007/978-1-4615-1619-4

Le Page-Degivry MT, Garello C (1992) In situ abscisic acid synthesis. Plant Physiol 98:1386–1390

Lei Y, Liu Q, Hettenhausen C, Cao G, Tan Q, Zhao W et al (2017) Salt-tolerant and -sensitive alfalfa (*Medicago sativa*) cultivars have large variations in defense responses to the lepidopteran insect *Spodoptera litura* under normal and salt stress condition. PLoS One 12(7):e0181589. https://doi.org/10.1371/journal.pone

Leymarie J, Robayo-Romero ME, Gendreau E, Benech Arnold RL, Corbineau F (2008) Involvement of ABA in induction of secondary dormancy in barley (*Hordeum vulgare* L.) seeds. Plant Cell Physiol 49:1830–1838

Li B, Foley ME (1994) Differential polypeptide patterns in imbibed dormant and after-ripened *Avena fatua* embryos. J Exp Bot 45:275–279

Liu A, Gao F, Kanno Y, Jordan MC, Kamiya Y, Seo M et al (2013) Regulation of wheat seed dormancy by after-ripening is mediated by specific transcriptional switches that induce changes in seed hormone metabolism and signaling. PLoS One 8:e56570. https://doi.org/10.1371/journal.pone.0056570

Loïc R, Duval M, Gallardo K, Catusse J, Bally J, Job C, Job D (2012) Seed germination and vigor. Annu Rev Plant Biol 63:507–533. https://doi.org/10.1146/annurev-arplant-042811-105550

Mares DJ (1983) Preservation of dormancy in freshly harvested wheat-grain. Aust J Agric Res 34: 33–38

Maria BR, Van Poucke K, Jan HW, De Veylder BL, Groot SPC, Inzé D, Engler G (2005) The role of the cell Cycle machinery in resumption of postembryonic development. Plant Physiol 137(1): 127–140. https://doi.org/10.1104/pp.104.049361

Matilla AJ (2000) Ethylene in seed formation and germination. Seed Sci Res 10:111–126

Matilla Angel J (2007) Ethylene in seed formation and germination. Sci Agric 60(3):601–606

Matilla Angel J (2020) Seed dormancy: molecular control of its induction and alleviation. *Plants* 9(10):1402

Mayer AM, Poljakoff-Mayber A (1989) The germination of seeds. Pergamon, Oxford. 270 p

Montague JL, Barrett SCH, Eckert CG (2008) Re-establishment of clinal variation in flowering time among introduced populations of purple loosestrife (*Lythrum Salicaria*, Lythraceae). J Evol Biol 21:234–245. https://doi.org/10.1111/j.1420-9101.2007.01456.x

Morrison DA, McClay KC, Rish PS (1998) The role of the lens in controlling heat-induced breakdown of testa-imposed dormancy in native Australian legumes. Ann Bot 82:35–40

Munns R, Schachtman D, Condon A (1995) The significance of a two-phase growth response to salinity in wheat and barley. Aust J Plant Physiol 22:561–569

Muscolo A, Sidari M, Mallamaci C, Attina E (2007) Changes in germination and glyoxylate and respiratory enzymes of *Pinus pinea* seeds under various abiotic stresses. J Plant Interact 2(4): 273–279. https://doi.org/10.1080/17429140701713795

Nadella D, Foley ME (2003) Seed dormancy: genetics of dormancy. In: Thomas B (ed) Encyclopedia of applied plant sciences. Elsevier, Amsterdam, pp 1323–1333. https://doi. org/10.1016/B0-12-227050-9/00062-4. ISBN 9780122270505. https://www.sciencedirect.com/ science/article

Nakabayashi K, Okamoto M, Koshiba T, Kamiya Y, Nambara E (2005) Genome-wide profiling of stored mRNA in Arabidopsis thaliana seed germination: epigenetic and genetic regulation of transcription in seed. Plant J 41:697–709

Nautiyal PC (2009) Seed and seedling vigour traits in groundnut (*Arachis hypogaea* L.). Seed Sci Technol 37:721–735

Nautiyal PC, Shono M (2010) Analysis of the role of mitochondrial and endoplasmic reticulum localised small heat-shock proteins in tomato (*Lycopersicon esculentum* Mill.) plant. Biol Plant 54(4):715–719

Nautiyal PC, Yadav SK (2019) Influence of physiological and environmental factors on groundnut seed development, quality and storage: an overview. Seed Res 47(1):1–14

Nautiyal PC, Bandyopadhyay A, Ravindra V (1993) Problems with defining seed dormancy characteristics of groundnut varieties. J Oilseeds Res 10:271–276

Nautiyal PC, Bandyopadhyay A, Zala PV (2001) In situ sprouting and regulation of fresh seed dormancy in Spanish type groundnut (*Archis hypogaea* L. ssp. *fastigiata* var. *vulgaris*). Field Crop Res 70:233–241

Nazar R, Iqbal N, Syeed S, Khan NA (2011) Salicylic acid alleviates decreases in photosynthesis under salt stress by enhancing nitrogen and sulfur assimilation and antioxidant metabolism differentially in two mung bean cultivars. J Plant Physiol 168:807–815

Neill S, Horgan JR, Rees AF (1987) Seed development and vivipary in *Zea mays* L. Planta 58:364

Nelson DC, Flematti GR, Riseborough JA, Ghisalberti EL, Dixon KW et al (2010) Karrikins enhance light responses during germination and seedling development in *Arabidopsis thaliana*. Proc Natl Acad Sci U S A 107:7095–7100

Ngoc PV, Kathare PK, Huq E (2018) Phytochromes and phytochrome interacting factors. Plant Physiol 176:1025–1038

Ordonez-Parra King CA (2022) The great diversity in kinds of seed dormancy. Seed Sci Res 31: 249. https://doi.org/10.1017/S096025852100026X

Palmiano E, Juliano BO (1972) Biochemical changes in the rice-grain during germination. Plant Physiol 49:751–756

Pandey R, Paul V, Dadlani M (2009) Mobilization of seed reserves and environmental control of seed germination. In: Singhal NC (ed) Seed science and technology. Kalyani Publishers, New Delhi, pp 84–114

Parihar SS, Dadlani M, Lal SK, Tonapi VA, Nautiyal PC, Basu S (2014) Effect of seed moisture content and storage temperature on seed longevity of hemp (*Cannabis sativa* L.). Ind J Agric Sci 84(11):1303–1309

Perino C, Come D (1991) Physiological and metabolic study of the germination phases in apple embryo. Seed Sci Technol 19:1–14

Perreira EP, Zanin GM, Castro HF (2003) Immobilisation and catalytic properties of lipase on chitosan for hydrolysis and etherification reactions. Braz J Chem Eng 20(4):343

Petrov VD, Van Breusegem F (2012) Hydrogen peroxide-a central hub for information flow in plant cells. AoB Plants 2012:pls014. https://doi.org/10.1093/aobpla/pls014

Preston J, Tatematsu K, Kanno Y, Hobo T, Kimura M, Jikumaru Y et al (2009) Temporal expression patterns of hormone metabolism genes during imbibition of *Arabidopsis thaliana* seeds: a comparative study on dormant and non-dormant accession. Plant Cell Physiol 50:1786–1800

Quettier AL, Eastmond PJ (2009) Storage oil hydrolysis during early seedling growth. Plant Physiol Biochem 47:485

Raboy V, Noaman MM, Taylor GA, Pickett SG (1991) Grain phytic acid and protein are highly correlated in winter wheat. Crop Sci 31:631–635

Rajjou L, Gallardo K, Debeaujon I, Vandekerckhove J, Job C, Job D (2004) The effect of alpha-amanitin on the *Arabidopsis* seed proteome highlights the distinct roles of stored and neosynthesised mRNAs during germination. Plant Physiol 134:1598–1613

Renu J (2018) Role of enzymes in seed germination. Int J Creat Res Thoughts 6(2): 1481–1485. ISSN: 2320-2882. www.ijcrt.org

Rodriguez María V, Barrero JM, Corbineau F, Gubler F, Benech-Arnold RL (2015) Dormancy in cereals (not too much, not so little): about the mechanisms behind this trait. Seed Sci Res 25:99–119. https://doi.org/10.1017/S0960258515000021

Sanchez-Linares L, Gavilanes-Ruíz M, Díaz-Pontones D, Guzman Chavez F, Calzada-Alejo V, Zurita-Villegas V, Luna-Loaiza V, Moreno-Sanchez R, Bernal-Lugo I, Sanchez-Nieto S (2012) Early carbon mobilisation and radicle protrusion in maise germination. J Exp Bot 63:4513–4526. https://doi.org/10.1093/jxb/ers130

Sedlakova V, Hanacek P, Grulichova M, Zablatzka L, Smykal P (2021) Evaluation of seed dormancy, one of the key domestication traits in chickpea. Agronomy 11:2292. https://doi.org/10.3390/agronomy11112292

Seshu DV, Dadlani M (1991) Mechanism of seed dormancy in rice. Seed Sci Res 1:187–194

Shin J, Kim K, Kang H, Zulfugarov IS, Bae G, Lee CH, Lee D, Choi G (2009) Phytochromes promote seedling light responses by inhibiting four negatively-acting phytochrome-interacting factors. Proc Natl Acad Sci U S A 106:7660–7665

Sławomir B, Ratajczak W, Ratajczak L (2015) Regulation of storage lipid metabolism in developing and germinating lupin (*Lupinus* spp.) seeds. Acta Physiol Plant 37:119. https://doi.org/10.1007/s11738-015-1871-2

Sohn SI, Pandian S, Kumar TS, Zoclanclounon YAB, Muthuramalingam P, Shilpha J, Satish L, Ramesh M (2021) Seed Dormancy and pre-harvest sprouting in rice—an updated overview. Int J Mol Sci 22:11804. https://doi.org/10.3390/ijms222111804

Steber CM, Cooney S, McCourt P (1998) Isolation of the GA-response mutant sly1 as a suppressor of ABI1-1 in *Arabidopsis thaliana*. Genetics 1496(1):509–521

Thongtip A, Mosaleeyanon K, Korinsak S et al (2022) Promotion of seed germination and early plant growth by KNO3 and light spectra in Ocimumtenuiflorum using a plant factory. Sci Rep 12:6995

Tomoko S (1997) Phytochrome regulation of seed germination. J Plant Res 110:151–161

Tully RE, Beevers H (1978) Protease and peptidases of castor bean endosperm. Enzyme characterisation and changes during germination. Plant Physiol 62:726–750

Vleeshouwers LM, Bouwmeester HJ, Karssen CM (1995) Redefining Seed dormancy: an attempt to integrate physiology and ecology. J Ecol 83(6):1031. https://doi.org/10.2307/2261184

Wang T, Sistrunk LA, Leskovar DI, Cobb BG (2011) Characteristics of storage reserves of triploid watermelon seeds: association of starch and mean germination time. Seed Sci Technol 39:318–326. https://doi.org/10.15258/sst.2011.39.2.05

Wang Z, Cao H, Sun Y, Li X, Chen F, Carles A, Li Y, Ding M, Zhang C, Deng X et al (2013) *Arabidopsis* paired amphipathic helix proteins snl1 and snl2 redundantly regulate primary seed dormancy via abscisic acid–ethylene antagonism mediated by histone deacetylation. Plant Cell, 25:149–166

Wang Z, Ren Z, Cheng C, Wang T, Ji H, Zhao Y, Deng Z, Zhi L, Lu J, Wu X et al (2020) Counteraction of ABA-mediated inhibition of seed germination and seedling establishment by ABA signaling terminator in *Arabidopsis*. Mol Plant 13:1284–1297

Wareing PF, Saunders PF (1971) Hormones and dormancy. Annu Rev Plant Physiol 22:261–288

Xu TM, Tian BQ, Sun ZD, Xie BJ (2011) Changes of three main nutrient during Oat germination. Nat Prod Res 23:534–537

Yamaguchi S, Smith MW, Brown RGS, Kamiya Y, Sun T (1998) Phytochrome regulation and differential expression of gibberellin 3β-hydroxylase genes in germinating *Arabidopsis* seeds. Plant Cell 106(1):2115–2126

Yamauchi Y, Takeda-Kamiya N, Hanada A, Ogawa M, Kuwahara A, Seo M, Kamiya Y, Yamaguchi S (2007) Contribution of gibberellin deactivation by AtGA2ox2 to the suppression of germination of dark-imbibed *Arabidopsis thaliana* seeds. Plant Cell Physiol 486(1):555–561

Zhao M, Zhang H, Yan H, Qiu L, Baskin CC (2018) Mobilisation and role of starch, protein, and fat reserves during seed germination of six wild grassland species. Front Plant Sci 9:234. https://doi.org/10.3389/fpls.2018.00234

Zhu JK (2003) Regulation of ion homeostasis under salt stress. Curr Opin Plant Biol 6:441–445

Seed Vigour and Invigoration

Sudipta Basu and Steven P. C. Groot

Abstract

Seed vigour is an important aspect of seed quality. It is a quantitative trait which is responsible for overall seed performance in terms of rate and uniformity of seed germination, seedling growth, emergence ability under unfavourable environments and post storage performance. Seed vigour is controlled by genetic factors, initial seed quality, production environments, harvesting and storage conditions. Seed vigour tests provide a more sensitive index of seed performance per se than the germination test. Efforts have been focused on developing novel or improving existing methods of vigour estimation in different crops. The vigour tests are tools routinely used for in-house seed quality control programs, especially for field and vegetable crops. Some treatments can improve seed vigour, although the treatment effects are more evident under sub-optimum than optimum growing conditions. This chapter deals with different aspects of seed vigour and its effects on plant growth and discusses physiological and biochemical parameters to understand underlying mechanisms.

Keywords

Crop production · Seed quality · Seed vigour · Seed vigour test · Seed invigouration

S. Basu (✉)
Division of Seed Science & Technology, ICAR-Indian Agricultural Research Institute, New Delhi, India

S. P. C. Groot
Wageningen University & Research, Wageningen Seed Centre, Wageningen, Netherlands
e-mail: steven.groot@wur.nl

M. Dadlani, D. K. Yadava (eds.), *Seed Science and Technology*,
https://doi.org/10.1007/978-981-19-5888-5_4

67

1   Introduction

Quality-assured seed is essential for sustainable and profitable crop production in both agricultural and horticultural crops. Seed quality is determined by its genetic and physical purity, health status and physiological quality. Among the physiological quality parameters, seed vigour is an important and complex trait which is determined by interaction of genetic and environmental factors (Finch-Savage and Bassel 2016). Unlike germination and viability, seed vigour is not a single measurable parameter but a quantitative attribute, which is controlled by several factors associated with overall seed performance that includes rate and uniformity of seed germination, seedling growth; emergence under unfavourable environmental conditions; and performance after storage. The high seed vigour status of a lot ensures a more rapid and uniform crop establishment and growth across diverse environmental conditions. The vigour differences among seed lots may be due to variability in genotype, environment, maturity at harvest, mechanical integrity, seed treatments and seed ageing. Under optimum conditions, seeds from different sources may germinate at comparable rates. However, under stressful conditions in the field, the seeds can exhibit variable performance due to differential vigour status. High vigour seed lots may store well for longer duration without evident loss in their germination capacity in comparison to the low vigour seed lots, which deteriorate faster under similar conditions. Freshly primed seeds, for instance, can emerge rapidly and uniform compared with the original non-primed seeds, but the shelf life of primed seeds is most often reduced. As a consequence, the primed seeds have a high vigour for germination speed, but often a lower vigour with respect to seed storage. Another example is the physical sanitation of seeds, aimed at the removal of seed-borne pathogens, but which may also result in damage to the seed which needs repair upon rewetting of the seeds.

1.1   Definitions of Seed Vigour

As early as in 1876, Friedrich Nobbe described vigour as 'Triebkraft' a German word meaning 'the driving force'. Isely (1957) defined it as 'the sum total of all seed attributes which favour stand establishment under unfavourable field conditions'. Seeds which perform well are termed as 'high vigour' seeds. A later definition of seed vigour is 'the sum total those properties of seed that determine the potential level of activity and performance of the seed during germination and seedling emergence' (Perry 1978). Here we define seed vigour as 'the sum of seed properties that determine the ability of viable seeds to germinate fast and uniform, and to produce healthy seedlings with rapid and uniform emergence under both optimal and suboptimal environmental conditions', combining the definitions formed by the Association of Official Seed Analysts (AOSA) and International Seed Testing Association (ISTA) (AOSA 1983; ISTA 2021).

2   Factors Affecting Seed Vigour

Modern agriculture emphasizes on technology and high inputs for obtaining high yield. Successful seedling establishment is the first critical step towards successful crop production and efficient use of all inputs. Plant stand establishment and performance are strongly influenced by seed vigour, therefore seed vigour has an indirect effect on yield (Tekrony and Egli 1997; Finch-Savage 1995). Low vigour seed lots provide seedlings with a lower leaf area index, dry matter accumulation and crop growth rate. Seed vigour is primarily influenced by genetic and environmental factors, including seed treatments. A major cause of the loss in vigour is attributed to deterioration during development, harvesting, drying and storage which commences when the seed becomes physiologically mature and continues during storage. In direct seeded crops, there is a relationship between the number of plants established per unit area and total yield. Efforts on inputs, agronomic practices and technological interventions in post establishment stages are not able to compensate for the initial setback in plant stand due to low vigour. This impact is higher in non-tillering crops or where gap filling is not feasible. Also in the transplanted crops, low seed vigour affects the plant stand and population but to a lesser extent due to wastage of seed/planting material, increased labour costs and reduced product quality due to low vigour. The impact of variation in seed vigour on both total and marketable yield varies among species and depends on the specific production practices and stage of crop harvest. The effect of vigour is highest in crops where the economic produce is harvested already at late seedling (micro greens and herbs) or vegetative stage (leafy vegetables and salad greens), followed by those crops which are harvested at fruiting (okra, beans, cucumber) and seed maturity (corn, cereals and pulses). Various factors have an impact on seed vigour.

2.1   Acquisition of Seed Vigour and Seed Maturity

In general, seed vigour increases during seed maturation and is at its peak at the time of natural shedding, which is later than what agronomists call physiological maturity defined as the stage where maximum dry weight is achieved at the end of seed filling. During the phase of late seed maturation (LSM), also called maturation drying, the developing seeds switch their metabolic activity from the production of storage reserves (oil, starch and proteins), to the synthesis of protection mechanisms (which including production of late embryogenesis abundant proteins (LEAs), heat shock proteins (HSPs), sugars for a glassy cytoplasm, and a change in the nuclear DNA confirmation) and degradation of chlorophyll. During this maturation drying phase, there is a gradual increase in seed vigour. Seed maturity at harvesting plays a key role in the seed vigour status. Shaheb et al. (2015) reported that seed quality parameters, such as seed germination and vigour indices, were significantly influenced by harvesting time. Seed development is not uniform throughout the inflorescence in Malvaceae, Brassicaceae and Umbelliferae species, therefore a seed lot harvested at any one time from the mother plant contains seeds at various stages of development (Bewley et al. 2013). Cabbage seed quality and maturity have been assessed by measuring the amount of chlorophyll fluorescence (CF) from intact seeds. Seeds with the lowest amount of CF gave the highest percentage of germination and normal seedlings (Jalink et al. 1998). An improvement in seed quality of soybean seeds was observed by removing green seeds using the CF sorting technique (Cicero et al. 2009).

Although in nature, seeds are most often dispersed from the mother plant at maturity, during domestication of our agricultural crops there has been a selection for genotypes that do not shed the seeds. When crop seeds remain on the mother plant after reaching full maturity, they can start deteriorating due to high temperatures, humidity or UV light, resulting in vigour reduction. Rain prior to harvesting can induce germination in the seeds when they are not dormant. To enable initiation of metabolic activity during germination, part of the protection is removed, which may result in reduced longevity of the seeds, or ultimately loss of desiccation tolerance when preharvest sprouting occurs.

2.2 Environment

Seed lots may differ substantially in their vigour status depending on the environment during seed development, harvesting and post harvest handling. Seed development under environmental stress (drought, frost, low/high temperatures, pest incidence) is reported to have lower vigour than those produced under congenial conditions. Fast drying during harvesting and processing can cause damage and cracking when the cells in the outer layers shrink much faster compared with those more inside. With grain legumes such as peas and beans, this will result in an increase in the proportion of abnormal seedlings. Elevated temperature during early seed development can result in decrease in seed size, number and fertility, and reduction in seed vigour in cereals, legumes and vegetable crops. Hence specific geographic locations with favourable climatic conditions are selected for commercial seed production. Variability in ambient temperature has been reported to increase seed dormancy and reduce germination rates in crops which require lower temperature for germination and seedling establishment (Reed et al. 2022).

2.3 Seed Size

Seed size can affect the rate of germination or emergence, total germination and seedling growth. In wheat, seed size is positively correlated with seed vigour wherein larger seeds tend to produce more vigourous seedlings (Cookson et al. 2001; Muhsin et al. 2021). Small muskmelon seeds had the lowest germination, emergence and seedling growth showing the association of seed physical parameters with vigour (Nerson 2002). With increased seed size, higher germination and emergence was observed in *Triticale* (Kaydan and Yagmur 2008) and maize (Mir 2010). Bolder seeds were capable of emerging from greater planting depth and exhibited enhanced ability to penetrate ground cover, weeds and surface hindrances in both wild and cultivated types of different species (Fenner 1991; Massimi 2018). In chick pea, large-sized seeds recorded maximum germination percentage, seedling vigour and protein content, dehydrogenase content, alpha-amylase activity (Anuradha et al. 2009). Large seeds also performed better under salinity stress (Mehmet et al. 2011). Plants grown from large seeds were vigourous and produced higher dry matter as compared to those grown from smaller seeds in groundnut (Steiner et al. 2019). The generally higher vigour of larger seeds may relate to their maturity status at harvesting, immature seeds being smaller. With spinach, however, larger seeds with a thicker pericarp result in slow and lesser germination under moist conditions and warmer temperatures, because of poor oxygen diffusion through a thicker pericarp (Magnée et al. 2020). However, smaller spinach seeds with higher chlorophyll levels also performed poorer compared to more mature low chlorophyll seeds from the same size. So, the seed performance is based on parameters more than just size. Hence, it is difficult to establish a consistent association between the seed size and vigour, the relationship being more species-specific and dependent on growing conditions.

2.4 Seed Reserves

Early seed vigour trait in some crops could be related to higher reserve utilization efficiency and mobilization. Low vigour seeds are sensitive to stress and their lower tolerance was associated with reduced lipid and protein content and increased amino acids, carbohydrates and phosphorus compounds in the embryo (Andrade et al. 2020). Among maize genotypes, sweet corn seeds have poor seed germination and vigour due to enhanced solute leakage, and insufficient energy supply during seed germination attributed to lower starch and higher sugar content (Styer and Cantliffe 1983).

2.5 Positional Effect

The position of the fruit can influence seed vigour both directly and indirectly through its relationship with variation in the progression of seed development. Seeds from the lower regions of the canopies, main branches and primary tillers can be more vigourous than those from the intermediate and upper plant positions, secondary or tertiary tillers, which again may relate to difference in seed maturity. With Pinto bean (*Phaseolus vulgaris*) the largest seeds with higher vigour were obtained from the lower canopy (Ghassemi-Golezani et al. 2011). Contrasting observations were made with okra, where fruits obtained from middle nodes pro- duced higher vigour seeds (Hedau et al. 2010). Seeds of pumpkin in middle and stylar segments had better seed vigour as compared to peduncular segment since the ovules at stylar and middle segments were first to get fertilized by high vigour pollen and have a temporal advantage in competing for resources from the mother plant

(Kumar et al. 2015). In cucumber, seeds from the top and peduncular fruit segments are delayed in reaching maximum quality compared with seeds from middle position, and this was correlated with a slower decline of chlorophyll fluorescence (Jing et al. 2000).

2.6   Seed Coat and Imbibition Damage

The outer cover or testa of a seed acts as a modulator between the internal seed structure and the environment, and is an important factor for vigour and field emergence, either by maintaining the integrity of internal seed components, protecting the embryo, allowing gas exchange between embryo and environment, and regulating the imbibition process. Imbibition damage has been shown to affect vigour in a range of temperate and tropical grain legumes (Powell 1985), but it may also occur in small seeded crops. Imbibition damage is caused by the rapid entry of water into the cotyledons during imbibition, which can result in disruption and disorganization of cell membranes, and leakage of sugars, amino acids and ions from the seeds. Especially under low temperature conditions the membrane is less flexible and more prone to disruption and imbibitional injury. The induction of imbibition damage is greater when seeds are very dry, creating a large difference in water potential upon imbibition and a strong flow of water through the membranes. Storing seeds overnight at a high humidity can reduce this problem. Fast drying of high moisture seeds can cause seed coat cracks that can increase the incidence of imbibition damage (Oliveira et al. 1984). The seed lots with extensive testa damage imbibed quickly with an induction of imbibition damage. These lots showed low vigour and emerged poorly in the field. Seed lots with little testa damage, on the other hand, imbibed slowly and exhibited lower imbibition damage, thus highlighting the importance of testa integrity in determining the vigour status in bold seeded legume. Measurement of electric conductance of seed leachate can be effective in identifying vigour differences among seed lots of such species. Equilibrating seeds to an acceptable range of moisture content is therefore suggested before performing EC test for seed vigour. Closure of hilar opening under conditions of excessive drying is a means of natural protection against imbibition damage, but this may significantly increase the proportion of so-called hard seeds, resulting in poor germination and field emergence.

Maternal inheritance also plays a role, as in the development of seed coat pigments and control of dormancy (Debeaujon et al. 2000). Sun et al. (2015) reported selection for a high permeability trait of the seed coat related to seed vigour during legume domestication.

2.7   Seed Ageing and Storage

Orthodox seeds deteriorate and age during storage under conditions of warm temperature, high seed moisture content and oxygen, resulting in a vigour decrease over time. Weathering of the seeds while still on the mother plant influences ageing to alter their vigour before harvest. Seed ageing or deterioration is the major cause of differences in seed vigour due to the accumulation of deleterious changes within the seed and ultimately the ability to germinate is lost. Reduced vigour of aged seeds can be linked to biophysical, biochemical and physiological changes associated with ageing (Bewley et al. 2013). During ageing under dry conditions, damage is induced by oxidation of DNA, RNA, proteins and membranes. DNA oxidation can result in strand breaks that need to be repaired before DNA replication. Limited DNA damage results in retardation of germination, while high levels of DNA damage could result in failure of germination or poor seedling quality. Oxidation of membrane lipids increases the risk of membrane leakage, hence aged seeds often show higher levels of electrical conductivity. Damage to the mitochondrial membranes, induced by oxidation, restricts the onset of aerobic respiration and thereby reduced seed vigour (Bewley et al. 2013). Vitamin E is the most important lipophilic anti-oxidant that protects the membranes. Seeds from Arabidopsis mutants deficient in vitamin E synthesis showed a relative very short shelf life (Sattler et al. 2004). Under the stressed storage environment (e.g., high temperature or relative humidity in uncontrolled storage) high vigour seed lots tolerate stress and lose vigour at a slower pace as compared to their low vigour counterparts. For an increasing number of crops, genetic variation in ageing tolerance has been observed under either dry or humid storage conditions. Sharma (2018) reported differential storage pattern in maize genotypes under ambient and controlled storage conditions wherein sweet corn and high protein maize showed greater decline in seed vigour and poor storability among genotypes compared.

2.8   Seed Processing

Seed processing is important in upgrading seed quality before use. However, processing seeds at high or very low moisture content may result in mechanical injuries on the seed coat and decline of seed vigour. Sweet corn seeds, for instance, are rather sensitive to mechanical damage by deshelling, especially after harvesting at relatively high seed moisture levels. During seed drying, moisture is removed initially from internal tissues of seed to seed surface and later from the seed surface into the atmosphere. High temperature or fast drying of seeds with high moisture content results in development of cracks on the surface, due to tissue shrinkage, which affects the seed vigour. Thus, selection of processing equipment, its slope and speed settings and seed moisture content during seed processing are important factors affecting seed quality.

2.9   Physical Sanitation Treatment

Some pathogens move with the seeds to the next generation interfering with the health of the new seedling. Sanitation treatments are performed to prevent seedling diseases. With increasing restrictions on the use of chemical disinfection, physical sanitation methods and the use of natural components are becoming more popular. While the aim of these treatments is to kill the pathogen, they may often have a deteriorating effect on seed vigour. Seed technologists have to balance between a treatment harsh enough to irradicate the pathogen or its pathogenicity, while mild enough to maintain seed vigour. However, the initial seed vigour also influences the sensitivity of the seeds to sanitation treatments. Less mature *Brassica oleracea* or *Daucus carota* seeds are more sensitive to hot water or aerated steam treatments compared with mature seeds (Groot et al. 2006). During seed priming part of the protection, that was induced during seed maturation, is removed to enable initiation of metabolic activity. Likely for that reason primed seeds are more sensitive to these physical sanitation treatments (Groot et al. 2008).

2.10   Genetic Variation

Seed vigour is a complex quantitative trait affected by multiple factors. With the use of high-throughput sequencing and genomic techniques quantitative trait loci (QTL) related to seed vigour have been identified in different crops (Reed et al. 2022). QTL analysis for seed vigour-related traits has been studied in different crops like *Arabidopsis*, rice, maize, *Brassica* sp. and lettuce. Li et al. (2017) and Shi et al. (2020) identified 19 and 26 QTLs respectively related to seed vigour in rice and wheat. Liu et al. (2011) identified QTLs for seed vigour-related traits which could provide information on early seedling vigour of maize. The genetic information on QTLs related to early vigour in different crops could be explored for developing breeding lines/genetic stocks with early vigour traits which could be utilized for developing high vigour lines through breeding. Morris et al. (2016) identified two genes related to seed vigour in *Brassica oleracea*. Alleles from the BolCVIG1 gene gave different splicing variants that coincided with variation in abscisic acid sensitivity, while the second gene BoLCVIG2 was a homologue of an Arabidopsis alternative-splicing regulator gene (AtPTB1).

3   Seed Vigour Assessment

Seed vigour is accepted as an important seed quality component, but official vigour testing is carried out for only a few crops. Numerous tests that have been attempted in different crops are not accepted universally due to variability in results. Still, development of reliable vigour tests for specific crop species for determining their planting value is desirable (Lopes et al. 2012; Marcos-Filho 2015; Reed et al. 2022).

Seed vigour tests aim to provide a more sensitive index of seed performance per se than the germination test. Efforts have been focused on developing novel methods or improving the existing methods of vigour estimation in different crops, by measuring the rate and uniformity of seed germination and seedling growth, emergence ability of seeds under adverse environmental conditions and performance after storage. Such vigour tests are routinely used for in-house seed quality control programmes, especially for field and vegetable crops.

While researchers have applied a large number of vigour tests in different crop species which were found useful for the assessment of seed vigour, there is lack of consistency and reproducibility of tests across species. However, the most commonly applied tests can be grouped based on physical attributes; germination and seedling growth; performance during or after subjecting to stress conditions; and physiological and biochemical parameters.

- **Physical tests:** based on seed density; seed size; seed colour; embryo size and image.
- **Performance tests:** based on measurable parameters of germination, such as rate of germination; radicle emergence; and vigour indices (VI) based on seedling growth (weight or length) and germination.
- **Stress tests:** based on seed germination after subjecting the seed to a defined condition of stress, such as low temperature; high temperature combined with high humidity or seed moisture; or a combination of multiple stresses.
- **Physiological and biochemical parameters**: based on the permeability of cellular membranes; changes in respiratory functions; key respiratory enzymes.

However, only some of these are used for official purposes as have been recommended by the ISTA and/or AOSA.

ISTA having a vision of 'uniformity in seed quality evaluation worldwide' provides a framework within which quality may be evaluated and compared by vigour tests. The ISTA seed vigour committee evaluates the performance of seed vigour test methods by the reproducibility of vigour method and the relationship between vigour test results and seedling emergence in the field.

The following vigour tests have been validated and recommended by ISTA for seed vigour estimation (for details please see ISTA Rules 2021):

- Conductivity test: *Cicer arietinum, Glycine max, Phaseolus vulgaris, Pisum sativum* (garden peas only, excluding petit pois varieties) and *Raphanus sativus*
- Accelerated ageing test: *Glycine max*
- Controlled deterioration test: *Brassica* spp.
- Radicle emergence test: *Zea mays, Brassica napus* (oilseed rape, Argentine canola), *Raphanus sativus, Triticum aestivum* L. subsp. *aestivum*
- Tetrazolium vigour test: *Glycine max*

AOSA also recommends most of the above-mentioned tests.

3.1   Seed Vigour Tests

3.1.1   Seed Size/Density
Seed vigour estimation may simply be based on seed size, and seed density. In general, the larger seeds with higher density have higher seed vigour than smaller and lighter seeds.

3.1.2   Performance-Based Tests
Rapid germination is an important component of the seed vigour concept since it usually corresponds to rapid seedling growth and emergence in the field. Tests based on seedling growth parameters are simple to perform and usually do not require special equipment besides those used for germination. Such vigour tests can be based on seedling performance including first count of the germination test, rate of germination or seedling emergence, mean germination time, seedling growth (length or dry weight), seedling vigour indices, and more recently the emergence rate of the primary root (radicle emergence).

1. **Radicle emergence test:** The slow rate of germination is an early physiological expression of seed ageing, the major cause of reduced vigour. The standard germination tests evaluate radicle emergence with an early and late count of normal seedlings at a fixed period after imbibition. The duration of the test can be long and may for some species extend beyond 1 month. The radicle emergence test is relatively rapid with an early count of emergence of radicle (2 mm) for vigour estimation. It is an accepted vigour test for maize (*Zea mays* L.), aubergine (brinjal), oilseed rape and radish (ISTA 2017). Differences in radicle emergence rates have been attributed to the length of the delay from the start of imbibition to radicle emergence (Matthews and Khajeh-Hosseini 2007), which is dependent upon the time required for damage repair before radicle emergence. Fast radicle emergence is indicative of high vigour.

3.1.3   Stress Tests
The following tests are stress tests which impose stress conditions on the seed and based on the performance of the seeds under stress conditions the quality of seed lot in terms of vigour is measured.

1. **Cool test:** The test is limited to measuring the effect of cool temperature (18 °C) on the germination of cotton seed and the growth rate of cotton seedlings (Hampton and TeKrony 1995). Since the seeds are subjected to germination at a temperature below its optimum of 25 °C, their ability to produce normal seedlings and speed of germination are taken as indication of vigour. Though not a recommended test, it may provide useful indication of vigour in some cases.
2. **Cold test:** Cold test estimates the ability of the seed to withstand the low temperature stress when the soil is humid and cold which hampers the growth of weak seedlings experienced during early spring planting. This test is commonly used for assessing seed vigour in maize (Caseiro and Marcos-Filho 2000).

The seeds are sown in soil collected from maize field (to simulate field conditions and microbial load) and exposed to low temperature (10 °C) for 7 days followed by assessment of seed germination. The seed lots exhibiting high germination after cold test are reported to be vigourous. This test was reported both by ISTA (Hampton and TeKrony 1995) and AOSA (1983), but due to difficulty in standardization of test procedure, this is not recommended officially, though used by seed companies for internal quality control in the maize growing belts of North America and Europe (Matthews and Powell 2009).

3. **Hiltner (Brick-gravel) test:** Perhaps the oldest test and developed by Hiltner in Germany in 1917. Weak seedlings are not able to generate enough force to overcome the mechanical stress imposed by the pressure of brick gravels. The higher the number of seedlings emerged through the brick gravel layer, the greater is the vigour of the seed lot. This method was suitable to study the vigour levels of cereal seeds. A paper piercing test (Fritz 1965) is in principle similar to the Hiltner brick gravel test. In this test the germination is measured in sand covered by a layer of moist paper, the vigourous seed lots emerge by piercing the paper.

4. **Accelerated aging test:** The accelerated ageing (AA) test is the most commonly used vigour test. The test was developed by Delouche and Baskin (1973) to assess the storage potential of a number of species, and recommended as a vigour test by McDonald and Phannendranath (1978). Seeds are exposed to high temperature (40–45 °C) and humidity (90–100% RH) for a specified time which causes metabolic inactivation by deterioration, especially in low vigour seeds. The lots showing higher germination percentage after AA have higher vigour as compared to the lots with lower vigour post AA test. ISTA has validated this vigour test for *Glycine max* only (ISTA 2015).

5. **Controlled deterioration test:** Controlled deterioration test (CD test) was developed by Matthews (1980) and predicts field planting potential of seed. This test is quite similar to the AA test, in subjecting seeds to rapid ageing at high m.c. and high temperature conditions. However, while in AA test the seed m.c. increases gradually from the start of test, in CD the seeds are first equilibrated to a high m.c. and then subjected to high temperature stress. In *Brassica* species seeds are brought to 20% moisture and then exposed to 45 °C for 24 h. The high vigour seeds retain high germination after deterioration. ISTA has validated this test (ISTA 2015). Comparable CD tests have been published for other crops, with some modifications in seed moisture content, temperature and duration, in which a positive relationship has been observed between the CD test tolerance and the field emergence for several crops (Kazim and Ibrahim 2005). The AA and CD test shows a poor correlation with storability under dry conditions, due to the difference in the physiological status of the seeds under humid conditions, with a liquid cytoplasm in contrast to a glassy cytoplasm under dry storage conditions. However, it may show a reasonable correlation with storage under humid tropical conditions at comparable relative humidity levels.

3.1.4 Physiological and Biochemical Tests

The physiological and biochemical tests measure the physio-chemical changes that occur during ageing providing indirect vigour estimation.

Tests based on respiratory parameters

1. **GADA test:** The glutamic acid decarboxylase activity (GADA) test was developed by Grabe (1964). GADA is a key respiratory enzyme during germination. Hence, the activity of this enzyme helps in the estimation of vigour.
2. **Tetrazolium vigour test:** Tetrazolium test is primarily used as a quick viability test. But Kittock and Law (1968) used it for the vigour estimation of seeds based on the colour intensity of stained embryos or cotyledons. The colourless 2,3,5-triphenyl tetrazolium chloride can pass intact membranes and turns into red-coloured formazan in living cells catalysed by dehydrogenase enzymes involved in aerobic respiration. In seeds the metabolically active cells with intact membranes stain pink or red. ISTA has validated the Tetrazolium test for the assessment of seed viability of a large number of crops (Leist et al. 2003) and for *Glycine max* as test for seed vigour (ISTA 2015).
3. **Respiration (RQ) test:** During the process of respiration, oxygen is taken up by seeds and carbon dioxide is released. The ratio of the volume of carbon dioxide evolved to the volume of oxygen consumed per unit time is called respiratory quotient (RQ). The RQ was found to be more often related to the vigour than oxygen uptake alone. This test was used for vigour estimation in maize (Woodstock 1988).
4. **Respiratory activity test:** The respiratory activity test measures the amount of carbon dioxide released from seeds during the respiration process, which results from the oxidation of organic substances in the cellular system after the start of the imbibition process. Thus, high respiratory rates are associated with vigourous seeds due to the oxidation of a large amount of reserve tissues present in their cells. This test is useful in vigour estimation of soybean (Dode et al. 2013) and okra (De Sousa et al. 2018). In GADA test, vigour assessment is based on estimation of glutamic acid decarboxylase enzyme activity whereas in this test the seed vigour is proportional to the respiration rate of the seeds, higher the respiration rate, higher the carbon dioxide release and thus the vigour.
5. **Conductivity test:** Conductivity test developed by Matthews and Bradnock (1967) evaluates the integrity of cell membranes and their ability to repair them during the 'soak period'. The test is based on the hypothesis that biomembranes undergo disintegration during seed ageing, rendering low vigour seeds with poor membrane integrity, and hence causing release of more solutes to the steep water, increasing the electrical conductivity of the solution. Seeds are soaked in de-ionized water for a stipulated time to allow seed leachates to come out in the soak water. The conductivity of steep water is inversely proportional to the seed vigour. Woodstock (1988) reported that the rate of electrolyte leakage in seeds was negatively correlated with membrane intactness. The extent of solute leakage during seed imbibition has been reported to be associated with the level of seed vigour by researchers (Wilson and Mohan 1998; Zhao and Wang 2005). ISTA

has recommended the conductivity test for vigour estimation in *Cicer arietinum*, *Glycine max*, *Phaseolus vulgaris*, *Pisum sativum* (garden peas only, excluding petit pois varieties) and *Raphanus sativus*.

3.2 Some Novel Methods of Vigour Assessment

1. **Ethanol test:** The ethanol test is a promising method that can be used to differentiate seed lots with different vigour levels. Decline in seed vigour due to ageing results in damage of the membranes, including in mitochondria, which impairs aerobic energy metabolism. Loss of mitochondrial integrity due to ageing directs the seeds to anaerobic respiration and ethanol release, which has been associated with seed vigour in various crop seeds (Buckley and Huang 2011). The test is based on the alcoholic fermentation theory, where the enzymes pyruvate decarboxylase and alcohol dehydrogenase act on pyruvate, producing ethanol and CO_2 as well as oxidizing NADH during this process. Ethanol release measured using a modified breath analyser was successfully used as a rapid test to rank the vigour of seed lots of canola (Buckley and Buckley 2009) and cabbage (Kodde et al. 2012).

2. **Single seed oxygen consumption** at the beginning of germination has been considered as an indicator of seed vigour (Reed et al. 2022) and is largely correlated to the activation of metabolic activity. Oxygen consumption is directly related to seed respiration and energy production. The Q2 machine, with oxygen-sensing technology, provides a fast and automatic measurement of oxygen consumption and respiration efficiency of individual seeds and has been found as a measure of seed vigour in several species, including *Beta vulgaris* L. and *Pinus massoniana* Lamb. (Zhao and Zhong 2012) and is recommended for seed vigour assessment, although this has been debated by Powell (2017).

3. **Volatile organic compounds production**: Seeds undergo deterioration due to lipid peroxidation leading to emission of volatile organic compounds (VOC). Since it has been established by many researchers that there is a significant difference between high and low vigour seeds with respect to quantity and profile of VOCs emitted, there is great potential for utilizing the VOC profile to obtain a quick and reproducible test of seed vigour (Umarani et al. 2020). Further research is needed to develop standard and reproducible protocols for fingerprinting of VOCs for seed vigour assessment and to identify the standard volatile biomarker (s) specific to crop species.

4. **Seed quality analysis by spectral properties:** Rapid discolouration and micro-organism contamination during production and storage is the main cause of seed discolouration. The colour sorter uses an electronic eye that can pick up different colours according to the way the machine is adjusted. As seed falls down a shoot, it passes in front of the electric eye. If the colour of the seed is different than the desired colour, the electric eye will activate a sudden burst of air that pushes that seed into a reject bin while the rest of the seed passes through to another bin.

Colour sorting improved germination and vigour in eastern Gamma grass (*Tripsacum dactyloides*) seeds (Klein et al. 2008).

5. **Seed maturity analysis by chlorophyll fluorescence (CF):** Seed within a lot that has not reached full maturity will most likely be less vigourous with impaired germination capabilities. Sometimes the seeds have finished the seed filling phase, but not the seed maturation phase. Such seeds are difficult to separate using normal density, size or shape separations. Chlorophyll fluorescence measures the amount of green chlorophyll present in each seed or the seed coat. The higher the chlorophyll in the seed the lesser mature and vigourous the seed (Jalink et al. 1999). The principle is used in commercial seed sorters.

6. **Seed separation by near-infrared (NIR):** Near-infrared light transmission through the seed gives off a signal that is linked to the internal components of the seed, which can be related to seed vigour and germination. Low vigour, dead and contaminated seed with fungal pathogen gives specific NIR signal which can be used to separate them from healthy and vigourous seeds. Near-infrared spectroscopy could predict seed germination and vigour status of soybean seeds (Al-Amery et al. 2018).

7. **Multispectral and X-ray imaging:** This could be used for rapid and non-destructive evaluation of seed quality, overcoming intrinsic subjectivities of seed testing. X-ray-based seed vigour estimation in tomato was undertaken with X-ray digital imaging (Bruggink 2017). These techniques help to sort broken, undersized, diseased seeds from healthy seeds. Bianchini et al. (2021) used multispectral and X-ray images for the characterization of *Jatropha curcas* L. seed quality and de Medeiros André et al. (2020) used FT-NIR spectroscopy and X-ray imaging for vigour estimation. Mahajan et al. (2018) reported the use of machine vision-based alternative testing approach for vigour testing of soybean seeds (*Glycine max*) wherein multispectral and X-ray imaging system rapidly and efficiently undertook non-destructive characterization of seed quality.

8. **Seed imaging systems:** The utilization of computers for seed vigour assessment had increased significantly in recent times. The SeedVigor Imaging System (SVIS®) developed by Sako et al. (2001) is being used successfully for vigour assessment in soybeans and corn (Marcos-Filho et al. 2009). The image analysis is performed on scanned images of seedlings, whose parts were identified and marked by developed software. After the computer image processing, data was obtained on the root length, hypocotyl and indices of vigour and uniformity of growth. Marcos-Filho (2015) reported that the vigour index, uniformity of growth and seedling length were consistently comparable with the results of recommended vigour tests. Another programme, SeedVigor Automated Analysis System (Vigor-S), was designed based on similar principles to the SVIS® to provide an efficient evaluation of seed physiological potential (Rodrigues et al. 2020).

4 Seed Invigoration

After harvest seeds are not always at highest level of vigour, and they may further deteriorate during storage. The seed invigoration treatments play a pivotal role in seed quality improvement where these treatments enhance the germinability, planting value, field performance, and yield via physiological and cellular repair. They may subsequently rejuvenate post storage seed quality. The selection of seed invigouration treatments is crop and problem specific and has variable applicability for enhancement of seed quality and planting value upgrading. In this chapter, we primarily put emphasis on 'seed priming' as the most widely reported invigouration treatment to reinvigourate low vigour seed lots. Seed invigouration treatments involving partial hydration of seeds are metabolically more advanced towards radicle protrusion (germination *sensu stricto*) than un-treated seeds. Therefore, these seeds are exhibiting rapid and uniform germination.

Rapid seed germination and stand establishment is critical for crop production especially under sub-optimum environmental conditions. Various seed priming treatments have been devised to improve the rate and uniformity of seedling emergence and crop performance. The priming treatments provide controlled hydration of seeds that trigger several metabolic, biochemical, and molecular alterations enabling seeds to germinate more efficiently, exhibit faster and synchronized germination, and impose vigour and stress tolerance to young seedlings (Varier et al. 2010).

4.1 Effect of Seed Priming

4.1.1 Biophysical and Structural Changes

Seed water status refers to the measurement of state of water in relation to seed. Water in seed is available in three forms, i.e. bound, free and adsorbed states. The free water is bound by weak hydrogen bond and is available for metabolic activities. The physical state of water is affected by hydrophilic compounds, viz. protein and sugars in plant tissues. Better performances of primed seeds are attributed to the modifications of seed water availability and binding properties (Nagarajan et al. 1993). Matsunami et al. (2021) studied the role of aquaporins (AQPs), transmembrane proteins that serve as water channels allowing rapid and passive movement of water in seed water metabolism. Among the AQPs, plasma membrane intrinsic proteins (PIPs) and the tonoplast intrinsic proteins (TIPs), expressed in the plasma and vacuole membranes respectively, are known to play a key role in transcellular and intracellular plant water transport.

Internal seed morphology of primed seed's free space (empty area between the embryo and the endosperm or between the seed contents and the integument within a seed) plays an important role in seed germination of primed seeds. The X-ray studies of primed tomato seeds reported that priming allowed greater water uptake by the embryo resulting in rapid growth and penetration of radicle tip. SEM observations on internal microstructural changes of osmoprimed okra seeds showed abundance of starch granules in primed seeds with pitted surface of starch granules and visible changes in their surface roughness, indicating activation of hydrolytic enzymes during the priming treatment. These seeds also showed an increased amount of amorphous material, probably proteins or other biopolymers that indicated the synthesis of protein bodies in primed seeds (Dhananjaya 2014).

4.1.2   Cellular and Metabolic Changes

Seed priming initiated activities of cell wall hydrolases as endo-β-mannanase that help in lowering mechanical constraints during the initial period of germination and radical protrusion (Toorop et al. 1998). The α and β tubulin subunits, a constituent of microtubules, are reported to help in the maintenance of the cellular cytoskeleton. An increase in the abundance of α and β subunits of tubulin has been recorded during priming in tomato seeds (de Castro et al. 2000). During priming under abiotic stresses, the cellular structures and proteins accumulated during the course of water uptake are known to be protected by specific proteins such as late embryogenesis abundant (LEA) protein and dehydrins (Chen et al. 2012). The expression of LEA proteins undergoes sequential changes with a decline during the imbibition phase, upregulation in the dehydration phase followed by degradation during the germination phase (Soeda et al. 2005).

The priming-induced enhancement in seeds is due to activation of DNA repair mechanisms, synchronization of the cell cycle in G_2 and preparation to cell division (Bewley et al. 2013). Cell division starts just after radical protrusion, thus, seed priming, which prolongs Phase II of seed germination and is finished just before Phase III (radicle protrusion), thus does not affect cell division in itself but prepares the seed for cell division (Varier et al. 2010). Seed priming also increases the ratio of cells in the G_2 phase to the G_1 phase, an expression of the beneficial effects of priming on seedling performance (Bino et al. 1992). Mir et al. (2021) reported the beneficial effect of priming on seedling performance due to the action of replicative DNA synthesis processes prior to seed germination in hydroprimed maize seeds.

4.1.3   Physiological and Biochemical Changes

Enhanced vigour and germination of primed seeds is attributed to a range of physiological and biochemical changes viz., damage repair, better mobilization of reserves into growing seedling, reduced time of imbibitions required for the onset of RNA and protein synthesis, polyribosome formation, increase in total RNA and total protein content, improved membrane integrity, control of lipid peroxidation, increased sugar content, protein and nucleic acid synthesis, removal of inhibitors like abscisic acid, efficient production and utilization of germination metabolites, increased activity of enzymes viz. alpha amylase, acid phosphatase, esterases, dehydrogenases, isocitrate lyase, protease, peroxidase, catalase, glutathione reductase, superoxide dismutase and ROS production (Pandita et al. 2007; Vashisth and Nagarajan 2010; Lutts et al. 2016).

The reactive oxygen species (ROS) functions as signalling molecules in plants thus regulating growth and development, programmed cell death and hormone signalling. Reactive oxygen species (ROS) such as superoxide radicals ($O_2^{\bullet-}$), hydrogen peroxide (H_2O_2) and hydroxyl radicals (OH) are generated as a result of redox reactions in seeds. Seed priming treatment strengthens the ROS mechanism in the seeds facilitating better performance of primed seeds (Apel and Hirt 2004).

4.1.4 Stress Resistance

Seed priming is an effective way to alleviate the inhibition of seed germination and seedling growth under stress conditions as it provides tolerance against abiotic stress (drought, heat, salinity, nutrient). Halopriming improved field emergence in chilli seeds under salinity stress (Khan et al. 2009) whereas ascorbic acid priming reduced aluminium stress in maize (Kussumoto et al. 2015). Hydropriming in paddy seeds enhanced resistance against carbon dioxide stress and oxidative damage, while osmopriming with $CaCl_2$ in wheat seeds provided resistance against drought stress (Nedunchezhiyan et al. 2020). Seed priming with cytokinins (plant growth substance) also imparted salinity and drought tolerance (Bryksova et al. 2020). Beneficial microorganisms or plant growth-promoting microorganisms used for seed-enhanced plant stress tolerance against drought was also recorded due to bacterial priming (*Bacillus* sp.) (Amruta et al. 2019; Singh et al. 2016). Magnetic field treatments imparted tolerance to biotic and abiotic stresses (Bhardwaj et al. 2012).

4.2 Storage of Primed Seeds

Priming often adversely affects the longevity of seeds. However, the extent of impairment depends on priming protocol, storage temperature, duration and moisture, storage period and crop species (Parrera and Cantliffe 1994). During priming, DNA replication is initiated in the embryonic axis and progresses from G_1 to S phase and subsequently to G_2 phase. When primed seeds are dried and stored, these nuclei get arrested in G_2 phase, and are vulnerable to cellular damage (Bewley et al. 2013). The longevity of primed *Digitalis purpurea* seeds could be preserved by slow drying at low temperature and moisture level followed by storing under low temperatures (Butler et al. 2009), while in commercial priming slow drying at a higher temperature is often used to improve shelf life (Bruggink et al. 1999). Gene expression studies showed a decline in the expression of LEA genes during *Brassica* seed priming and a reinduction of expression of LEA genes during slow and warm drying of the primed seeds (Soeda et al. 2005). Notwithstanding the decreased longevity of primed seeds in many species, priming offers high commercial value by practising storage at low temperature, controlled and slow/fast drying post priming or other chemical methods.

5 Conclusion

Seed vigour is an important trait of seed quality assessment which ensures high performance under variable environmental conditions. The early vigour trait is important for optimum field establishment. Numerous tests have been attempted in different crops but most are not universally accepted due to high variability in their results. Novel and innovative methods of seed vigour estimation have been developed for reliable assessment of seed quality which needs further refinement. Seed invigoration treatments are effective for improving the vigour status of low vigour seeds but post-treatment storability is a cause of concern.

Acknowledgement Dr. Sudipta Basu acknowledges the contribution of Mr. Mallanna Malagatti, Ph.D. Scholar of ICAR-Indian Agricultural Research Institute, New Delhi for helping in the preparation of the draft of the chapter.

References

Al-Amery M, Geneve RL, Sanches MF, Armstrong PR, Maghirang Elizabeth B, Lee C, Vieira RD, Hildebrand DF (2018) Near-infrared spectroscopy used to predict soybean seed germination and vigour. Seed Sci Res 28:245–252

Amruta N, Kumar MP, Kandikattu HK, Sarika G, Puneeth ME, Ranjitha HP, Vishwanath K, Manjunatha C, Pramesh D, Mahesh HB, Narayanaswamy S (2019) Bio-priming of rice seeds with novel bacterial strains for management of seedborne *Magnaporthe oryzae* L. Plant Physiol Rep 24(4):507–520

Andrade CMM, Coelho GC, Virgílio GU (2020) Modelling the vigour of maize seeds submitted to artificial accelerated ageing based on ATR-FTIR data and chemometric tools (PCA, HCA and PLS-DA). Heliyon 6(2):e03477. https://doi.org/10.1016/j.heliyon.2020.e03477

Anuradha R, Balamurugan P, Srimathi P, Sumathi P (2009) Influence of seed size on seed quality of chick pea (*Cicer arietinum* L.). Legume Res Int J 32:133–135

Apel K, Hirt H (2004) Reactive oxygen species: metabolism, oxidative stress and signal transduction. Annu Rev Plant Biol 55:373–399

Association of Official Seed Analysts (AOSA) (1983) Seed vigor testing handbook. Contribution no. 32 to the AOSA handbook on seed testing. AOSA, Springfield, IL, pp 1–93

Bewley JD, Bradford KJ, Hilhorst HWM, Nonogaki H (2013) Seeds: physiology of development, germination and dormancy. Springer, New York, NY

Bhardwaj J, Anand A, Nagarajan S (2012) Biochemical and biophysical changes associated with magnetopriming in germinating cucumber seeds. Plant Physiol Biochem 57:67–73

Bianchini VJM, Mascarin GM, Silva LCAS, Valter A, Carstensen JM, Boelt B, Silvada BC (2021) Multiple spectral and X ray images for characterization of *Jatropha curas* L. seed quality. Plant Methods 17:9. https://doi.org/10.1186/s13007-021-00709-6

Bino RJ, Vries JND, Kraak HL, Pijlen JGV (1992) Flow cytometric determination of nuclear replication stages in tomato seeds during priming and germination. Ann Bot 69:231–236

Bruggink H (2017) X-ray based seed analysis and sorting, Paper presented at ISTA conference

Bruggink GT, Ooms JJJ, Van der Toorn P (1999) Induction of longevity in primed seeds. Seed Sci Res 9:49–53

Bryksova M, Hybenova A, Hernandiz AE, Novak O, Pencik A, Lukas S, Diego de N, Dolezal K (2020) Hormonal priming to mitigate abiotic stress effects: a case study of N^9 substituted cytokinin derivatives with a florinated carbohydrate moiety. Front Plant Sci 11:599228. https://doi.org/10.3389/fpls.2020.599228

Buckley WT, Buckley KE (2009) Low molecular weight volatile indicators of canola seed deterioration. Seed Sci Technol 37:676–690. https://doi.org/10.15258/sst.2009.37.3.15

Buckley WT, Huang J (2011) A fast ethanol assay to detect seed deterioration. Seed Sci Res 39: 510–526. https://doi.org/10.1017/S0960258511000274

Butler H, Hay FR, Ellis RH, Murray RD (2009) Priming and re-drying improve the survival of mature seeds of *Digitalis purpurea* L. during storage. Ann Bot 103:1261–1270. https://doi.org/10.1093/aob/mcp059

Caseiro RF, Marcos-Filho J (2000) Alternative methods for evaluation of corn seed vigor by the cold test. Sci Agric 57:459–466. (in Portuguese, with abstract in English)

Chen K, Fessehaie A, Arora R (2012) Dehydrin metabolism is altered during seed osmopriming and subsequent germination under chilling and desiccation in *Spinacia oleracea* L. cv. Bloomsdale: possible role in stress tolerance. Plant Sci 183:27–36

Cicero SM, Schoor van der R, Halink H (2009) Use of chlorophyll fluorescence sorting to improve soybean seed quality. Revista Brasileira de Sementes 31:145–151

Cookson WR, Rowarth JS, Sedcole JR (2001) Seed vigour in perennial ryegrass (*Lolium perenne* L.): effect and cause. Seed Sci Technol 29:255–270

de Castro RD, van Lammeren AAM, Groot SPC, Bino RJ, Hilhorst HWM (2000) Cell division and subsequent radicle protrusion in tomato seeds are inhibited by osmotic stress but DNA synthesis and formation of microtubular cytoskeleton are not. Plant Physiol 122:327–336

de Medeiros André D, da Silva LJ, João OR, Kamylla Ca F, Jorge TFR, Abraão AS, da Silva C (2020) Machine learning for seed quality classification: an advanced approach using merger data from FT-NIR spectroscopy and X-ray imaging. Sensors 20:4319. https://doi.org/10.3390/s20154319

De Sousa LM, Salvador BT, De Freitas RMO, Narjara WN, Leite TDS, De Paiva E (2018) Vigor determination of Okra seeds by respiratory activity. Biosci J Uberlândia 34:1551–1554

Debeaujon I, Leon-Kloosterziel KM, Koornneef M (2000) Influence of the testa on seed dormancy, germination, and longevity in *Arabidopsis*. Plant Physiol 122:403–414

Delouche JC, Baskin CC (1973) Accelerated aging techniques for predicting the relative storability of seed lots. Seed Sci Technol 1:427–452

Dhananjaya P (2014) Studies on seed quality enhancement in okra (*Abelmoschus esculentus* L. *Moench*), Doctoral dissertation. ICAR-Indian Agricultural Research Institute, New Delhi

Dode JS, Meneghello GE, Timm FC, Moraes DM, Peske ST (2013) Teste de respiraçãoemsementes de soja para avaliação da qualidadefisiológica. Ciência Rural, Santa Maria 43:193–198. http://www.scielo.br/pdf/cr/v43n2/a3613cr5997

Fenner M (1991) The effects of the parent environment on seed germinability. Seed Sci Res 1:75–84

Finch-Savage WE (1995) Influence of seed quality on crop establishment, growth and yield. In: Basra AS (ed) Seed quality: basic mechanisms and agricultural implications. Haworth Press, New York, NY, pp 361–384

Finch-Savage WE, Bassel GW (2016) Seed vigour and crop establishment: extending performance beyond adaptation. J Exp Bot 67:567–591. https://doi.org/10.1093/agriculturalimplications

Fritz T (1965) Germination and vigour tests of cereals seed. Int Seed Test Assoc Proc 30:923–927

Ghassemi-Golezani K, Zafarani-Moattar P, Raey Y, Mohammadi M (2011) Response of pinto bean cultivars to water deficit at reproductive stages. J Food Agric Environ 8:801–804

Grabe DF (1964) The GADA test for seed storability. Seed Technology Papers. 217

Groot SPC, Birnbaum Y, Rop N, Jalink H, Forsberg G, Kromphardt C, Werner S, Koch E (2006) Effect of seed maturity on sensitivity of seeds towards physical sanitation treatments. Seed Sci Technol 34:403–413

Groot SPC, Birnbaum Y, Kromphardt C, Forsberg G, Rop N, Werner S (2008) Effect of the activation of germination processes on the sensitivity of seeds towards physical sanitation treatments. Seed Sci Technol 36:609–620

Hampton JG, TeKrony DM (1995) Handbook of vigour test methods, 3rd edn. International Seed Testing Association, Zurich

Hedau NK, Gyanendra S, Mahajan V, Singh SRK, Anita G (2010) Seed quality and vigour in relation to nodal position and harvesting stage of okra under mid hill of North-Western Himalayas. Ind J Hortic 67:251–253

Isely D (1957) Vigor tests. Proc Assoc Off Seed Anal 47:176–182

ISTA (2015) International rules for seed testing. International Seed Testing Association, Basserdorf

ISTA (2017) International rules for seed testing. International Seed Testing Association, Bassersdorf, p 296

ISTA (2021) International rules for seed testing, vol 1. International Seed Testing Association, Bassersdorf, p i-19-18. https://doi.org/10.15258/istarules.2021

Jalink H, van der Schoor R, Frandas A, van Pijlen JG, Bino RJ (1998) Chlorophyll fluorescence of Brassica oleracea seeds as a non-destructive marker for seed maturity and seed performance. Seed Sci Res 8(4):437–443. https://doi.org/10.1017/S0960258500004402

Jalink H, van der Schoor R, Birnbaum YE, Bino RJ (1999) Seed chlorophyll content as an indicator for seed maturity and seed quality. Acta Hortic 504:219–228. https://doi.org/10.17660/ActaHortic.1999.504.23

Jing C, Hai Jan HW, Bergervoet JH, Klooster M, Sheng-Li D, Raoul JB, Henk WM, Hilhorst GSPC (2000) Cucumber (Cucumis sativus L.) seed performance as influenced by ovary and ovule position. Seed Sci Res 10:435–445

Kaydan D, Yagmur M (2008) Germination, seedling growth and relative water content of shoot in different seed sizes of triticale under osmotic stress of water and NaCl. Afr J Biotechnol 7:2862–2868. https://doi.org/10.5897/AJB08.512

Kazim M, Ibrahim D (2005) Controlled deterioration for vigour assessment and predicting seedling growth of winter squash (Cucurbita maxima) seed lots under salt stress. N Z J Crop Hortic Sci 33:193–197. https://doi.org/10.1080/01140671.2005.9514349

Khan HA, Pervez MA, Ayub CM, Ziaf KR, Balal M, Shahid MA, Akhtar N (2009) Hormonal priming alleviates salt stress in hot Pepper (Capsicum annuum L.). Soil Environ 28:130–135

Kittock DL, Law AG (1968) Relationship of seedling vigour to respiration and tetrazolium reduction in germinating wheat seeds. Agron J 60:268–288

Klein JD, Wood LA, Geneve RL (2008) Hydrogen peroxide and color sorting improves germination and vigor of eastern gamagrass (Tripsacum dactyloides) seeds. Acta Hortic 782:93. https://doi.org/10.17660/ActaHortic.2008.782.8

Kodde J, Buckley WT, Groot CC, Retiere M, Zamora AM, Groot SPC (2012) A fast ethanol assay to detect seed deterioration. Seed Sci Res 22:55–62. https://doi.org/10.1017/S0960258511000274

Kumar V, Tomar BS, Dadlani M, Singh B, Kumar S (2015) Effect of fruit retention and seed position on the seed yield and quality in pumpkin (Cucurbita moschata) cv Pusa Hybrid 1. Indian J Agric Sci 85:1210–1213

Kussumoto AB, Machemer-Noonan K, Júnior FGS, Ricardo AA (2015) Dry priming of maize seeds reduces aluminum stress. PLoS One 10:e0145742. https://doi.org/10.1371/journal.pone.0145742

Leist N, Krämer S, Jonitz A (2003) ISTA working sheets on tetrazolium testing. International Seed Testing Association, Bassersdorf

Li J, Wan H, Wei H, Wang Q, Zhou Y, Yang W (2017) QTL mapping for early vigor related traits in an elite wheat-breeding parent Chuanmai 42 derived from synthetic hexaploid wheat. Pak J Agric Sci 55:33–45

Liu J, Fu Z, Xie H, Hu Y, Liu Z, Duan L, Xu S (2011) Identification of QTLs for maize seed vigor at three stages of seed maturity using a RIL population. Euphytica 178:127–135

Lopes MM, Barbosa RM, Vieira RD (2012) Methods for evaluating the physiological potential of scarlet eggplant (Solanum aethiopicum) seeds. Seed Sci Technol 40:86–94

Lutts S, Paolo B, Lukasz W, Szymon KS, Roberta P, Katzarina L, Muriel Q, Malgorzata G (2016) Seed priming: new comprehensive approaches for an old empirical technique. In: Araujo S, Balestrazzi A (eds) New challenges in seed biology - basic and translational research driving seed technology. IntechOpen, London, pp 1–46. https://doi.org/10.5772/64420

Magnée KJ, Scholten OE, Postma J, van Lammerts BET, Groot SP (2020) Sensitivity of spinach seed germination to moisture is driven by oxygen availability and influenced by seed size and pericarp. Seed Sci Technol 48:117–131. https://doi.org/10.15258/sst.2020.48.1.13

Mahajan S, Mittal SK, Das A (2018) Machine vision based alternative testing approach for physical purity, viability and vigour testing of soybean seeds (*Glycine max*). J Food Sci Technol 55: 3949–3959. https://doi.org/10.1007/s13197-018-3320-x

Marcos-Filho J (2015) Seed vigor testing: an overview of the past, present and future perspective. Sci Agric 72:363–374. https://doi.org/10.1590/0103-9016-2015-0007

Marcos-Filho J, Kikuti ALP, Lima LB (2009) Procedures for evaluation of soybean seed vigor, including an automated computer imaging system. Revista Brasileira de Sementes 31:102–112. (in Portuguese, with abstract in English)

Massimi M (2018) Impact of seed size on seed viability, vigor and storability of *Hordeum vulgare* L. Agric Sci Dig 38:62–64. https://doi.org/10.18805/ag.A-293

Matsunami MH, Hayashi M, Murai-Hatano J, Ishikawa-Sakurai S (2021) Effect of hydropriming on germination and aquaporin gene expression in rice. Plant Growth Regul 97:263. https://doi.org/10.1007/s10725-021-00725-5

Matthews S, Bradnock WT (1967) The detection of seed samples of wrinkle-seeded peas (*Pisum sativum* L.) of potentially low planting value. Proc Int Seed Test Assoc 32:553–563

Matthews S (1980) Controlled deterioration: a new vigour test for crop seeds. In: Hebblethwaite PD (ed) Seed production. Butterworths, London, UK, pp 647–661

Matthews S, Khajeh-Hosseini M (2007) Length of the lag period of germination and metabolic repair explain vigour differences in seed lots of maize. Seed Sci Technol 35:200–212

Matthews S, Powell A (2009) Seed viability and vigour. In: Singhal NC (ed) Seed science and technology. Kalyani Publishers, New Delhi, pp 115–146

McDonald MB, Phannendranath BR (1978) A modified accelerated aging vigor test procedure. J Seed Technol 3:27–37

Mehmet KD, Suay B, Gamze K, Oguzhan U (2011) Seed vigour and ion toxicity in safflower (*Carthamus tinctorius* L.) seedlings produced by various seed sizes under NaCl stress. Arch Biol Sci 63:723–729

Mir HR (2010) Studies on physical, physiological and biochemical changes associated with seed enhancements in maize. Graduate dissertation. Indian Ageicultural Research Institute, New Delhi

Mir HR, Yadav SK, Ercisli S, Al-Huqail AA, Dina SA, Siddiqui MH, Alansi S, Sangita Y (2021) Association of DNA biosynthesis with planting value enhancement in hydroprimed maize seeds. Saudi J Biol Sci 28:2634–2640. https://doi.org/10.1016/j.sjbs.2021.02.068

Morris K, Barker GC, Walley PG, Lynn JR, Finch-Savage WEC (2016) Trait to gene analysis reveals that allelic variation in three genes determines seed vigour. New Phytol 212:964–976. https://doi.org/10.1111/nph.14102

Muhsin M, Nawaz M, Khan I, Chattha MB, Khan S, Aslam MT, Chattha MU (2021) Efficacy of seed size to improve field performance of wheat under late sowing conditions. Pak J Agric Res 34:247–253

Nagarajan S, Chahal SS, Gambhir PN, Tiwari PN (1993) Relationship between leaf water relation parameter in three wheat cultivars. Plant Cell Environ 16:87–92

Nedunchezhiyan V, Velusamy M, Subburamu K (2020) Seed priming to mitigate the impact of elevated carbon dioxide associated temperature stress on germination in rice (*Oryza sativa* L.). Arch Agron Soil Sci 66:83–95. https://doi.org/10.1080/03650340.2019.1599864

Nerson H (2002) Relationship between plant density and fruit and seed production in muskmelon. J Am Soc Hortic Sci 127:855–859

Oliveira M, Matthews S, Powell AA (1984) The role of split seed coats in determining seed vigour in commercial seed lots of soyabean as measured by the electrical conductivity test. Seed Sci Technol 12:659–668

Pandita VK, Anand A, Nagarajan S (2007) Enhancement of seed germination in hot pepper following presowing treatments. Seed Sci Technol 35:282–290

Parrera CA, Cantliffe DJ (1994) Pre-sowing seed priming. Hortic Rev 16:109–141

Perry DA (1978) Report of the vigour test committee 1974–1977. Seed Sci Technol 6:159–181

Powell AA (1985) Impaired membrane integrity—a fundamental cause of seed quality differences in peas. In: Hebblethwaite Heath MC, Dawkins TCK (eds) The pea crop. Butterworths, London, pp 383–395

Powell AA (2017) A review of the principles and use of the Q2 seed analyser. International Seed Testing Association, Bassersdorf. https://www.seedtest.org/api/rm/9DD4DNNYFVH4ZV6/areviewoftheprinciplesanduseoftheq2seedanalyser1.pdf

Reed RC, Bradford KJ, Khanday I (2022) Seed germination and vigor: ensuring crop sustainability in a changing climate. Heredity 128:1–10. https://doi.org/10.1038/s41437-022-00497-2

Rodrigues M, Junior FGG, Marcos-Filho J (2020) Vigor-S: system for automated analysis of soybean seed vigor. J Seed Sci 42:e202042039. https://doi.org/10.1590/2317-1545v42237490

Sako Y, Mc Donald MB, Fujimura K, Evans AF, Bennett MA (2001) A system for automated seed vigour assessment. Seed Sci Technol 29:625–636

Sattler SE, Gilliland LU, Magallanes-Lundback M, Pollard M, Della Penna D (2004) Vitamin E is essential for seed longevity and for preventing lipid peroxidation during germination. Plant Cell 16(6):1419–1432. https://doi.org/10.1105/tpc.021360

Shaheb MR, Islam MN, Nessa A, Hossain MA (2015) Effect of harvest times on the yield and seed quality of French bean. SAARC J Agric 13:1–13

Sharma V (2018) Seed composition and its relation with vigour and storage behaviour in maize. Doctoral Dissertation. Indian Agricultural Research Institute, New Delhi

Shi H, Guan W, Shi Y, Wang S, Fan H, Yang J, Chen W, Zhang W, Sun D, Jing R (2020) QTL mapping and candidate gene analysis of seed vigor-related traits during artificial aging in wheat (*Triticum aestivum*). Sci Rep 10:22060–22063

Singh A, Gupta R, Pandey R (2016) Rice seed priming with picomolar rutin enhances rhizospheric Bacillus subtilis CIM colonization and plant growth. PLoS One 11:e0146013

Soeda Y, Maurice CJM, Konings Oscar V, van Houwelingen AMML, Geert M, Stoopen CA, Maliepaard JK, Bino RJ, Groot SPC, van der Geest AHM (2005) Gene expression programs during brassica oleracea seed maturation, osmopriming, and germination are indicators of progression of the germination process and the stress tolerance level. Plant Physiol 137:354–368

Steiner F, Zuffo AM, Busch A, Sousa TDO, Zoz T (2019) Does seed size affect the germination rate and seedling growth of peanut under salinity and water stress? Pesquisa Agropecuária Tropical 49:1

Styer RC, Cantliffe DJ (1983) Changes in seed structure and composition during development and their effects on leakage in two endosperm mutants of sweet corn. J Am Soc Hortic Sci 108:717–720

Sun L, Miao Z, Cai C (2015) GmHs1-1, encoding a calcineurin like protein, controls hard-seededness in soybean. Nat Genet 47:939–943

Tekrony DM, Egli DB (1997) Accumulation of seed vigour during development and maturation. In: Ellis RH, Black M, Murdoch AJ, Hong TD (eds) Basic and applied aspects of seed biology. Kluwer, Dordrecht, pp 369–384

Toorop PE, van Aelst AC, Hilhorst HWM (1998) Endosperm cap weakening and endo-β-mannanase activity during priming of tomato (*Lycopersicon esculentum* cv. Moneymaker) seeds are initiated upon crossing a threshold water potential. Seed Sci Res 8:483–492. https://doi.org/10.1017/S0960258500004451

Umarani R, Bhaskaran M, Vanitha C, Tilak M (2020) Fingerprinting of volatile organic compounds for quick assessment of vigour status of seeds. Seed Sci Res 30:112–121. https://doi.org/10.1017/S0960258520000252

Varier A, Vari AK, Dadlani M (2010) The subcellular basis of seed priming. Curr Sci 99:450–456

Vashisth A, Nagarajan S (2010) Effect on germination and early growth characteristics in sunflower (*Helianthus annuus*) seeds exposed to static magnetic field. J Plant Physiol 167:149–156. https://doi.org/10.1016/j.jplph.2009.08.011

Wilson DO Jr, Mohan SK (1998) Unique seed quality problems of sh2 sweet corn, vol 20. Seed Technology, pp 176–186

Woodstock LW (1988) Seed imbibition: a critical period for successful germination. J Seed Technol 12:1–15

Zhao G, Wang J (2005) Seed vigour test of sweet corn (*Zea mays* L. *saccharata* Sturt) and evaluation of its field survival ability. Plant Physiol Commun 41:444–448

Zhao G, Zhong T (2012) Improving the assessment method of seed vigor in *Cunninghamia lanceolata* and *Pinus massoniana* based on oxygen sensing technology. J For Res 23:95–101. https://doi.org/10.1007/s11676-012-0238-4

Seed Longevity and Deterioration

Umarani Ranganathan and Steven P. C. Groot

Abstract

The fundamental deteriorative processes that lead to loss of seed viability contrastingly vary between desiccation insensitive (orthodox) and desiccation sensitive seeds (recalcitrant). Orthodox seeds which undergo maturation drying are bestowed with protective mechanisms which guard the seeds against deterioration. They include the accumulation of antioxidants, non-reducing sugars, protective proteins such as late embryogenesis abundant proteins, heat-shock proteins, lipocalins, hormones and chemical protectants (raffinose family oligosaccharides, flavonoids, lignins, vitamin E). The nuclear DNA is packed denser and chlorophyll is degraded. Besides, the cytoplasm is capable of transitioning between liquid and glassy state depending on the moisture content of the seeds aiding in the maintenance of seed viability potential. In the dry seeds, the glassy state of the cytoplasm ensures the stabilization of cellular components by arresting cell metabolism. However, even with low moisture content and a glassy state of cytoplasm, reactive oxygen species generated due to the presence of oxygen in the storage atmosphere may cause the ageing of seed. As the seed moisture content increases, mitochondrial respiration gets activated, also leading to increased production of reactive oxygen species, owing to inefficient mitochondrial activity. The reactive oxygen species lead to the oxidation of essential molecules such as DNA, RNA, proteins and lipids. Further, mitochondrial membranes also get oxidized, leading to reduced aerobic respiration potential. When the damage is not substantial, orthodox seeds are capable of repairing the molecular damages that accumulate during storage, enabling the seeds to partially

U. Ranganathan (✉)
Seed Centre, Tamil Nadu Agricultural University, Coimbatore, India

S. P. C. Groot
Wageningen University & Research, Wageningen Seed Centre, Wageningen, Netherlands
e-mail: steven.groot@wur.nl

© The Author(s) 2023
M. Dadlani, D. K. Yadava (eds.), *Seed Science and Technology*,
https://doi.org/10.1007/978-981-19-5888-5_5

91

overcome the damages and extend their longevity. This includes activation of repair of cell membranes, DNA, RNA, proteins and mitochondria as the seeds imbibe water.

Unlike the orthodox seeds, the recalcitrant seeds are largely devoid of protective mechanisms which guard the seeds against rapid deterioration. The recalcitrant seeds are shed from the mother tree at high moisture content while they are metabolically active. After dispersal, the seeds undergo deteriorative changes during drying due to the damage to the cytoskeleton (physical damage), besides reactive oxygen species-induced damage due to lack of antioxidant activity (metabolism-induced damage). Even when maintained under high moisture content, seeds exhibit dysfunction of the cell organelles and extensive vacuolization predisposing the seeds to deterioration. Thus, recalcitrant seeds are prone to deterioration either under low or high moisture content.

Keywords

Orthodox longevity behaviour · Reactive oxygen species · Seed deterioration · Seed ageing · LEA proteins · Glassy state · Recalcitrant seed

1   Introduction

Seed, the basic unit of propagation, serves a crucial role in the evolution and survival of higher plants. Acquisition of desiccation tolerance during the late maturation stage bestows upon orthodox seeds a unique property of remaining viable for a prolonged period in a dry quiescent state. However, as all living organisms, seeds also undergo ageing, gradually losing vigour and viability, until they eventually die. Seed longevity refers to the period from maturation to the loss of seed viability in dry storage. Seed storage potential is highly related to the environmental conditions of storage, the physiological status of seeds as well as genetic factors. Certain plant species have shown a remarkably long longevity period for their seeds, e.g. lotus (*Nelumbo nucifera*) seeds have remained viable for nearly 1300 years (Shen-Miller et al. 2002) and a *Phoenix dactylifera* seed could germinate even after 2000 years (Sallon et al. 2008). The variation in the shelf life of seeds may be due to differences in the biochemical components of these seeds, which may be attributed to their genetic constitution. Seeds of certain species have a very short storage life. For instance, seeds of trees belonging to *Dipterocarpaceae* as those from *Shorea robusta* remain viable only for 7–10 days (Saha et al. 1992).

A characteristic and contrasting seed storage behaviour among plant species has led to the classification of seeds into two prominent types viz., recalcitrant and orthodox seeds (Bewley et al. 2013). The recalcitrant seeds are 'desiccation-sensitive'. On maturation these seeds do not undergo drying or undergo drying only to a limited extent, therefore they are shed from the trees with relatively high moisture content of 0.4–4.0 g water/g and often high metabolic activity. For that reason, the post-harvest life of the recalcitrant seed is very short, extending from just few days to few months, depending on the species and storage conditions. Some recalcitrant seeds from temperate regions, as those of *Quercus suber* (cork oak) can be stored for about a year at temperatures around 0 °C, due to slowed down metabolic activity, but eventually die due to depletion of reserve carbohydrates (SPC Groot, personal observations). Recalcitrant seeds from tropical species do not withstand such cold storage and consequently have a very short shelf life. On the other hand, the orthodox seeds, which are desiccation-insensitive undergo maturation drying to reach a moisture content of <15%, before being shed from the mother plant. They can generally be stored at ambient conditions for a very long period ranging from months to many years. A third category with intermediate seed storage behaviour has been identified for seeds that can withstand desiccation but not storge at sub-zero temperatures (Ellis et al. 1990).

The loss of seed viability during storage occurs due to the onset of deterioration processes. Seed ageing or seed deterioration is commonly described as an irreversible, cumulative and inexorable process (McDonald 1999) that can cause the build up of cellular damages, and result in delayed seedling emergence, reduced ability to withstand stresses, and ultimately loss of viability (Zhang et al. 2021). Limited damage can often be repaired, but as this will take energy and time, it will result in a slowing down of average germination rate, accounting for protrusion of the radicle. Substantial deterioration may still allow this embryonic root protrusion but may result in seedling abnormalities and in case of extreme deterioration of seeds, no radicle protrusion and seedling formation occurs.

The process of seed deterioration distinctly varies between the orthodox and recalcitrant seeds. Wide variations of longevity are observed among species, genotypes and populations (Clerkx et al. 2004; Walters et al. 2005; Probert et al. 2009), which to a large extent are attributed to the differences in their biochemical or biophysical characteristics (Walters et al. 2010). Irrespective of the inherent storage potential of the seeds, all the orthodox seeds show extended viability under cool and dry storage, while at elevated temperatures and relative humidity, the seed storability is greatly reduced (Ellis and Hong 2006). Seed longevity duration is governed mainly by two factors, these are the genetic constitution—which is manifested in the chemical constitution and structural features of the seed parts, and the environment in which the seed is stored. Among the environmental factors, relative humidity (RH) of the storage air which directly influences the seed moisture content and the temperature are the two primary factors influencing seed longevity besides the gaseous composition of the storage environment (Roberts 1972). Hence, seed longevity depends on environmental and seed factors.

2 Seed Factors

The storage potential of the orthodox seeds has been attributed to cellular protective mechanism. At the end of seed development, orthodox seeds undergo the maturation processes, with corresponding decline in moisture level and suspension of their metabolic activities. The dry state is called as quiescent state which aids in the seed storage potential. The protective mechanisms imposed during late seed maturation include the accumulation of antioxidants, non-reducing sugars, and protective proteins such as late embryogenesis abundant (LEA) proteins, heat shock proteins (HSPs) and lipocalins. The nuclear DNA is packed denser and chlorophyll is degraded. Seed storage proteins can act as primary targets of oxidation, thereby helping in buffering of reactive oxygen species (ROS) generated during dry storage. Since chlorophyll levels can be measured very sensitively using its fluorescent properties, chlorophyll fluorescence levels of individual seeds can be used as a marker for seed maturity for those seeds that contain chlorophyll during development, as most do. The lower the level of fluorescence, higher the level of seed maturity.

2.1 Role of Chemical Protectants in Various Tissues

The main damaging factor during seed storage is oxidative stress induced by ROS. Molecular and enzymatic antioxidants are therefore essential in seed longevity. Under dry conditions, enzymes cannot access the ROS due to restricted molecular mobility within the cytoplasm (Gerna et al. 2022). Under these dry conditions, seeds rely on low molecular weight antioxidants as tocochromanols (tocopherols and tocotrienols), ascorbate (vitamin C) and glutathione. Seeds are often rich in tocopherols, including vitamin E (alfa-tocopherol), which are lipophilic antioxidants, that play an important role in preventing oxidation of storage and membrane lipids. Vitamin E-deficient *Arabidopsis* mutants have a considerably reduced seed longevity (Sattler et al. 2004). Ascorbate and glutathione are the main water soluble antioxidants present in seeds. At the low water content of dry seeds, enzymatic regeneration of the antioxidants is not possible and eventually the antioxidant pool will get exhausted (Groot et al. 2012; Gerna et al. 2022). Exposing seeds to higher RH environment, or by imbibition, will allow enzymatic ROS scavenging activity by glutathione-reductase, superoxide dismutase, peroxidases and catalases. These enzymes also play a role in regeneration of molecular antioxidants. A study with barley seeds showed that tocopherol and glutathione levels decrease during seed ageing, both under dry genebank storage and during controlled deterioration at 45 °C and 75% RH (Roach et al. 2018).

During seed maturation, the cells accumulate sucrose and raffinose family oligosaccharides (RFOs), which are raffinose, stachyose and verbascose. These sugars have been suggested to play a role in the formation of the glassy state and thereby improving seed longevity (Bernal-Lugo and Leopold 1992; Vandecasteele et al. 2011). Galactinol is the direct precursor of raffinose and a positive genetic correlation between seed galactinol content and longevity has been shown for Arabidopsis, tomato and cabbage, while its role was confirmed by shorter longevity shown by seeds from Arabidopsis galactinol synthase mutants (de Souza Vidigal et al. 2016). In that study Arabidopsis and tomato seed longevity was tested at 85% RH and 40 °C, while the cabbage seeds had been stored in paper bags at 20 °C without RH control. Galactinol has also been shown to enable protection against oxidative stress in Arabidopsis leaves (Nishizawa et al. 2008).

Late embryogenesis abundant proteins (LEAs) and heat shock proteins (HSPs) are synthesized at the end of seed maturation. They play a role in seed longevity, by stabilization of the glassy cytoplasm, protecting structural proteins, condensation of chromatin and dismantling of thylakoids in chloroplasts (Sano et al. 2015; Ballesteros et al. 2020).

The seed coat is a maternal tissue which surrounds the embryo and nutritive tissues, forming both a physical and biochemical layer of protection. The seed coat cells become dead at the end of the seed development. Metabolites accumulated during seed development determine the composition and structure of the seed coat and influence the chemical and mechanical protection to the seed and the longevity potential. The polyphenols found in the seed coat are flavonoids, lignins and lignans. During initial seed development polymeric colourless compounds accumulate in the vacuoles of the innermost layer or endothelium cells. Later during desiccation, they are oxidized into a brown pigment by polyphenol oxidase to form flavonoids called flavonols. The flavonoids act as antioxidants and scavenge the ROS eventually reducing the oxidative stress. In rapeseed (*Brassica napus*) the dark-pigmented seeds survive longer under accelerated ageing conditions (Zhang et al. 2006). Peroxidation of flavonoids accumulated in seed coats may cause browning and reduced water permeability of the seed coat. The proanthocyanidins (PAs) (also known as condensed tannins) present in the seed coat can also have antimicrobial properties, thereby providing a chemical barrier against infections by fungi. In cowpea (*Vigna unguiculata*), PAs were also found to be poisonous to bruchid larvae and prevented their infestation (Lattanzio et al. 2005).

Lignin is a polymer of monolignol units which are rich in flax seeds. It is known to protect the seeds against mechanical stress (Capeleti et al. 2005) besides having antioxidant properties. Defence-related proteins which accumulate in testa of Arabidopsis and soybean (*Glycine max*) are polyphenol oxidases, peroxidases and chitinases (Moïse et al. 2005; Pourcel et al. 2005).

2.2   Role of Hormones

Hormones such as abscisic acid, auxin and gibberellic acid are reported to be involved in acquisition of seed longevity. In Arabidopsis, the loss of functional mutants of ABI3 (ABSCISIC ACID INSENSITIVE 3) and LEC1 (LEAFY COTY-LEDON 1) produced seeds that lost their viability within few weeks of harvest (Sugliani et al. 2009), implying that these LAFL transcription factors influenced the longevity of seeds. ABI3 is also shown to regulate seed de-greening by induction of *STAYGREEN* (Delmas et al. 2013) and induce acquisition of seed longevity. ABI3 and LEC1 were reported to directly or indirectly regulate proteins such as HSP and LEA besides storage proteins which were linked with seed longevity. The alternative splicing of *ABI3* mRNA is expected to be associated with acquisition of seed longevity in Arabidopsis and many other species because the relative abundance of transcripts of full-length *ABI3i* decreased as the seeds entered maturation drying.

ABI5 is yet another important regulator of longevity in some plants such as *Medicago truncatula* and pea (Zinsmeister et al. 2016). The seeds of *abi5* mutants of these species showed 40–60% reduction in longevity compared to wild-type, in dry storage. ABI5 regulates LEA proteins and raffinose family oligosaccharide (RFO) synthesis which plays a role in conferring stability to the seeds in dry state. It is also known to regulate antioxidant levels in seeds during storage, especially tocopherol profile (Zinsmeister et al. 2016).

ABI5 acts downstream of ABI3 and regulates AtEM gene expression (Lopez-Molina et al. 2002) suggesting that interactions existed between ABI3 and PYR/PP2C/SnRK2 signalling pathways. ABA regulates aquaporins which are involved in transport of water and other small molecules including H_2O_2. ABA may control water relations and H_2O_2 accumulation via ABI3 modulation of aquaporins, thereby contributing to seed longevity (Sano et al. 2015).

Auxin is also involved in the regulation of seed longevity (Righetti et al. 2015; Carranco et al. 2010). Exogenous application of auxin during maturation had increased the seed longevity period. In the embryo, auxin signalling was found to be associated with ABA signalling pathway. Auxin is proven to induce the expression of *ABI3* and its LEA protein target EARLY METHIONE1 (EM1) since their expression was deregulated in the auxin biosynthesis mutants. Auxin could also regulate the genes involved in longevity directly (Pellizzaro et al. 2019).

Gibberellins offer resistance against seed deterioration by reinforcing the seed coat (Bueso et al. 2014). Increased tolerance of seeds to accelerated ageing was observed when there was an overexpression of ARABIDOPSIS THALIANA HOMEOBOX 25 (ATBH25) leading to increased accumulation of gibberellins and transcripts of the gibberellin biosynthesis gene GIBBERELLIN 3—OXIDASE 2 (Bueso et al. 2014). However, since mutants with defective gibberellin synthesis had also recorded high seed longevity as that of wild-type, the role of gibberellin seemed to be inconclusive (Clerkx et al. 2004).

Brassinosteroids (BR) may also positively influence seed longevity since it has been reported that the BR-deficient mutants *cyp85at/a2* and *det2* recorded significantly longer longevity period than the wild type (Sano et al. 2017).

3   Storage Factors

3.1   Moisture Content, Water Activity or Equilibrium Relative Humidity

The main driver for seed deterioration during storage of orthodox seeds is humidity. Water is important for most chemical and enzymatic reactions. Oxidation of lipids, proteins and nucleic acids, important building blocks of living organisms, is stimulated by moisture, oxygen and temperature. In the oily or lipophilic part, deterioration is due to oxidation of the unsaturated fatty acids in the oil bodies and membranes. Whereas in the non-oily, or hydrophilic part, deterioration is mainly due to oxidation of proteins, DNA and RNA and cross-linking of macromolecules. An initial rule of the thumb for the quantitative effects of humidity on seed ageing was formulated by Harrington (1972), stating that when seed moisture content is between 5% and 14%, each 1% decrease in seed moisture doubles the shelf life of the seeds.

While studying the effect of moisture on seed ageing, a clear distinction should be made between seed moisture content and water activity or storage relative humidity (RH). Traditionally seeds were characterized by data on their moisture content concerning the seed trade. Also, seed technologists were used to defining the humidity level of the seeds in moisture content, either on a fresh or dry-weight basis. But the seed moisture content does not define the availability of water in the non-oily part of the seeds, and hence the deterioration processes and rates at which these reactions are taking place. Imagine castor bean (*Ricinus communis*) seeds containing 50% oil and 10% moisture content on a wet basis, which will mean a moisture content of 20% in the non-oily part. In comparison, seeds from a common bean (*Phaseolus vulgaris* L.) that contain only 2% oil will experience at the same total seed moisture content only about 10% moisture in their non-oily part. As a consequence, the physiological conditions of both seeds will differ, despite their similar seed moisture content. Seed oil content does not only vary between crops but can also vary between varieties (Yao et al. 2020) and production conditions. In the food industry, it is common to use water activity (a_w) as a measure of the moisture status of commodities, including seeds. When in equilibrium with the humidity of the surrounding air, the a_w is more or less linearly related to the relative humidity RH, be it that a_w is expressed between 0 and 1.0 and the RH in percentages. At the first Seed Longevity Workshop of the International Society of Seed Science (Wernigerode, Germany, 5–8 July 2015) it was concluded that for studies on seed ageing it is better to compare seeds based on their a_w or equilibrium RH (eRH) instead of their moisture content.

According to the Seed Viability Equation, seed survival prolongs with decreasing seed moisture contents, but there is a limit to this. In fact at very low moisture levels, under so-called ultra-dry storage conditions, which equals eRH levels below around 15–20%, seed deterioration could be faster (Chai et al. 1998). This is likely because of a faster rate of lipid oxidation, as has been described in food science (Labuza and Dugan 1971).

3.1.1 Glassy or Liquid Cytoplasm

Non-reducing sugars such as sucrose and raffinose family oligosaccharides (RFOs) are stored in seeds as energy sources during the seed development process. Raffinose and stachyose are efficient inhibitors of sucrose crystallization. During late seed maturation, drying leads to transformation of cell cytoplasm from a fluid state to the viscosity of glass, thereby disrupting the normal crystal matrices (Koster and Leopold 1988). The glassy state has an extremely low molecular mobility that enables arresting of cell metabolism and stabilization of cellular components. This in turn reduces the deteriorative processes and results in the extension of seed longevity (Sun 1997). At 25 °C for at least four species with varying seed oil content, the transition from a cytoplasmic glass phase to a more viscous cytoplasm is reported to occur at an RH between 44% and 49%, which is lower than the RH levels that leads to loss of viability in recalcitrant seeds (Buitink and Leprince 2008). Glass formation is therefore not the mechanism for desiccation tolerance, but it enables the orthodox seeds to withstand long term survival under dry conditions.

Upon incubation at higher RH levels or imbibition, the seeds absorb moisture and the cytoplasm can first become viscous, also called rubbery and subsequently liquid, allowing enzyme activity and resumption of metabolic activities. With increasing temperatures, this transition occurs even at a lower RH. The rubbery phase is a damaging state as the hydration level increases the rate of deleterious chemical processes, while it is still too low for the active repair of the cellular damage (Bewley et al. 2013). After transition towards a liquid cytoplasm, enzymes can become active. At higher levels of eRH, more enzymes may become active. With storage of peanuts at different RH levels in hermetically closed containers, oxygen levels significantly start to decline from an RH of 70% onwards, but the oxygen consumption considerably speeds up with a further increase of the RH, indicating onset of aerobic respiration activities from eRH of around 70%, which accelerates with the increase in RH (Groot et al. 2022).

The glass phase transition is crucial for the physiological status of the seeds and the response to environmental conditions. A common error made in the studies on seed ageing is that the analyses of enzyme activity are performed on liquid extracts obtained from dry stored seeds. Extracts made from dry stored seeds can indeed have some enzyme activity, but that only shows that these enzymes present in these seeds can be activated in a watery extract. These enzymes may include those involved in the scavenging of reactive oxygen species (ROS), repair of damage or respiration activity providing cellular energy. However, this does not mean that they are active in the dry seed. Only in the liquid cytoplasmic state these enzymes can become active in situ, while their actual activity still depends on the available water. Some enzymes will require full hydration during imbibition to express maximum activity.

3.2   Temperature

Seeds deteriorate faster at higher temperature conditions, as these accelerate the rate of chemical oxidation. For this reason, gene banks are recommended to dry and store their valuable germplasm at sub-zero temperatures (Rao et al. 2006). More expensive seeds of horticultural species are generally stored in temperature and humidity-controlled warehouses, set at 15 °C and 30% RH. As mentioned above, increasing the temperature decreases the a_w at which glass phase transition occurs. A second rule of thumb formulated by Harrington stated that the storage life approximately doubles for every 5 °C decrease in temperature (Harrington 1972). Ellis et al. (1989) suggested that chemical reactions that take place in the stored seeds are dependent mainly on moisture content of the seed as well as the storage temperature. Indeed it has been shown for many desiccation-tolerant seeds that by lowering the storage temperature, seed longevity can be improved. Genes and Nyomora (2018), for example, reported that the seeds of *Escoecaria bussei* could retain the seed viability for a period of 9 months when stored at 15 °C, while they lost the germination potential within 3 months, if stored at 30 °C. Strelec et al. (2010) also observed that wheat seeds stored at 40 °C had a greater decrease in germination and vigour compared to seeds stored at 25 °C. Seeds with an intermediate seed storage behaviour can withstand desiccation but are sensitive to storage at sub-zero temperatures (Ellis et al. 1990). An example is seeds from oil palm (*Elaeis guineensis* Jacq.) (Ellis et al. 1991).

3.3   Oxygen

Most organisms need oxygen to survive and allow metabolic activity, but seeds in a state of glassy cytoplasm under dry conditions are not metabolically active and they do survive without the need of oxygen (O_2) (González-Benito et al. 2011). Only above a water potential of about -14 MPa providing oxygen becomes essential to sustain respiration (Roberts and Ellis 1989). Under dry storage conditions, molecular oxygen is a main source of ROS. The internal environment of seeds is rich in metal ions such as Fe^{2+}, Cu^{2+} and Zn^{2+}, needed for metabolism in the emerging seedling as cofactors for enzymes, including those involved in respiration and photosynthesis. Interaction of molecular oxygen with these metal ions results in the formation of ROS (Fig. 1) (Hayyan et al. 2016). These ROS can react with organic molecules, resulting in oxidation of lipids, proteins and nucleic acids. Therefore, the reason for seed deterioration under dry conditions is oxidation. The higher the oxygen concentration, the faster the ageing of seeds (Groot et al. 2012). In the food industry it is common practice to limit food oxidation, mainly lipid oxidation that would otherwise result in a rancid taste, by packaging under controlled atmosphere with low oxygen levels or even anoxia. Indeed, low oxygen or anoxia storage under dry conditions can extend seed longevity considerably (González-Benito et al. 2011; Schwember and Bradford 2011; Gerna et al. 2022).

To a lesser degree, ROS can also be generated in dry seeds by autocatalytic lipid oxidation. Under humid storage conditions, when seeds possess liquid cytoplasm,

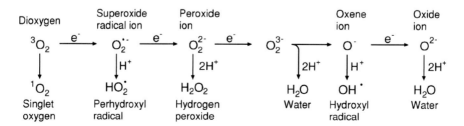

Fig. 1 Generation of ROS by energy transfer or sequential reduction of triplet oxygen. (Original picture: Apel and Hirt 2004)

mitochondrial respiration activity can start, and ROS may also be generated by the inefficient mitochondrial activity.

3.4   Pests and Pathogens

In addition to the direct influence of RH, temperature and oxygen on seed deterioration, these also play an important role in the survival and multiplication of pests and pathogens, which impacts the loss of seed viability. Fungi can severely deteriorate seeds during storage. They can grow in and on the seeds at about 70% RH with an increased rate at higher humidity levels, but they are considerably reduced in their survival at temperatures below zero (Roberts 1972). Although storage under reduced oxygen levels can strongly inhibit fungal growth, fungi like *Aspergillus* spp. can still proliferate at 0.5% oxygen, be it at a lower rate (Hall and Denning 1994).

Insects as the rice weevil and cowpea bruchid can induce considerable damage to storage grains and pulses. Some insects can survive and propagate under rather dry conditions, till about 30% RH (Roberts 1972), but insects are sensitive to sub-zero temperatures and hypoxia. Both conditions are recommended as alternative for chemical fumigation (Kalpana et al. 2022; Kandel et al. 2021). However, it should be considered that the eggs or pupae can have, related to their low metabolic activity, a greater tolerance compared to the adult insects and modified atmospheric treatments can take some weeks, also depending on the temperature.

4   Modelling Seed Ageing

Ageing of seed lots should be considered on a population level, in which the frequency of seeds capable of radicle protrusion decreases over storage time, with a sigmoidal curve. When this curve is plotted on a probit scale, the decline in frequency of surviving seeds will show a straight line (Ellis and Roberts 1980). The angle of this line may vary between species. Variation in storage environments will also change the angle of this line, decreasing in case of less deteriorating conditions. Plotting survival curves of seed lots using probit transformation can aid in estimating the p50 value, the storage time resulting in a 50% decline in viability, to characterize the effect of genetics or seed treatments on the tolerance to ageing under specific conditions (Hay et al. 2018). According to Ellis and Roberts (1980), the viability of the seed population after a certain storage period can mathematically be defined by the equation

$$v = K_i - p/\sigma,$$

The percentage viable seeds (v) are equal to the initial viability (K_i, or intercept with the Y-axis) decreased by the seed death expected during the storage period (p) according to the slope ($1/\sigma$) of the line. To estimate the survival of seeds from specific species under different storage conditions, Ellis and Roberts (1980) have developed the Seed Viability Equation:

$$\log \sigma = K_{\mathrm{E}} - C_{\mathrm{W}} \log m - C_{\mathrm{H}} T - C_{\mathrm{Q}} T^2,$$

Here K_{E} is a species constant that accounts for the differences in storability among species, m is the seed moisture content (in percentage on fresh weight basis), and T is the storage temperature (°C). C_{W}, C_{H} and C_{Q} are species-specific constants, empirically determined by storing seeds at different moisture levels in hermetically sealed pouches at different temperatures. Related to the large efforts and time needed for these experiments, it should be considered that these constants have been determined for only limited number of species and most often for only a single seed lot of that species. A comparison of eight species indicated that the temperature constants C_{H} and C_{Q} are rather similar and the constant C_{W} for moisture content is common when expressed in relation to the equilibrium RH instead of absolute moisture content (Bewley et al. 2013).

The Seed Information Database maintained by the Royal Botanic Gardens, Kew, UK (http://data.kew.org/sid/viability/) has compiled available data on these species-specific constants, along with tools to calculate estimated seed longevities, under different seed moisture contents and temperature conditions, and the seed moisture content at a certain RH, based on the seed oil content.

Limited data have been published on the quantitative effect of oxygen on seed ageing, but very strong shelf life-extending effects have been reported for seeds of *Brassica* species where seed viability had not or hardly declined after almost 40 years storage under a combination of ultra-dry storage and anoxia (González-Benito et al. 2011). Preliminary data of experiments with dry storage of primed celery seeds under different oxygen levels and temperatures shows that halving the oxygen level during storage increases shelf life with a factor of 1.6, and this effect was independent of the storage temperature (Groot et al. unpublished data). It was observed that reducing the oxygen level from 21% (air) to 1% extends shelf life almost eight times.

5 Estimating Seed Longevity

Estimating the potential longevity or storability of commercial seed lots is of great significance in the seed trade. The viability equations and nomographs based on these were attempts to provide a reliable estimate of the shelf life of seed lots based on their initial germination (%), and specified regimes of seed moisture content or storage RH (%) and storage temperatures. Though useful to some extent, these tools could not differentiate between the lots of similar germination but variable vigour status. For managing the commercial seed inventory, it is important to estimate the shelf life of all seed lots of a given species or determine the efficacy of seed treatments on subsequent storability. To provide answers in short periods of time, experiments are performed to estimate storability at high humidity and temperature conditions. Such 'accelerated ageing' (AA) and 'controlled deterioration' (CD) tests are often performed under humidity levels between 75% and 100% RH and at temperatures around 40 or 50 °C. When seed lots showing large differences in tolerance to ageing are compared, they generally show similar rankings in these tests. However, for more subtle differences in tolerance, these tests seem to fail in showing a reasonable relationship with the shelf life of seeds under commercial storage conditions. Such deviations are more prominent in the case of primed seeds, which exhibit a fast decline in longevity under dry storage. The main reason for this lack of correlation is that commercial seed storage is performed under dry conditions, with a glassy cytoplasm, while in the accelerated ageing and controlled deterioration tests the cytoplasm is in a liquid state. Seed storage experiments performed at high RH levels can possibly provide more reliable information on the storage behaviour of high moisture seeds. When the same seeds are stored under dry conditions with a glassy cytoplasm, their relative ageing pattern can be quite different.

Since oxygen is a third factor associated with accelerating seed ageing under dry conditions, an alternative seed ageing treatment has been developed by storing seeds under an elevated partial pressure of oxygen (EPPO) in pressure tanks under 200 times the normal air pressure, without increasing the seed water content or storage temperature (Groot et al. 2012). This test has successfully been employed to analyse genetic variation in seed longevity with seeds from Arabidopsis (Buijs et al. 2020) and barley (Nagel et al. 2016). However, more extensive testing is needed with seeds of contrasting morphological, chemical and physiological characteristics to validate the robustness of this test.

6   Types and Causes of Seed Deterioration

Damage induced during storage of seed with a glassy cytoplasm is due to reactive oxygen species (ROS) that can induce oxidation of essential molecules, such as DNA, RNA, proteins and lipids. DNA oxidation results in strand breaks, that can only be repaired upon imbibition. Protein oxidation, or carbonylation, occurs mainly with the storage proteins, that are less likely to be protected against oxidation compared to structural proteins and enzymes. As such storage proteins may act as ROS scavengers. Oxidation of the unsaturated membrane phospholipids results in increasing their melting point and cross-linking, both make the cell and organellar membranes less flexible. This creates more risks on leakage when water passes the membranes upon imbibition, especially at lower temperatures or at large differences in water potential. Because of the latter, it is recommended to humidify aged seeds by storing them overnight at 100% RH before sowing. Oxidation of the mitochondrial membranes results in reduced aerobic respiration potential and the need for anaerobic respiration to sustain the supply of energy. Ethanol production during seed imbibition is inversely related to seed vigour (Woodstock and Taylorson 1981). Indeed, seeds stored dry for prolonged periods in the presence of oxygen show more anaerobic respiration during early germination, compared to seeds stored for a shorter period or under less deteriorating conditions (Kodde et al. 2012).

7   Repair Mechanisms

Protective mechanisms involved in repair of damage to the cell membranes, DNA, RNA, proteins and mitochondria are rapidly activated in the seeds once they imbibe water during the early hours of seed germination.

With respect to RNA, pre-existing or stored mRNAs are involved in resumption of metabolic activity in seeds immediately after imbibition (Bewley et al. 2013). Reduction of total RNA content and RNA integrity are associated with loss of germination ability of seeds. RNA is more vulnerable to oxidation damage compared to DNA since it possesses only a single strand. Damaged mRNA blocks translation, and this loss of translational activity in imbibed seeds is correlated with loss of seed longevity (Sugliani et al. 2009). The molecular mechanisms which are associated with repair of RNA have not yet been elucidated (Sano et al. 2015).

Similar to DNA and RNA, spontaneous damage to proteins, such as oxidations or covalent modifications, take place during seed ageing, which culminates in loss of protein function (Sugliani et al. 2009). The major damage of protein is due to oxidation of methionine to methionine sulfoxide by the ROS resulting in damage of protein (Stadtman 2006). The repair of oxidized proteins happens due to reduction of methionine sulfoxide by methionine sulfoxide reductase (MSR) (Weissbach et al. 2005). Therefore, MSR repair system plays a major role in preservation of seed longevity. Spontaneous covalent modification of proteins results in conversion of L-aspartyl or asparaginyl residues to abnormal Zisoaspartyl residues. L-Isoaspartyl O-methyl transferase (PIMT) is capable of repairing these residues. High PIMT activity has been reported with germinating sacred lotus seeds which show remarkable seed longevity duration (Shen-Miller et al. 2002). Thus, repair of protein appears to play a significant role in extending the longevity of seeds under dry storage.

8   Storability of Recalcitrant Seeds

Recalcitrant seeds possess high moisture content and are metabolically active when shed from the mother plant on maturation. They are largely devoid of seed protective mechanism which can enable the seeds to withstand desiccation and induce greater longevity potential. Delahaie et al. (2013) reported that critical LEA proteins, which contribute to the desiccation tolerance and extended storability of seeds, were absent in the recalcitrant seeds of *Castanospermum australe*. However, most recalcitrant seeds reportedly do accumulate LEA proteins and respond to ABA, drought or temperature (Farrant et al. 1996), although they may not be effective at very low seed water contents wherein seed would have already lost viability. Further, the organelles of a mature recalcitrant seeds were found to be in a highly differentiated state and they also undergo specific ultra-structural changes resembling the modifications that occur during the early stages of seed germination (Berjak and Pammenter 2000). The changes include development of mitochondria and deposition of starch, dense material in the plastids, mitochondrial development and appearance of golgi bodies in root primordia and strong development of polysomes. This reveals that 'dessication sensitive' recalcitrant seeds lack the ability to "switch off" the metabolic activity by undergoing maturation drying after attaining maturity, unlike the 'desiccation insensitive' orthodox seeds, eventually leading to poor seed storage potential.

The physiological basis of viability loss in recalcitrant seeds has been reviewed by Umarani et al. (2015). During dehydration the cytoskeleton of these seeds suffer physical or mechanical damage due to lack of intracellular support, which does not reassemble upon rehydration, resulting in viability loss. These seeds are also more prone to '*metabolism-induced damage*' because the embryos of seeds, when shed from the mother tree, are metabolically active with high respiration rate. An uncontrolled activity of ROS coupled with concurrent failure of antioxidant system also leads to generation of free radicals and destruction of metabolic system of the cells. Further, even if maintained with high seed moisture content, dysfunction of the cell organelles, extensive vacuolization, consumption of reserves, dysfunction of mitochondria (Pammenter and Berjak 2000), presence of unstacked golgi bodies, dilated endoplasmic reticulum cisternae and abnormal vacuoles become rampant (Motete et al. 1997). The confluence of all these changes results in acute damage to cell metabolism, culminating in the loss of viability of recalcitrant seeds.

References

Apel K, Hirt H (2004) Reactive oxygen species: metabolism, oxidative stress, and signal transduction. Annu Rev Plant Biol 55:373–399

Ballesteros D, Pritchard HW, Walters C (2020) Dry architecture: towards the understanding of the variation of longevity in desiccation-tolerant germplasm. Seed Sci Res 30(2):142–155

Berjak P, Pammenter N (2000) What ultrastructure has told us about recalcitrant seeds. Rev Bras Fisiol Veg 12:22–55. (Edicao Especial)

Bernal-Lugo I, Leopold AC (1992) Changes in soluble carbohydrates during seed storage. Plant Physiol 98(3):1207–1210

Bewley JD, Bradford KJ, Hilhorst HWM, Nonogaki H (2013) Seeds. Physiology of development, germiantion and dormancy, 392 pp. Springer, New York–Heidelberg–Dordrecht–London 2013978-1-4614-4692-7. Seed Sci Res 23(4):289–289

Bueso E, Muñoz-Bertomeu J, Campos F, Brunaud V, Martínez L, Sayas E, Ballester P, Yenush L, Serrano R (2014) *Arabidopsis thaliana* HOMEOBOX25 uncovers a role for gibberellins in seed longevity. Plant Physiol 164(2):999–1010

Buijs G, Willems LA, Kodde J, Groot SPC, Bentsink L (2020) Evaluating the EPPO method for seed longevity analyses in Arabidopsis. Plant Sci 301:110644

Buitink J, Leprince O (2008) Intracellular glasses and seed survival in the dry state. C R Biol 331(10):788–795

Capeleti I, Bonini EA, Ferrarese MDLL, Teixeira ACN, Krzyzanowski FC, Ferrarese-Filho O (2005) Lignin content and peroxidase activity in soybean seed coat susceptible and resistant to mechanical damage. Acta Physiol Plant 27(1):103–108

Carranco R, Espinosa JM, Prieto-Dapena P, Almoguera C, Jordano J (2010) Repression by an auxin/indole acetic acid protein connects auxin signaling with heat shock factor-mediated seed longevity. Proc Natl Acad Sci 107(50):21908–21913

Chai JF, Ma RY, Li LZ, Du YY (1998) Optimum moisture contents of seeds stored at ambient temperatures. Seed Sci Res 8:23–28

Clerkx EJ, Blankestijn-De Vries H, Ruys GJ, Groot SPC, Koornneef M (2004) Genetic differences in seed longevity of various Arabidopsis mutants. Physiol Plant 121(3):448–461

Delahaie J, Hundertmark M, Bove J, Leprince O, Rogniaux H, Buitink J (2013) LEA polypeptide profiling of recalcitrant and orthodox legume seeds reveals ABI3-regulated LEA protein abundance linked to desiccation tolerance. J Exp Bot 64:4559–4573. https://doi.org/10.1093/jxb/ert274

Delmas F, Sankaranarayanan S, Deb S, Widdup E, Bournonville C, Bollier N, Samuel MA (2013) ABI3 controls embryo degreening through Mendel's I locus. 1. Proc Natl Acad Sci 110(40): e3888–e3894

Ellis RH, Hong TD (2006) Temperature sensitivity of the low-moisture-content limit to negative seed longevity–moisture content relationships in hermetic storage. Ann Bot 97(5):785–791

Ellis RH, Roberts EH (1980) Improved equations for the prediction of seed longevity. Ann Bot 45(1):13–30

Ellis RH, Hong TD, Roberts EH (1989) A comparison of low moisture content limit to the logarithmic relation between seed moisture and longevity in 12 species. Ann Bot 63:601–611. https://doi.org/10.1093/oxfordjournals.aob.a087788

Ellis RH, Hong TD, Roberts EH (1990) An intermediate category of seed storage behaviour? I. Coffee. J Exp Bot 41(9):1167–1174

Ellis RH, Hong TD, Roberts EH, Soetisna U (1991) Seed storage behaviour in Elaeis guineensis. Seed Sci Res 1(2):99–104

Farrant JM, Pammenter NW, Berjak P, Farnsworth EJ, Vertucci CW (1996) Presence of dehydrin like proteins and levels of abscicic acid in recalcitrant (desiccation-sensitive) seeds may be related to habitat. Seed Sci Res 6:175–182

Genes FE, Nyomora AMS (2018) Effect of storage time and temperature on germination ability of Escoecaria bussei. Tanzania J Sci 44(1):123 133

Gerna D, Ballesteros D, Arc E, Stöggl W, Seal CE, Marami-Zonouz N, Na CS, Kranner I, Roach T (2022) Does oxygen affect ageing mechanisms of *Pinus densiflora* seeds? A matter of cytoplasmic physical state. J Exp Bot 73(8):2631–2649

González-Benito ME, Pérez-García F, Tejeda G, Gómez-Campo C (2011) Effect of the gaseous environment and water content on seed viability of four Brassicaceae species after 36 years storage. Seed Sci Technol 39(2):443–451

Groot SPC, Surki AA, De Vos RCH, Kodde J (2012) Seed storage at elevated partial pressure of oxygen, a fast method for analysing seed ageing under dry conditions. Ann Bot 110(6): 1149–1159

Groot SPC, Van Litsenburg MJ, Kodde J, Hall RD, de Vos RC, Mumm R (2022) Analyses of metabolic activity in peanuts under hermetic storage at different relative humidity levels. Food Chem 373:131020

Hall LA, Denning DW (1994) Oxygen requirements of *Aspergillus* species. J Med Microbiol 41(5): 311–315

Harrington JF (1972) Seed storage and longevity. In: Kozlowski TT (ed) Seed biology, insects, and seed collection, storage, testing and certification. Academic Press, New York, pp 145–245

Hay FR, Valdez R, Lee J-S, Sta Cruz PC (2018) Seed longevity phenotyping: recommendations on research methodology. J Exp Bot 70(2):425–434

Hayyan M, Hashim MA, Al Nashef IM (2016) Superoxide ion: generation and chemical implications. Chem Rev 116(5):3029–3085

Kalpana, Hajam YA, Kumar R (2022) Management of stored grain pest with special reference to *Callosobruchus maculatus*, a major pest of cowpea: a review. Heliyon 8(1):e08703

Kandel P, Scharf ME, Mason LJ, Baributsa D (2021) Effect of hypoxia on the lethal mortality time of adult *Sitophilus oryzae* L. Insects 12(10):952

Kodde J, Buckley WT, de Groot CC, Retiere M, Zamora AMV, Groot SPC (2012) A fast ethanol assay to detect seed deterioration. Seed Sci Res 22(1):55–62

Koster KL, Leopold AC (1988) Sugars and desiccation tolerance in seeds. Plant Physiol 88(3): 829–832

Labuza TP, Dugan LR Jr (1971) Kinetics of lipid oxidation in foods. Crit Rev Food Sci Nutr 2(3): 355–405

Lattanzio V, Terzano R, Cicco N, Cardinali A, Venere DD, Linsalata V (2005) Seed coat tannins and bruchid resistance in stored cowpea seeds. J Sci Food Agric 85(5):839–846

Lopez-Molina L, Mongrand S, McLachlin D, Chait B, Chua NH (2002) ABI5 acts downstream of ABI3 to execute an ABA-dependent growth arrest during germination. Plant J 32(3):1–12

McDonald MB (1999) Seed deterioration: physiology, repair and assessment. Seed Sci Technol 27: 177–237

Moïse JA, Han S, Gudynaitę-Savitch L, Johnson DA, Miki BL (2005) Seed coats: structure, development, composition, and biotechnology. In Vitro Cell Dev Biol Plant 41(5):620–644

Motete N, Pammenter NW, Berjak P, Frédéric JC (1997) Response of the recalcitrant seeds of *Avicennia marina* to hydrated storage: events occurring at the root primordia. Seed Sci Res 7(2): 169–178

Nagel M, Kodde J, Pistrick S, Mascher M, Börner A, Groot SPC (2016) Barley seed ageing: genetics behind the dry elevated pressure of oxygen ageing and moist controlled deterioration. Front Plant Sci 7:388

Nishizawa A, Yabuta Y, Shigeoka S (2008) Galactinol and raffinose constitute a novel function to protect plants from oxidative damage. Plant Physiol 147(3):1251–1263

Pammenter NW, Berjak P (2000) Evolutionary and ecological aspects of recalcitrant seed biology. Seed Sci Res 10(3):301–306

Pellizzaro A, Neveu M, Lalanne D, Ly Vu B, Kanno Y, Seo M, Leprince O, Buitink J (2019) A role for auxin signaling in the acquisition of longevity during seed maturation. New Phytol 225(1): 284–296

Pourcel L, Routaboul JM, Kerhoas L, Caboche M, Lepiniec L, Debeaujon I (2005) TRANSPARENT TESTA10 encodes a laccase-like enzyme involved in oxidative polymerization of flavonoids in Arabidopsis seed coat. Plant Cell 17(11):2966–2980

Probert RJ, Daws MI, Hay FR (2009) Ecological correlates of *ex-situ* seed longevity: a comparative study on 195 species. Ann Bot 104(1):57–69

Rao NK, Hanson J, Dulloo ME, Ghosh K, Nowell A (2006) Manual of seed handling in genebanks, vol 8. Bioversity International, Rome

Righetti K, Vu JL, Pelletier S, Vu BL, Glaab E, Lalanne D, Pasha A, Patel RV, Provart NJ, Verdier J, Leprince O, Buitink J (2015) Inference of longevity related genes from a robust coexpression network of seed maturation identifies zegulators linking seed storability to biotic defense-related pathways. Plant Cell 27(10):2692–2708

Roach T, Nagel M, Börner A, Eberle C, Kranner I (2018) Changes in tocochromanols and glutathione reveal differences in the mechanisms of seed ageing under seedbank conditions and controlled deterioration in barley. Environ Exp Bot 156:8–15

Roberts EH (1972) Storage environment and the control of viability. In: Roberts EH (ed) Viability of seeds. Springer, Dordrecht, pp 14–58

Roberts EH, Ellis RH (1989) Water and seed survival. Ann Bot 63(1):39–39

Saha PK, Bhattacharya A, Ganguly SN (1992) Problems with regard to the loss of seed viability of *Shorea robusta* Gaertn. F. Indian For 118:70–75

Sallon S, Solowey E, Cohen Y, Korchinsky R, Egli M, Woodhatch I, Simchoni O, Kislev M (2008) Germination, genetics, and growth of an ancient date seed. Science 320(5882):1464–1464

Sano N, Ono H, Murata K, Yamada T, Hirasawa T, Kanekatsu M (2015) Accumulation of long-lived mRNAs associated with germination in embryos during seed development of rice. J Exp Bot 66(13):4035–4046

Sano N, Kim JS, Onda Y, Nomura T, Mochida K, Okamoto M (2017) RNA Seq using bulked recombinant inbred line populations uncovers the importance of brassinosteroid for seed longevity after priming treatments. Sci Rep 7(1):8075

Sattler SE, Gilliland LU, Magallanes-Lundback M, Pollard M, Della Penna D (2004) Vitamin E is essential for seed longevity and for preventing lipid peroxidation during germination. Plant Cell 16(6):1419–1432

Schwember AR, Bradford KJ (2011) Oxygen interacts with priming, moisture content and temperature to affect the longevity of lettuce and onion seeds. Seed Sci Res 21(3):175–185

Shen-Miller J, Schopf JW, Harbottle G, Cao RJ, Ouyang S, Zhou KS, Southon JR, Liu GH (2002) Long-living lotus: germination and soil γ-irradiation of centuries-old fruits, and cultivation, growth, and phenotypic abnormalities of offspring. Am J Bot 89(2):236–247

de Souza Vidigal D, Willems L, van Arkel J, Dekkers BJ, Hilhorst HW, Bentsink L (2016) Galactinol as marker for seed longevity. Plant Sci 246:112–118

Stadtman ER (2006) Protein oxidation and ageing. Free Radic Res 40(12):1250–1258

Strelec I, Popović R, Ivanišić I, Jurković V, Jurković Z, Ugarčić-Hardi Z, Sabo M (2010) Influence of temperature and relative humidity on grain moisture, germination and vigor of three wheat cultivars during one year storage. Poljoprivreda 16(2):20–24

Sugliani M, Rajjou L, Clerkx EJ, Koornneef M, Soppe WJ (2009) Natural modifiers of seed longevity in the Arabidopsis mutants *abscisic acid insensitive 3-5* (*abi3-5*) and *leafy cotyledon 1-3* (*lec1-3*). New Phytol 184(4):898–908

Sun WQ (1997) Glassy state and seed storage stability: the WLF kinetics of seed viability loss at T>Tg and the plasticization effect of water on storage stability. Ann Bot 79(3):291–297

Umarani R, Aadhavan EK, Faisal MM (2015) Understanding poor storage potential of recalcitrant seeds. Curr Sci 108:2023–2034

Vandecasteele C, Teulat-Merah B, Morère-LePaven M-C, Leprince O, Ly Vu B, Viau L, Ledroit L, Pelletier S, Payet N, Satour P, Lebras C, Gallardo K, Huguet T, Limami AM, Prosperi J-M, Buitink J (2011) Quantitative trait loci analysis reveals a correlation between the ratio of sucrose/raffinose family oligosaccharides and seed vigour in *Medicago truncatula*. Plant Cell Environ 34(9):1473–1487

Walters C, Wheeler LM, Grotenhuis JM (2005) Longevity of seeds stored in a genebank: species characteristics. Seed Sci Res 15(1):1–20

Walters C, Ballesteros D, Vertucci VA (2010) Structural mechanics of seed deterioration: standing the test of time. Plant Sci 179(6):565–573

Weissbach H, Resnick L, Brot N (2005) Methionine sulfoxide reductases: history and cellular role in protecting against oxidative damage. Biochim Biophys Acta, Proteins Proteomics 1703(2): 203–212

Woodstock LW, Taylorson RB (1981) Ethanol and acetaldehyde in imbibing soybean seeds in relation to deterioration. Plant Physiol 67(3):424–428

Yao Y, You Q, Duan G, Ren J, Chu S, Zhao J, Li X, Zhou X, Jiao Y (2020) Quantitative trait loci analysis of seed oil content and composition of wild and cultivated soybean. BMC Plant Biol 20(1):1–13

Zhang XK, Yang GT, Chen L, Yin JM, Tang ZL, Li JN (2006) Physiological differences between yellow-seeded and black-seeded rapeseed (*Brassica napus* L.) with different testa characteristics during artificial ageing. Seed Sci Technol 34(2):373–381

Zhang K, Zhang Y, Sun J, Meng J, Tao J (2021) Deterioration of orthodox seeds during ageing: influencing factors, physiological alterations and the role of reactive oxygen species. Plant Phys Biochem 158:475–485

Zinsmeister J, Lalanne D, Terrasson E, Chatelain E, Vandecasteele C, Vu BL, Dubois-Laurent C, Geoffriau E, Le Signor C, Dalmais M, Gutbrod K, Dörmann P, Gallardo K, Bendahmane A, Buitnk J, Leprince O (2016) ABI5 is a regulator of seed maturation and longevity in legumes. Plant Cell 28(11):2735–2754

Principles of Quality Seed Production

Sanjay Kumar, K. V. Sripathy, K. Udaya Bhaskar, and Banoth Vinesh

Abstract

Plant breeding and seed technology are the two arms of crop improvement programmes. The ultimate goal of any plant breeding programme is to make available quality seeds of new improved varieties in adequate quantity to farmers. Modern plant breeding techniques have enabled the development of crop varieties with desired traits at a much higher pace than ever before, addressing the challenges of food and nutritional security. A large number of new crop varieties are being bred continuously to address specific needs viz. productivity, quality, tolerance against abiotic and biotic stresses, cropping intensity, etc. However, to the farmers, all these scientific achievements would be of little use unless they have access to seeds of these varieties, which are genetically pure, physiologically sound (germination, vigour), free from physical impurities and seed-borne diseases. The pace of progress in food production largely depends upon the speed with which a country is able to multiply quality seeds of high-yielding varieties. Hence, the quality seed supply chain must be supported by desired policy and technically sound systems both in national and international domains. In order to achieve this, every country needs a well-established infrastructure for seed production, quality assurance, storage and marketing. Similarly, the regulatory framework for the variety testing, release and notification, and regulation of the seed market are also important to preserve the interests of the farming community. Successful seed production requires establishing variety identity, adherence to maintaining variety purity, and the application of good

S. Kumar (✉) · K. V. Sripathy · B. Vinesh
ICAR-Indian Institute of Seed Science, Maunath Bhanjan, Uttar Pradesh, India
e-mail: director.seed@icar.gov.in; sripathy.v@icar.gov.in; banoth.vinesh@icar.gov.in

K. Udaya Bhaskar
ICAR-Indian Institute of Seed Science, Regional Station, Bengaluru, India
e-mail: udaya.kethineni@icar.gov.in

© The Author(s) 2023
M. Dadlani, D. K. Yadava (eds.), *Seed Science and Technology*,
https://doi.org/10.1007/978-981-19-5888-5_6

109

farming practices along with careful management of crops, following widely
accepted production technologies, and quality standards.

Keywords

Variety release · Seed quality · Isolation distance · Field inspection · Roguing ·
Genetic purity · Seed conditioning · Seed marketing

---

1   Introduction

Quality seed is the most important input for enhancing crop production. The use of
quality seed, along with the standard package of practices, enhances the crop yield
by 15–20% or more (Prasad et al. 2017). Profit maximization is achieved when
productivity augmentation is realized per unit of inputs used (seed, water, fertilizer,
pesticides and manpower). The basic objective of a seed production programme is to
supply quality planting material at the right time, and at affordable prices. Low seed
quality can potentially decrease the rate of germination and seedling emergence
(Finch-Savage and Bassel 2016) leading to poor stand establishment in the field and
consequently yield loss in many crops such as direct sown rice (Rahman and Ellis
2019), corn (García et al. 1995; Moreno-Martinez et al. 1998), wheat (Ganguli and
Sen-Mandi 1990), cotton (Iqbal et al. 2002), barley (Samarah and Al-Kofahi 2008)
and garden pea (Hampton and Scott 1982). The quality of seed depends on how and
when it was grown, and whether the seed producer was fully acquainted with the
genetic constitution of the variety, its flowering and pollination behaviour, and other
basic principles of quality seed production. Using cleaned grain as seed, and using
farm-saved seed repeatedly cause deterioration of genetic constitution due to cross-
contamination, as well as poor germination, vigour and seed health due to improper
management, that result in overall decline in performance, yield penalty,
non-uniform harvest and poor quality (Fig. 1).

  Hence, strict measures need to be followed during seed production, harvesting
and post-harvest operations to maintain the desired varietal purity. Similarly, seed
agronomy, which may vary from general crop agronomy; care in the selection of a
proper site for seed production (based on the previous crop history); seedbed
preparation; timely completion of tillage operations; irrigation; supplementary pollination measures, if required; weed control; pest and disease control measures;
identification and removal of the contaminants, off-types, obnoxious weeds, objectionable crop plants, diseases with seed-borne nature, etc. differentiate seed production from general crop husbandry.

  Policy support from institutions, support from the regulatory system, and state-of-
the-art facilities for production, processing, quality assurance, storage and distribution are considered necessary for undertaking a well-orchestrated seed programme
for ensuring the supply of quality seed to the farmers. This chapter discusses variety
release procedures and limited generation schemes followed for the seed production
of crop varieties; causes for genetic deterioration and measures to safeguard the

Fig. 1 Effect of repeated use of farm-saved seed on crop quality. Roadside view at Phurlak village, Haryana (India) (Courtesy: Dr. Rakesh Seth, ICAR-IARI, RS, Karnal)

genetic identity and purity in crop varieties, and the general agronomic and seed technology principles that are to be followed for production of quality seed.

2   Variety Release and Seed Certification

Development and release of new crop varieties for commercial cultivation by crop breeders from research institutions and private seed companies is a continuous process. In most countries, notification for release or registration of varieties is a mandatory requirement for seed certification. For being eligible to registration, candidate varieties need to undergo multi-location agronomic evaluation to confirm their value in cultivation and use (VCU) before release for commercial cultivation. When plant breeders have a promising candidate for release, they submit the release proposal to the appropriate review committee, which contains comprehensive information on the candidate variety, viz. breeding history, description of the variety, and its performance data for two or more years. Variety identification and release may consist of two or more steps during which after the initial evaluation of a candidate variety, it is submitted to higher committee for consideration. The first group provides closer knowledge of the candidate variety, while the second group observes uniformity of release proposals and provides a more objective and comparative evaluation of the candidate variety with other checks and justification for its release based on one or more important criteria. In the USA, for the purpose of certification, state seed certification agencies are aided in determining the eligibility of varieties for certification by a national review board established by the Association of Official Seed Certifying Agencies (AOSCA). The board evaluates the new variety and advices AOSCA on their acceptability for certification purpose (Copeland and McDonald 2001) (see chapter 'Role of Seed Certification in Seed Quality Assurance' for more). Under Organization for Economic Cooperation and Development (OECD) seed certification system, a variety which is part of 'OECD list of varieties' accepted by any of the National Designated Authorities (NDA) are eligible for certification in accordance with rules of the OECD seed schemes. Seed production can be taken up in any country, if a given variety is part of OECD list of varieties.

In India, new varieties are tested for VCU under All India Coordinated Research Project (AICRP) trials at multi-locations following a two-tier system referred as the initial varietal trials (IVT) and advanced varietal trials (AVT). After a comprehensive testing for three years, superior entries are identified by varietal identification committee (VIC) constituted under the AICRP for a specific crop/crop group. The proposal of the qualifying entries identified by the VIC is forwarded to the central sub-committee on crop standards, notification and release of varieties for evaluation. The latter recommends to the Central Seed Committee for Release and Notification of Variety (Chand et al. 2020). The notification of approved varieties is published in official gazette of India. This is followed by the notification of the variety by the Seeds Division, Department of Agriculture and Farmers Welfare, Government of India, along with its area of adaptation and standard descriptors. Only notified crop varieties are eligible for certification process as per The Seeds Act, 1966.

3   Generation Scheme of Seed Multiplication

The generation scheme of seed multiplication is integral to the concept of certification, and permits only limited seed classes to be produced from a given lot of breeder or pre-basic seed. The guiding principle followed in generation system recognizes that maintaining the desired levels of variety purity is difficult in one-step large-scale seed production. Hence, stringent methods are followed for maintaining the highest purity in the first step of multiplication, which then is taken to the subsequent levels increasing the scale of multiplication. In most systems prevailing in the world, seed multiplication follows four generations, though these may be designated and used differently. The AOSCA follows a (four-generation) system that includes breeder seed, foundation seed, registered seed and certified seed. The breeder seed is produced under the control of a plant breeder and is labelled with white tag. The foundation seed is produced from the breeder seed under the contract with the foundation seed-producing organizations authorized by the plant breeder, and is also identified by a white tag. The registered seed is produced from the foundation seed by the registered seed producers and labelled with a purple tag. The certified seed is produced from the registered seed by the certified seed producers/growers, and marked with a blue tag. The first three generations are for non-commercial use (only for the purpose of controlled multiplication), while the certified seed is marketed for crop production (McDonald and Copeland 1997).

Similarly, under the OECD seed schemes, though three generations are recognized viz. pre-basic, basic and certified seed, a successive generation of certified seed is also permitted. The pre-basic and basic seed are controlled by the official maintainer or designated authorities and produced by the recognized institutions only. Pre-basic seed carries a white label with violet stripes, while basic seed carries a white label (without any stripes). Certified seed is produced by the certified seed growers under the administrative control of the designated authorities. First generation of the certified seed carries a blue label, whereas successive generations of certified seed are identified by a red label. The pre-basic and basic seed are used for the multiplication of succeeding classes, while certified seed is used for the commercial purpose of cultivation. In addition to the above-mentioned seed classes, two more seed classes exist under OECD seed schemes, viz. not finally certified seed and standard seed. Not finally certified seeds are the ones which are to be exported from the country of production after field approval, but have not been finally certified, which is mostly done by the recipient country after testing the other seed quality parameters. It carries a grey label. Standard seeds are only pertinent to vegetable seed scheme, where seed is declared by the supplier as 'true to the variety'. These conform to the other conditions in the scheme, and carry a dark yellow label.

In India and other SAARC countries, a three-generation system of seed multiplication is followed (Huda and Saiyed 2011). Breeder seed is exclusively produced by the originating breeder or by a sponsored institution in India. While the foundation and certified seed are produced by any of the State Seeds Corporations, National Seeds Corporation, seed cooperatives, public sector undertakings, non-governmental organizations, private seed companies and the farmers' producer organizations. As per the Indian Minimum Seed Certification Standards (IMSCS), only three generations are allowed beyond the breeder seed (Trivedi and Gunasekaran 2013). Seed quality pertinent to the foundation and certified seed classes are assured by the concerned third party (state seed certification agencies).

4 Genetic Deterioration in Crop Varieties

Several factors are responsible for the deterioration of a variety over a period of repeated use (Kadam 1942; Laverack 1994; Singhal 2001; Singhal 2016) which are summarized here with some modifications.

4.1 Genotypic Constitution and Pollination Behaviour

Cross-pollinated and often-cross-pollinated species are more prone to outcrossing and genetic contamination than the self-pollinated species. Similarly, maintenance of the genetic constitution of the synthetic and composite varieties needs more care than pure-line varieties. Residual heterozygosity and occurrence of aneuploidy (as in bread wheat, an allohexaploid species) can cause occurrence of off-types and genetic deterioration even in highly self-pollinated species (Atwal 1994).

4.2   Developmental Variation

Modern crop varieties are mostly bred to perform in a specific environmental condition. In such a scenario, if seed production is undertaken in the changed environmental backdrop (different soil fertility conditions, altered photo and thermo periods, relative humidity and elevations), developmental variations in the form of dissimilar plant phenotype may arise sometimes due to differential growth responses. To minimize the chance for such shifts to occur in the varieties, it is advisable to grow the seed crops in their areas of adaptation for the variety and in recommended growing seasons.

4.3   Mechanical Mixtures

It is a physical process by which the seeds of other varieties may get mixed inadvertently and deteriorate the genetic purity of the given variety (Fig. 2). This may happen through the seed drill during sowing, through wind carrying the harvested crop from one field to another, in the threshing yard where many varieties are kept together, during processing operations, etc., and also through rodents or other interferences. Hence, care is needed during production (roguing), harvesting, threshing and further handling to avoid mechanical admixtures. Proper hygiene needs to be maintained at stores and seed processing equipment needs to be thoroughly cleaned after operating each seed lot. In case of hybrid seed production, the male parent rows are to be harvested first and removed from the field before harvesting the seed crop (female parent rows).

Fig. 2 Level of varietal mixtures in paddy as a result of mechanical mixture during seed production and post-harvest operations (Source: ICAR-IARI, RS, Karnal)

4.4   Natural Out-Crossing

This is the major source of contamination in sexually propagated plant species through the flow of genes from dissimilar genotypes (which are cross-compatible). The extent of contamination depends upon the extent of the natural cross-fertilization with off-types, and diseased plants. To overcome the natural crossing, seed crop should be sufficiently isolated from the contaminants. The extent of contamination also depends on the intensity and direction of the wind and activity of the pollinators. Highly cross-pollinated species belonging to genus brassica, cucurbits, etc., need to be sufficiently isolated from fields of dissimilar varieties to avoid natural outcrossing.

4.5   Influence of Pests and Diseases

In case of foliar diseases, the size of the seed gets affected due to the poor supply of photosynthates from the infected plant parts. In case of seed and soil-borne diseases, use of infected seeds can cause widespread disease occurrence in the commercial crop (Fig. 3). Control of diseases through the use of healthy seed, and in some cases pesticide-treated seeds provide an effective means minimizing the pesticide use.

Fig. 3 Bakane disease caused by *Giberella fujikuroi* resulting in the abnormal elongation of culm in paddy (Source: ICAR-IARI, RS, Karnal)

Similarly, vegetative propagules deteriorate fast, if infected by viral, fungus or bacterial diseases. Hence, during seed production, plant protection measures need to be deployed at the right stages to check the incidence of pest and diseases.

4.6   Genetic Drift

In cross-pollinated crop species and multi-line varieties, the population is represented by a group of plants with distinct individual genetic constitution. Hence, during the maintenance of such varieties, if few gene combinations are missed out due to improper sampling, the resultant generation will not be representing the entire set of gene combinations present in genotype of the original variety. The genetic equilibrium that was supposed to be present in such population is lost suddenly (Nagel et al. 2019). This is known as genetic drift. This can be reduced to near-zero level by maintaining sufficiently large plant populations and following proper sampling methods during the maintenance of varieties (OPVs of cross-pollinated crops, synthetics, composites and multi-lines).

4.7   Minor Genetic Variations and Pre-mature Release of Varieties

Often, some small proportions of genetic heterogeneity may exist in the variety appearing phenotypically uniform and homogenous at the time of release. This may result in different plant types arising during repeated multiplications. Thus, pre-mature release of a variety may lead to quicker varietal deterioration. For instance, if a variety is bred for disease resistance and the gene(s) conferring this trait is not sufficiently fixed at the time of release, it may segregate producing susceptible and resistant plants during subsequent reproduction cycles. This not only impacts yields and other agronomic traits, but also poses serious problems in the process of certification with respect to the occurrence of off-types in the seed plots (Fig. 4). Post-control grow out testing of the breeder seed, and production of the nucleus seed after every few years following stringent measures are recommended to avoid such genetic variations. This type of genetic inconsistency is more common in cross-pollinated and often cross-pollinated species needing more care during variety maintenance.

5   Principles of Quality Seed Production

During quality seed production, utmost care is needed for ensuring genetic purity, physical and physiological quality and seed health. This may be attained by implementing a defined set of interventions and corrective measures, which are broadly classified under genetic, agronomic and seed technology principles as discussed in the following sections (for more see chapter 'Role of Seed Certification in Seed Quality Assurance').

Fig. 4 Occurrence of early flowering trait in paddy cv. Pusa 44 due to presence of residual heterozygosity (Courtesy: Dr. Rakesh Seth, ICAR-IARI, RS, Karnal)

5.1 Genetic Principles of Seed Production

Production of genetically pure seed is a challenging task and requires high technical know-how, skill and comparatively high financial investment (Agrawal 1994). During seed production, due attention is essential for maintenance of varietal purity and one should have a detailed understanding of underlying genetic principles. Varied interventions, depending on flowering pattern, determinate or indeterminate; floral structure and pollination behaviour—self- and cross-pollinated; genetic constitution of the variety—pure line, multi-line or composite, or hybrid; photo- and thermo-sensitivity and need for special stimulus with respect to floral initiation, are to be followed to avoid genetic deterioration of a crop variety. Following are the safeguards for maintaining genetic purity during seed production.

5.1.1 Maintenance Breeding

Maintenance breeding is the backbone of a quality seed production programme. Varietal maintenance (used synonymously as maintenance breeding) is a simple, but key technique for purification and stabilization of released genotypes. Though based on the basic principles of genetic constitution, it has a profound role in varietal spread, popularization and life of a variety. Maintenance procedures are the extension of normal breeding process, but selection is mild and aims not to improve the variety, but only to keep the genetic constitution unchanged (Peng et al. 2010). Based upon the original characteristics of a variety, minute deviations or poor-performing lines are discarded and seeds only from the uniform 'true-to-type' plants are pooled to form the pre-basic seed, which is the basis for generation system of seed multiplication under OECD seed schemes and other established systems of seed quality assurance (see chapter 'Principles of Variety Maintenance for Quality Seed Production' for more).

5.1.2 Confirmation of the Seed Source

For raising a certified seed crop, the initial seed should be of appropriate class and from the approved source. In a generation scheme of seed multiplication, specific class of seed is used for multiplication of the ensuing seed class. For instance, under the OECD seed scheme, pre-basic seed is used for the production of basic seed, which in turn is employed for the production of certified seed class. The source of the seed can be ascertained through the labels attached to the containers or bags used for seed production purpose (Table 1).

5.1.3 Previous Cropping History

The primary objective of this step is to avoid any genetic contamination through volunteer plants (grown from self-sown seeds of the previous crop). Seed production in related or similar crops in rotation may be followed to address other issues such as plant nutrition, maintenance of soil physical condition and minimizing the risk of soil-borne pathogens and weeds common to a particular group of crops (George 2011). Similarly, the dormant seeds from previous crop may lead to genetic contamination in planned seed crop. Under OECD seed schemes, previous crop requirement for crucifer species, grass species and legume species are five, two and three years, respectively, for production of basic and certified seeds.

5.1.4 Isolation Requirement

Isolation is required to avoid natural crossing with cross-compatible species and undesirable types from nearby fields, mechanical mixtures (at the time of sowing,

Table 1 Minimum varietal purity standards for non-hybrids under OECD seed schemes

Crop	Basic seed	Certified seed first generation	Certified seed second generation
Triticum aestivum	99.9%	99.7%	99.0%
Oryza sativa	99.7%	99.0%	98.0%
Zea mays	99.5%	99.0%	99.0%
Helianthus annuus	99.7%	99.0%	98.0%
Arachis hypogaea	99.7%	99.5%	99.5%
Glycine max	99.5%	99.0%	99.0%
Pisum sativum	99.7%	99.0%	98.0%

(OECD Seed Schemes Rules and Regulations, 2022, https://www.oecd.org/agriculture/seeds/documents/oecd-seed-schemes-rules-and-regulations-2022.pdf)

threshing, and processing) and contamination due to seed-borne diseases from adjoining fields. Protection from these sources of contamination is necessary for maintaining desired genetic purity. Three types of isolation can be achieved during seed production.

Isolation in Time

The planting time of two varieties of the same crop or two cross-compatible varieties of related crops can be staggered over time. The time isolation should be a minimum of 15 to 20 days or more depending on the flowering habit of the crop. As the plants will not blossom at the same time, cross-pollination becomes impossible (McDonald and Copeland 1997). Time isolation is only applicable to crops having determinate flowering habit. Time isolation in the case of paddy can be achieved by providing a time of over 25 days (flowering stage of other varieties over a 100 m range should be 25 days earlier or later as compared to variety in seed production plot) (Fig. 5).

Isolation by Distance

This type of isolation is based on the concept that if a seed crop is sufficiently distant from any other cross-compatible crop then the adverse pollen contamination will be negligible. The distance isolation is efficient and practical type to achieve in most of the crops. The isolation distance recommended in regulations for specific crop take into account the method of pollination (self or cross-pollinated) and mode of pollination (insect or wind). However, in practice, it is impossible to completely prevent foreign compatible pollen reaching a crop, because the wind or pollen-carrying insects can transfer pollen grains over relatively long distances. Under OECD seed schemes, distance isolation can be disregarded when there is sufficient protection from undesirable pollen sources (Table 2).

Isolation by Barrier

The isolation can also be achieved through physical barriers. It may include natural means like the use of such land for seed production, which are physically isolated by

Fig. 5 Isolation achieved in (**a**) time and (**b**) through a barrier crop (sugarcane) in the background in paddy seed production plot (Source: ICAR-IISS, Mau)

Table 2 Minimum isolation distance prescribed under the AOSCA seed certification (in North Dakota State, USA as an example)

Crop	Foundation seed (in feet)	Registered seed (in feet)	Certified seed (in feet)
Wheat, barley & oats	5	5	5
Millets (self-pollinating)	5	5	5
Mustard	1320	–	600
Rapeseed	600	–	300
Hybrid canola & rapeseed	2640	–	2640
Soybean, chickpea & lentils	5	5	5

(North Dakota Legislative Branch, Title 74, Seed Commission, Article 74-03, Seed Certification Standards; https://www.ndlegis.gov/information/acdata/html/74-03.html)

mountains, forests and rivers between the cross-compatible crops or growing taller crops like sorghum, maize, pearl millet, sugarcane, sesbania, etc., in between them (minimum 30 meters wide) or by artificially erecting large sleeves that surround an entire seed crop (up to two meters height) (Fig. 5). If the barrier crop is used, it should be planted in such a manner that it minimizes the flow of contaminating pollen from nearby sources to seed plot. Raising barrier crops is commonly practised in hybrid seed production plots of paddy, sunflower and pearl millet.

5.1.5 Compact Area Approach

Under this approach, seed production of single variety is allowed in a wide stretch of area. By allowing only one variety to be grown in a specified area or zone, the chances of undesirable cross-pollination are minimized. In the USA, compact area approach is followed for seed production of sweet corn in Idaho State (Delouche 1980). This approach is also advantageous in seed production of hybrid seed by making available sufficient pollen mass in a compact area.

5.1.6 Discarding the Peripheral Strip

In wind-pollinated crops, the pollen concentration in the air over a field is higher at the windward end and tends to decline towards the leeward end (Dark 1971). The marginal strip is important in the production of genetically pure seeds. When a cloud of contaminant pollen passes over the field, it is possible that some quantity of pollen grains drop out at random. Those falling over the centre of the plot will compete with the relatively high concentration of the crop's own pollen source and have a negligible chance of fertilizing, whereas those falling on the marginal areas will not have so much competition and will therefore have a higher chance of fertilization. The seeds from a five-meter-wide strip around the perimeter of the plot are harvested separately and these can either be destroyed or placed in a lower seed category. The bulk of the seed harvest will come from the inner area of the plot (George 2011).

5.2   Agronomic Principles of Seed Production

Agronomic interventions in a seed crop vary from that of commercial crop production. The basic principles are aimed at economically viable seed production practices.

5.2.1   Selection of the Agro-Climatic Region
A crop variety employed for seed production in a given area must be adapted to the photoperiod and temperature conditions prevailing in that area. Regions of moderate rainfall and humidity are more suited for seed production than the regions with high rainfall and humidity. Most crops require a dry sunny period and moderate temperature for induction of flowering and pollination (Neenu et al. 2013). Excessive rainfall during flowering will reduce the seed set and may lead to incidence of diseases in onion and other crops. Too high temperatures during flowering may result in pollen abortion. In general, regions with extreme temperatures should be avoided for seed production, unless particular crop is especially adapted to grow and produce seed under these conditions.

5.2.2   Field Preparation
Field preparation consists of eliminating any weeds and volunteer plants and making flat or raised seed beds, as required. Well-tilled seed bed help in improved germination, good stand establishment and removal of potential weeds during tillage operations. To avoid contamination of site with other crop or weed seed, equipment employed during field operations need to be cleaned of soil, residual weed or crop seeds before entering into the site. Stale seedbed technique, in which the seed beds are formed about a week before it is to be sown, can be adopted. This allows the weed seeds to germinate first, so these can be removed to minimize the weed competition in the seedbed before the seed crop is sown (George 2011).

5.2.3   Selection of a Variety
For certified seed production, the variety must be authentic, duly released/registered and must be clearly identifiable by a set of stable characteristics, which can be employed for field inspection by the certification agency. However, for quality declared seed (QDS) class or truthfully labelled seed (TLS) class, production of a non-registered variety is also taken by seed production agencies. A good market demand for the variety is desirable for successful seed business. Generally, the improved varieties having resistance against major pests and diseases and tolerance against extreme weather conditions are more in demand.

5.2.4   Seeding and Stand Establishment
Though the seeding method and seed rate vary from crop to crop, a desired plant stand is achieved by adopting good agricultural practices. However, the row-to-row distances are generally kept wider in a seed crop than in a commercial crop, to facilitate field inspection. The sowing of seed crops in wide-spaced rows also helps in conducting effective plant protection measures and roguing operations. In case of

hybrid seed production, suitable planting ratio needs to be followed for obtaining optimum seed yield. Besides, to allow synchrony in the flowering of the parental lines, staggered sowing may be followed (see chapter 'Hybrid Seed Production Technology' for details). The seed crops should be sown during the most favourable season, though, depending upon the incidence of diseases and pests, some adjustments could be made, if necessary.

5.2.5 Roguing of the Seed Crop

Roguing is the selective removal of undesirable plants from a seed crop on the basis of distinct morphological characteristics in order to improve one or more parameters (genetic purity, free from diseases and noxious weeds) of seed quality (Laverack and Turner 1995). Rogues include off-types, diseased plants, objectionable weeds and other crop plants (Parimala et al. 2013). Adequate and timely roguing is important in seed production. Roguing of off-types at flowering stage is more important than at vegetative stages. The undesirable plants, which are often not distinguishable at vegetative stage, should be identified by the distinctiveness and be removed soon after the emergence of the earheads or tassel or flowers, to avoid any genetic contamination. If cytoplasmic genic male sterile (CGMS) system is used, special attention is required to remove the pollen shedders in the female population. Diseased plants are also rogued out from the seed crop to avoid the spread of pathogens. Many off-type plants exhibit variations only after they mature, hence roguing at the maturity stage is also crucial (Table 3).

5.2.6 Weed Control

Presence of weeds in seed crop not only reduce the yield by competing for space, nutrients, moisture and sunlight, but also lowers the quality standard. Weed plants in the seed field or nearby areas may also serve as the host to a number of diseases. Presence of prohibited weed (objectionable/obnoxious weed) seed may result in the rejection of an entire seed plot/lot. For instance, by the presence of more than 1 and 2 wild rice plants (*Oryza sativa* L. var. *fatua* Prain) (Syn. *O. sativa* L.f. *spontanea* Rosch.) in a population of 10,000 plants of rice (*O. sativa*) crop of foundation and

Table 3 Specific field standards (wheat, barley, oats and rye) under the AOSCA seed certification scheme (in North Dakota State, USA as an example)

Factors	Maximum tolerance		
	Foundation seed	Registered seed	Certified seed
Other varieties[a]	1:10,000	1:5000	1:2000
Inseparable other crops	1:30,000	1:10,000	1:5000
Prohibited noxious weeds[b]	None	None	None

(North Dakota Legislative Branch, Title 74, Seed Commission, Article 74-03, Seed Certification Standards, Chap. 74-03-02; https://www.ndlegis.gov/information/acdata/pdf/74-03-02.pdf)
[a]Other varieties include plants that can be differentiated from the variety being inspected, but shall not include variants which are characteristic of the variety
[b]The tolerance for prohibited or objectionable weeds, or both, in the field will be determined by the inspector

certified seed plots, respectively, will be rejected. Hence, weed control is a critical aspect of seed production. Weed management strategy begins with the selection of a clean site and continues till the seed crop is harvested. Best agronomic practices should be followed for effectively managing the weeds.

5.2.7   Disease and Insect Control

Successful disease and pest control is another important aspect of raising a healthy seed crop. Apart from the reduction of yield, the quality of seed from diseased and insect-damaged plants is invariably poor. There are number of diseases which are systemic and seed-borne in nature. If not checked, the seed thus produced will carry the spores of the pathogens (inoculum) and produce diseased plants in the next generation. Production of disease-free seed can be achieved by using disease-free planting material, producing seed in isolated and disease-free zones, and using recommended plant protection measures. Insects can be managed in the field by the use of insecticides and in storage by proper sanitation, fumigation and seed treatment. Insecticide-impregnated seed packaging material is also effective in managing cross-infestation during storage (Agarwal et al. 2018). Some diseases are identified by the national authorities, as in India, as designated diseases. As per the IMSCS, loose smut in wheat, ashy stem blight in cowpea, halo blight in green gram and downy mildew in sunflower are designated as objectionable diseases during field inspection (Fig. 6).

5.2.8   Soil Fertility and Plant Nutrition

There is no direct association between soil fertility and seed quality although soil fertility and seed yield are positively correlated. However, soils deficient in minor elements may cause seed quality issues. In field beans, cotyledonary discolouration is observed due to deficiency of calcium and boron (McDonald and Copeland 1997). Hypocotyl necrosis in germinating seed of groundnut is common in crop grown in calcium-deficient soils. Nitrogen, phosphorus, potassium and several other elements play an important role for proper development of plants and seeds (White and Brown

Fig. 6   Seed discolouration in soybean due to influence of disease: (**a**) purple staining of seed due to *Cercospora kikuchii*, (**b**) seed discolouration due to soybean mosaic virus

2010). It is, therefore, mandatory that the soil health is tested, and the nutritional requirements of seed crops are met through the application of required fertilizers in adequate quantity.

5.3 Seed Technology Principles

Quality seed production is a function of not only genetic and agronomic principles but also seed technology principles. Aspects such as selection of field free from volunteer plants, strategies for enhancing the seed set, time and method of harvest, seed extraction, drying and other post-harvest operations constitute seed technology principles crucial for obtaining seeds of highest quality.

5.3.1 Selection of Field

The plot selected for seed production must be levelled and should have an assured source for irrigation. The field must be free from volunteer plants, weed plants, soil-borne diseases and have good soil texture and fertility. It should be feasible to isolate the plot as per the requirements of seed certification. In case space isolation is not possible for some reason, time and barrier isolation may be deployed. For instance, in case of maize, time isolation could be provided and barrier isolation may be achieved by planting rows of a tall barrier crop or additional border rows. Fields that have produced seed crops of small-seeded forage legume (e.g., red clover) in the preceding season should not be used to produce seed of another inseparable legume seed crop (e.g., alfalfa). Seed of such crops tends to remain viable in the soil and continue to germinate and contaminate subsequent crops.

5.3.2 Supplementary Pollination

Pollination occurs naturally without human interference either through the wind or pollinators. Pollen availability on the stigma determines the seed set and ultimately the seed yield. This can be augmented through human intervention or by supplementing the pollinator activity. Various kinds of bees (honey bee, leaf-cutting bee and alkali bee) are common and effective. Seed set and quality in berseem can be enhanced significantly by maintaining three to five honey bee hives in close proximity to seed fields (Prasad et al. 2014). However, safe isolation distance needs to be ensured in such cases to avoid genetic contamination by pollinators. Hand pollination is a commonly used strategy in hybrid seed production of sunflower where pollen is collected from the heads of male plant and applied gently over receptive stigma of female plants. In hybrid seed production (please see chapter 'Hybrid Seed Production Technology' for details) of rice, rope pulling or beating the male parent with stick is generally practised to release the pollen grains from the male parent (Fig. 7).

5.3.3 Harvesting

The development of seed is characterized by two distinct stages of maturity viz. physiological (or mass) maturity and harvest maturity. Physiological maturity is the end of seed filling period (Harrington 1972; Tekrony and Egli 1997), whereas, harvest maturity is the point of time that coincides with the maturation drying

Fig. 7 Supplementary pollination in hybrid seed production of paddy: (**a**) by beating the male parental lines with wooden stick and (**b**) by polling the rope over male parental lines. (Courtesy: Dr. S.K. Chakrabarthy, ICAR-IARI, New Delhi)

when it is harvested. The germination and vigour of the seed are at peak when the seed attains physiological maturity (Ghassemi-Golezani et al. 2011). However, as the seed moisture is quite high at this stage, seed is harvested only once it attains a safe moisture level that allows safe processing and storage (Ellis 2019). In plants with determinate flowering habit, seed maturity is uniform, whereas in crops with indeterminate flowering habit (carrot, sugarbeet, etc.), harvesting needs to be timed to obtain maximum seed yield and quality. Method of harvesting also influences seed quality, hence selecting the right method of crop-appropriate harvesting is important. Harvesting and threshing equipment must be thoroughly cleaned before harvesting each variety to avoid mechanical mixtures.

5.3.4 Drying and Storage of Raw Seed

Drying seeds to a safe level is critical to maintain the seed viability and vigour during the storage, and to keep seeds free from pests and disease incidence. Drying should be done using an optimum combination of temperature and airflow maintaining a temperature that does not adversely impact seed quality. In warmer environments, and natural drying by spreading the seeds in thin layer under the sun, or forced air drying at ambient air temperature can be performed. In case of mechanical drying, care should be taken to avoid any mechanical admixture. Pre-processing sheds and bags containing pre-processed seed also need to be well-cleaned (see chapter 'Seed Storage and Packaging' for more).

5.3.5 Seed Conditioning and Upgradation

After seed has been harvested and before it is dried and stored, it must be cleaned. Seed as it comes from the field, contains varying quantities of physical impurities such as trash, dried leaves, weed seeds, other crop seeds, etc. (McDonald and Copeland 1997). The purpose of conditioning is to remove these physical impurities, as well as to upgrade the seed quality and appearance. Satisfactory conditioning requires a specific sequence through several operations. Raw seed is initially conditioned by pre-cleaner to remove impurities such as crop debris and soil particles. It also removes bigger and smaller sized seeds to large extent. Later, pre-cleaned seed is conditioned through air-screen machine (seed grader) and quality upgradation is done through machines viz. indented cylinder separator, gravity separator, fractioning aspirator, roll mill, spiral separator, buckhorn machine and inclined belt separator. Specialized machineries such as huller-scarifiers (to scarify hard seeds) and debearders (to remove seed appendages like awns, beards or glumes) are also employed. The choice of operation and machinery depends on the kind of seed, the nature and type of contaminants in seed lot and quality standard need to be achieved after seed conditioning and upgradation (see chapters 'Seed Processing for Quality Upgradation' and 'Seed Quality Enhancement' for more).

5.3.6 Seed Treatment

Seed treatment promotes the planting value mainly by ensuring good seedling establishment and control of seed and soil-borne pathogens. The application of seed treatment is a specialized operation and last step in the conditioning of seed before the bagging. Range of contact and systemic fungicides, and insecticides are available for the purpose. Choice of chemicals depends upon the nature of protection needed. Treated seed should be clearly distinguishable and seed container properly labelled by the statement indicating that seed is treated and these are not suitable for food, feed and oil (see chapters 'Seed Health: Testing and Management' and 'Seed Quality Enhancement' for more).

5.3.7 Seed Packaging and Storage

After seed processing and treatment, seeds are ready for packaging into the containers of specified weight based on seed rate per unit area in various crops. The packaging materials should protect quality of seed and should have sufficient tensile strength, bursting strength and tearing resistance to withstand the handling stresses (Walters 2007). Such materials may not always protect the seeds against either insect pests or moisture regain. Based on the nature, seed packaging materials are classified as moisture vapour permeable container (freely permeable to water vapour and gases, e.g., jute bag, cloth bag, paper bag, multiwall paper bag), moisture vapour resistant container (materials resistant to the passage of moisture but, over a long period of time, there will be a slow passage of water vapour tending to equilibrate the relative humidity inside with surrounding environment, e.g., jute bag laminated with thin polythene film) and moisture vapour proof container (material is completely moisture and vapour impermeable and hence, seeds should be dried to low moisture levels before packaging and they can be hermetically sealed with altered gaseous content inside the package, e.g., tin can, polythene bags, aluminium foil pouches, glass bottles). Two simple rules say that for every 1% decrease in moisture content, storage life of the seed is doubled and for every 5 °C decrease in storage temperature, storage life is also doubled. The ideal temperature range for insect and fungal activity is 21 °C to 27 °C. Seed storage godown should be well ventilated and has provision for prevention of entry of rodents. Stored seeds need to be monitored at regular intervals for insect infestation and fumigation may be done if need arises. The stacks of bags should not be made directly on the floor. These should be arranged on the wooden platform in dry, cool, clean and rat proof godown (see chapter 'Seed Storage and Packaging' for more).

5.3.8   Seed Certification

Seed certification is a legally sanctioned system for quality assurance of seed multiplication and production. Certification programme is necessary (or sometimes obligatory) for the seed trade. OECD seed schemes are globally accepted seed certification system, whereas AOCSA system is widely followed in North America. In India, the state seed certification agency is the legally authorized body to manage and monitor the seed quality during multiplication. Seeds which are certified under the certification schemes have to meet both general and crop-specific field and seed standards. Producing high-quality seeds of the crop varieties and making them available to the farmers are the prime aim of any seed certification system (see chapter 'Role of Seed Certification in Seed Quality Assurance' for more).

5.3.9   Seed Certification Procedures

Application

Seed producer needs to submit an application along with the requisite fee to the designated seed certification authority, requesting for certification. The fee is for one season for a single variety and for an area as specified for one seed plot, which is mostly up to ten hectares. The official tag of the source seed should also be submitted (e.g., breeder seed tag in case of foundation seed production; and foundation seed tag in case of certified seed production) along with the application.

Field Inspection

The field inspections are performed by the concerned certification agency on all fields for which applications are received. The objective of the field inspection is to verify that proper care is taken to check the factors that may affect genetic purity and physical health of seeds during multiplication. A number of field inspections differ from crop to crop and certification schemes. Generally, field inspections would be carried out for a minimum of two or more times during pre-flowering, flowering, pre-harvest and harvest stages. During field inspection, isolations are verified; presence of off types, other crops, weed contaminations and diseased plants are checked and seed growers are guided to undertake necessary corrective measures, if required. The fields, which do not conform to the prescribed standards, and if there is enough evidence to prove that contamination has already occurred, shall be rejected for certification.

Seed Sampling, Testing and Tagging

The purpose of sampling is to draw a representative sample from a seed lot (of 10 t or more) of a size suitable for conducting quality testing, in which the probability of a constituent being present is same as its proportion present in the seed lot. Seed sample thus drawn by the authorized persons is sealed, labelled and submitted to the certification agency, and tested for quality parameters in an official seed testing laboratory. Seed lots which meet the prescribed seed standards (pure seed, germination, weed seeds, other crop seeds, and diseased seeds) alone will be eligible for allotment of seed certification tag. Under the OECD seed schemes, the results of seed

Table 4 Specific seed standards for chickpea, soybean and lentil under the AOSCA seed certification scheme (in North Dakota State, USA as an example)

Factor	Foundation seed	Registered seed	Certified seed
Pure seed (minimum)	98.0%	98.0%	98.0%
Total weed seed (maximum)	None	1 per pound	2 per pound
Other varieties (maximum)[a]	0.1%	0.2%	0.2%
Other crop seeds (maximum)			
Soybean and chickpea	None	1 per 2 pounds	1 per pound
Lentil	1 per 2 pounds	1 per pound	3 per pound
Inert matter	2.0%	2.0%	2.0%
Prohibited noxious weed seeds	None	None	None
Objectionable weed seeds[b]	None	None	None
Germination and hard seeds	85.0%	85.0%	85.0%

[a]Other varieties shall not include variants characteristic of the variety
[b]Objectionable weed seeds are dodder, hedge bindweed (wild morning glory), wild oats, buckhorn, hoary alyssum, quackgrass, wild vetch, giant foxtail, wild radish, nightshade species, and cocklebur
Seed label shall have the results of an *ascochyta* test performed on the harvested seeds of each seed lot

testing should, whenever possible, be given on the orange international seed lot certificate issued under the rules of the International Seed Testing Association (ISTA). Most of the agencies have adopted a two-tag system, in which seed analysis tag and certification tag are different. Certification tag is issued by the seed certification authority and seed analysis information is printed on seed label and affixed on seed containers separately (Table 4).

5.3.10 Marketing

Generally, the seed companies, both in the public and private sectors produce seed through the contract seed growers/certified seed growers, where production is taken up in the farmers' fields and raw seed is procured, processed, tested, packed and sold through the network of dealers (Chauhan et al. 2016). Public sector seed production agencies may also produce seed in their own farm and sell it through their own outlets. Effective promotion, branding, attractive packaging, product mix, established market channels, seed quality, seed price etc. play important roles in seed marketing.

6   Conclusion

Seed production is a series of well-defined specialized activities, requiring rigorous criteria to be followed by the seed producers at each of the stages to ensure that high-quality seed is produced and marketed. Continuous flow of new improved varieties with steady augmentation of variety replacement and seed replacement rates are going to be the key for future food and nutritional security of the countries. The genetic, agronomic and seed technology principles of quality seed production discussed broadly hold good for current scenario. However, with threats of climate change in future, alternate procedures may need to be adopted with respect to growing locations and agronomic practices. Seed production of high-value seeds, especially hybrid vegetables under controlled conditions of polyhouse/greenhouse/ net house, is a viable option to combat the threats of abiotic and biotic stresses.

The varietal spectrum across the cropping system is also expected to shift more towards early maturing and climate-resilient varieties in the times to come. Modified agronomic practices or growing seed crops in protected environments may take forefront for narrowing the ill effects of changing climate on seed production. Likewise, the prescribed field standards may need to be redefined in cross and often cross-pollinated species. Future researches will also need to address the management of some diseases and pests, which are of lesser economic significance today, but threaten to become major problems for seed production in the coming years.

References

Agarwal, D. K., Koutu, G. K., Yadav, S. K., Vishnuvat, K., Amit Bera, Ashwani Kumar, Vijayakumar, H. P., Sripathy, K. V., Ramesh, K. V., Jeevan Kumar, S. P., UdayaBhaskar, K. & Govind Pal (2018). Proceedings of joint group meeting of AICRP-National Seed Project (crops) & ICAR seed project- Seed Production in Agricultural Crops held during 09-11 May, 2018 at PAJANCOA & RI, Karaikal

Agrawal RL (1994) In seed technology, 2nd edn. Oxford and IBH publishing company Pvt. Ltd., New Delhi

Atwal SS (1994) Genetic purity standard for breeder seed of wheat. Seed Res 22:168–169

Chand S, Chandra K, Khatik CL (2020) Varietal release, notification and denotification system in India. In: In Plant Breeding-Current and Future Views. IntechOpen, London

Chauhan JS, Prasad SR, Pal S, Choudhury PR, Bhaskar KU (2016) Seed production of field crops in India: quality assurance, status, impact and way forward. Indian J Agric Sci 86(5):563–579

Copeland LO, McDonald MB (2001) In seed certification. In: Principles of seed science and technology. Springer, Boston, MA. https://doi.org/10.1007/978-1-4615-1619-4_14

Dark SOS (1971) Experiments on the cross-pollination of sugar beet in the field. J Natl Inst Agric Bot 12:242–266

Delouche JC (1980) Environmental effects on seed development and seed quality. Hort Sci 15:13–18

Ellis RH (2019) Temporal patterns of seed quality development, decline, and timing of maximum quality during seed development and maturation. Seed Sci Res 29:135–142

Finch-Savage WE, Bassel GW (2016) Seed vigour and crop establishment: extending performance beyond adaptation. J Exp Bot 67(3):567–591

Ganguli S, Sen-Mandi S (1990) Some physiological differences between naturally and artificially aged wheat seeds. Seed Sci Technol 18(3):507–514

García FC, Jiménez LF, Vázquez-Ramos JM (1995) Biochemical and cytological studies on osmoprimed maize seeds. Seed Sci Res 5(1):15–23

George RAT (2011) Agricultural seed production. CABI, pp 1–194

Ghassemi-Golezani K, Sheikhzadeh-Mosaddegh P, Shakiba MR, Mohamadi A, Nasrollahzadeh S (2011) Development of seed physiological quality in winter oilseed rape (Brassica napus L.) cultivars. Notulae Botanicae Horti Agrobotanici Cluj-Napoca 39(1):208–212

Hampton JG, Scott DJ (1982) Effect of seed vigour on garden pea production. N Z J Agric Res 25(3):289–294

Harrington JF (1972) Seed storage and longevity. In: Kozlowski TT (ed) Seed biology, vol III. Academic Press, New York, USA, pp 145–245

Huda MN, Saiyed IM (2011) Quality seed in SAARC countries. SAARC Agriculture Centre, Dhaka, Bangladesh, p 492

Iqbal N, Basra SMA, Rehman K (2002) Evaluation of vigour and oil quality in cotton seed during accelerated ageing. Int J Agric Biol 4:318–322

Kadam BS (1942) Deterioration of varieties of crops and the task of the plant breeder. Indian J Genet Plant Breed 2:159–172

Laverack GK (1994) Management of breeder seed production. Seed Sci Technol 22:551–563

Laverack GK, Turner MR (1995) Roguing seed crops for genetic purity: a review. Plant Varieties Seeds 8:29–46

McDonald MF, Copeland LO (1997) In seed production: principles and practices. Springer Science & Business Media, Berlin

Moreno-Martinez E, Vazquez-Badillo ME, Rivera A, Navarrete R, Esquivel-Villagrana F (1998) Effect of seed shape and size on germination of corn (Zea mays L.) stored under adverse conditions. Seed Sci Technol 26(2):439–448

Nagel R, Durka W, Bossdorf O, Bucharova A (2019) Rapid evolution in native plants cultivated for ecological restoration: not a general pattern. Plant Biol 21(3):551–558

Neenu S, Biswas AK, Rao AS (2013) Impact of climatic factors on crop production-a review. Agric Rev 34(2):97–106

Parimala, K., Subramanian, K., Mahalinga Kannan, S., & Vijayalakshmi, K. A. (2013). A manual on seed production and Certification. Centre for Indian Knowledge Systems (CIKS), seed node of the Revitalising rainfed agriculture network pp 1–19

Peng S, Huang J, Cassman KG, Laza RC, Visperas RM, Khush GS (2010) The importance of maintenance breeding: a case study of the first miracle rice variety-IR8. Field Crop Res 119(2–3):342–347

Prasad S.R., Vilas Tonapi, Bhale M.S., Amit Bera, Natarajan S., Sripathy K.V., Ramesh, K.V., UdayaBhaskar, K., Kamble U.R. and Govind Pal (2014). Proceedings of annual group meeting of AICRP-national Seed Project (Crops) held during 14-16 April, 2014 at SKUAST, Srinagar

Prasad, S. R., Chauhan, J. S. and Sripathy, K. V. (2017).An Overview of national and international seed quality assurance systems and strategies for energizing seed production chain of field crops in India. Indian J Agric Sci 87 (3), 287–300

Rahman SMA, Ellis RH (2019) Seed quality in rice is most sensitive to drought and high temperature in early seed development. Seed Sci Res 29(4):238–249

Samarah NH, Al-Kofahi S (2008) Relationship of seed quality tests to field emergence of artificial aged barley seeds in the semiarid Mediterranean region. Jordan J Agric Res 4(3):217–230

Singhal NC (2001) Concept of variety maintenance. Seed Tech News 3(1):3–5

Singhal NC (2016) Variety: definition, characteristics and maintenance. In: Singhal NC (ed) Seed science and technology, 2nd edn. Kalyani Publishers, New Delhi, pp 215–232

Tekrony DM, Egli DB (1997) Accumulation of seed vigour during development and maturation. In: Basic and applied aspects of seed biology. Springer, Dordrecht, pp 369–384

Trivedi RK, Gunasekaran M (2013) Indian minimum seed certification standards. The Central Seed Certification Board, Department of Agriculture and Cooperation, Ministry of Agriculture, Government of India, New Delhi, pp 401–402

Walters C (2007) Materials used for seed storage containers: response to Gómez-Campo [Seed Science Research16, 291–294 (2006)]. Seed Sci Res 17(4):233–242

White PJ, Brown P (2010) Plant nutrition for sustainable development and global health. Ann Bot 105(7):1073–1080

Vegetable Seed Production

Vinod K. Pandita, P. M. Singh, and Nakul Gupta

Abstract

Unlike field crops where the cultural practices of raising seed crops are mostly similar to the commercial crops, in case of the vegetables, not only the seed crops are grown for a much longer duration than the crops raised for the vegetable purpose, they often also have critical requirements of photoperiods, temperatures, humidity and precipitation. Moreover, compared to field crops, most of vegetable seeds are considered high value and low volume, and hence these require specific care in pre-harvest stages, i.e. isolation and roguing, during harvesting and extraction, and post-harvest operations such as drying, processing, packaging and storage. Due to their vast diversity in growth pattern, induction of flowering and pollination behaviour, seed production procedures need to be followed for each group, e.g. cole crops, cucurbits, leafy vegetables, root crops, solanaceous, malvaceous crops, etc. General principles and standard practices in vegetable seed production are provided in this chapter.

Keywords

Vegetable seed · Roguing · Pollination · Maturity indices · Vernalisation · Wet extraction

V. K. Pandita (✉)
ICAR-Indian Agricultural Research Institute, Regional Station, Karnal, India

P. M. Singh
Vegetable Improvement Division, ICAR-Indian Institute of Vegetable Research, Varanasi, India

N. Gupta
ICAR-Indian Institute of Vegetable Research, Varanasi, India

© The Author(s) 2023
M. Dadlani, D. K. Yadava (eds.), *Seed Science and Technology*,
https://doi.org/10.1007/978-981-19-5888-5_7

1   Introduction

Vegetables are important source of nutrients in human diets, especially for vitamins, minerals, antioxidants, etc.; hence are essential for nutritional security of a nation, and health and well-being of its people. With the advancement of breeding technologies, mechanisation and the practice of precision agriculture in open as well as controlled growing conditions have revolutionised vegetable farming industry with ~3% annual growth rate. This has been possible due to the growth of a vibrant vegetable seed industry making available sufficient quantity of vegetable seed of varieties suitable for different production systems. The public research institutions together with the private seed industry have contributed in this. Vegetable seeds are high value and low volume in nature hence, possess a much larger share in the global seed market in economic terms. Among vegetable seed, the biggest share is of solanaceous crops.

Once a new strain or variety of a vegetable crop is developed, only a small quantity of seed, known as the nucleus seed, is available with the breeder for multiplication. To meet the demands of the vegetable growers, this small quantity is to be multiplied rapidly in a manner that not only maintains the genetic purity of the variety, but also produces seed of good planting value. Seed is the initiating point of the majority of the vegetable crops including the root and bulb crops viz., carrot, radish and onion, except those which can be propagated vegetatively or from tissue-cultured plants. Based on the duration of growing season(s) for seed production, vegetables are classified as:

- Annuals: Seed-to-seed cycle is completed in one crop season, viz. solanaceous crops, e.g., tomato, chilli, brinjal, peas and beans, okra, and majority of the cucurbits.
- Biennials: In these crops vegetative growth is completed in the first season, and the seed is produced in the next season fulfilling a low-temperature requirement (vernalization) for inducing flowering. Cole crops (e.g. cauliflower, cabbage, broccoli, knol khol and others), carrot and radish are some examples of this group.
- Perennials: Vegetables which survive for more than two years but complete their reproductive cycle annually, viz. asparagus, artichoke and pointed gourd (*Trichosanthes dioica*).

Seed production of vegetables is different from raising a commercial crop in many ways (Table 1). It is a specialized activity, and somewhat different from the seed production of the grain crops, not only because the seed crop of vegetables is harvested much after the main crop, but also because it often requires supplementary pollination by keeping the beehives, and specialized treatments to induce flowering and seed set. Therefore, a thorough knowledge of crop biology, pollination techniques and climatic requirements are the prerequisite for undertaking seed production of vegetable crops.

Table 1 Differences between seed production and crop production

Seed production	Crop production
Seed production should be taken up as per edaphic and environmental requirements, in its area of adaptation	Can be grown in any area of general adaptation
Needs isolation from other varieties	Isolation is not required
Requires technical skill for maintenance of purity and quality	Needs skill for raising the crop
Roguing is compulsory	General weeding is performed
Harvesting should be done at physiological/harvestable maturity	Harvested at commercial maturity
Special treatment may be required to induce flowering and to supplement pollination	Not required
Maturity indices need to be considered for good seed quality at harvest	Marketable quality is considered for harvesting

2 Seed Quality Parameters

The essential quality parameters are genetic purity, physical purity, germination, vigour and seed health. A systematic seed production approach has to be followed for production of quality vegetable seed. Before the seed reaches the farmer, it is multiplied through several stages. The actual nomenclature of the seed lot at each stage depends on the system of multiplication adopted, and the legislation of each country.

The testing procedure for these quality parameters is well defined in ISTA Rules (ISTA 2015), whereas the minimum requirements with respect to germination, genetic purity, physical purity and maximum permissible limit for diseased seed, other crop seed, weed seed and moisture content are maintained as per the applicable legislation viz., IMSCS and/or OECD, etc.

2.1 Genetic Purity

In order to maintain the genetic potential of a genotype, be it a pure line variety or the parental lines in case of a hybrid, genetic purity of seed needs to be maintained at the highest possible level, or not below the prescribed standards during the course of seed multiplication. In India, as well as in other countries, a four-generation seed multiplication system is followed for maintaining maximum genetic purity with good seed yield. The stages in the seed multiplication chain in India are described below.

2.1.1 Breeder Seed

The breeder seed [equivalent to the pre-basic seed under OECD seed scheme] is produced from the nucleus seed (including the vegetatively propagating material, such as tubers), directly under the supervision and control of the originating or sponsoring plant breeder/breeding institution and maintaining safe isolation distance to avoid any out-crossing. There are no prescribed quality standards for the breeder seed; however, care is taken so that the genetic purity of the breeder seed shall be maintained to guarantee that the subsequent progeny of the seed, that is the Foundation seed (basic seed in OECD) meets the prescribed standards of genetic purity. The germination and purity standards shall be indicated on the label on actual basis and shall not be below that of the foundation seed.

2.1.2 Foundation Seed

Foundation seed is produced from the breeder seed, the production of which is undertaken following the specific requirements of season, previous crop history of the plot and maintaining prescribed isolation from all possible contaminants. The seed crop is grown following the best agronomic practices and is inspected by the personnel of the certification agency at critical growth stages for checking the presence of off-types, obnoxious weeds and occurrence of seed-borne diseases. The seed crop is accepted and harvested only if it meets the prescribed norms of variety and genetic purity and so handled to maintain specific standards during the stages of harvest and post-harvest handling. This class of seed is equivalent to the basic seed of the OECD schemes, and is used to produce the commercial seed of certified grade (C1 andC2). In case of the hybrids, parental line seeds are advanced through the breeder seed and foundation seed stages, while the hybrid seed is produced as certified seed only. In order to meet large seed demands, two generations of foundation seed (F1 and F2) may be produced with the permission of the competent authority, and restricting to only one generation of certified seed (C1).

2.1.3 Certified Seed

The certified seed is produced from the foundation seed, which is used for commercial crop production. Its production is also undertaken following the prescribed norms of field selection, isolation, field inspections and removal of the off-types, weeds and diseased plants and harvested only if it fulfils the standards of genetic purity and is free from diseases and pests. It shall be so handled as to maintain specific genetic identity and purity according to the standards prescribed for the crop being certified. This is equivalent to C1 (also C2) of the OECD schemes. Certified seed of hybrids is produced using foundation seed of the parental lines. Post-control plots are grown to confirm the genetic purity of variety/hybrid seed in a grow-out test.

2.1.4 Labelled Seed

In India, the USA, the UK, the EU and many other countries, non-certified but labelled seed of vegetables is permitted for commercial sale and cultivation. A large proportion of vegetable seeds in commerce falls under this category, the quality assurance of this class of seed is solely the responsibility of the seed-producing agency. Seed is produced following the norms of field inspections, harvesting and processing as prescribed for the certified class of seed. Similarly, the quality parameters, following the recommended procedures of testing, must meet the standards prescribed for CS.

3   Seed Certification

Seed certification schemes have been devised to produce high-quality seed, in which the varietal purity and quality standards of the seed/propagating material of registered/notified varieties or kinds are maintained following the recommended practices and meeting the quality standards prescribed for the given species. Certification by a designated agency ensures that the genetic purity and testing in a notified/designated Seed Testing Laboratory (STL) confirms to the acceptable levels of physical purity and germination potential. The detailed procedure of seed certification is discussed under Chap. 12. Notwithstanding some variations in the operational system, the following norms are followed in most of the countries):

- Seed certification in India is conducted by the designated authority (certification agency) notified under the Seeds Act, 1966.
- Seeds of those varieties which are notified/registered shall be eligible for certification.
- Certification agency verifies the seed source, class and other requirements of the seed for producing the seed crop.
- Seed plots are inspected by the designated/certification personnel, to verify those factors which might irreversibly affect the genetic purity or health of the seed.
- Seed crop meeting field standards for certification are harvested, threshed, and transported under the guidance of certification agency to the seed processing plant.
- After processing, samples are taken by the designated personnel of the certification agency and sent for analysis to the designated/notified seed testing laboratory.
- Once the seed lot is tested and found to meet the prescribed standards of purity, germination and moisture, the certification agency ensures packaging, tagging, sealing and issuance of the certificate.
- The germination (%), purity (%), moisture content (%) and date of seed testing are declared on the tag along with the name of the crop (kind) and variety.

4   Seed Production Technology

The following points are to be kept in mind for obtaining economic yields of quality seed of vegetable crops during the seed multiplication process:

4.1   Environmental Requirements

Vegetable seed production should be undertaken in the best agronomic conditions preferably in the area of adaptation, recommended during the multi-location trials (or VCU trials). Best seed yield and quality is obtained when the crop is grown in the right season and in the area for which the crop is adapted. The regions with abundant sunshine, dry and moderate temperature during maturity and harvest should be preferred for seed production. In many crops flowering is controlled by the photoperiod, e.g. lettuce and spinach require long-day conditions for flowering and seed setting (Waycott 1995; Kim et al. 2000; Pennisi et al. 2020), whereas majority of temperate vegetables like cabbage, cauliflower, beetroot, European type radish and carrot require a low-temperature stimulus to initiate flowering (vernalization). High rainfall areas are not considered suitable for seed production, as it may reduce seed yield, germination and storability and increase the risk of pests and pathogens and hence the need for artificial drying. Excessive wind on the other hand increases rapid water loss, adversely affects the activity of pollinators, carries pollen through wind over long distances, and increases seed shattering. Coyne (1969) reported distinct off-types in *Phaseolus vulgaris* L. under particular environment and recommended that maintenance should be carried out in relatively cool climates. Barker et al. (1984) recommended that maintenance should be carried out preferably at the location of variety development.

4.2   Land Requirements and Planting

The field for raising a vegetable seed crop should be free from 'volunteer' plants and soil-borne disease inoculum. Volunteer plants are the plants originating from the seed/plant material of the previous crop of the same species but different variety or of the same variety of a different seed class, which might remain dormant and germinate with the seed crop. In vegetables volunteer plants are common in spinach, tomato, etc. Besides, the field should be properly levelled with appropriate drainage and should contain sufficient organic matter and nutrients. Row spacings and planting populations within the rows of the seed crops may vary from those for commercial production (Singh et al. 2010). Adequate spacing is maintained not only for flower development, pollinator activity and ease of mechanical operations, but also for undertaking field inspections of the seed crop at different growth stages. In some cases, e.g. eggplant, pepper, tomato, melons, etc., the spacings used for seed production may be the same as those for commercial production (McDonald and Copeland 1997). Most of the cultural practices of annual vegetables for seed production are similar to those for the commercial crop with proper isolation and roguing to maintain varietal purity, and adequate measures for the control of diseases, pests and weeds to be continued over an extended period.

4.3   Pollination Requirements

Vegetable crops like garden pea, cowpea, French bean, tomato, etc. are self-pollinated, and the contamination rate due to unwanted natural crossing is very low. In vegetatively propagated vegetable crops like potato, the maintenance of identity and uniformity is usually easy (Mastenbroek 1982). The majority of other vegetable crops are cross-pollinated like okra, chilli, cucurbits and brassicas, and

Table 2 Extent of natural cross-pollination and pollinating insect in vegetable crops[a]

Vegetable name	Occurrence of natural cross-pollination (%)	Pollinating insect
Brinjal	0.70–15.00	Insects
Cabbage	73.0	Honey bees/bumble bees
Capsicum	7.00–37.00	Honey bees/insects
Carrot	97.6–98.90	Insects/bees
Cauliflower	40.0–50.0	Honey bees/bumble bees
Cucumber	65.0–70.0	Honey bees/solitary bees
Muskmelon	85.0–95.0	Honey bees
Onion	95.0–100.0	Insects
Potato	0.00–20.00	Bumble bees
Radish	Highly CP	Bumble bees/honey bees
Tomato	0.00–5.00	Honey bees/solitary bees

[a](Mastenbroek 1982)

hence require longer isolation for maintenance of genetic purity. The extent of natural cross-pollination, as shown for some vegetable crops in Table 2, determines the degree of isolation required. However, environment also has a major effect on pollination behaviour and subsequent consequences in variety maintenance (Rick 1950).

Normally, the natural insect population under open conditions is sufficient to take care of satisfactory pollination, but in case of excessive application of insecticides and high plant populations grown for seed crop, the natural insect population may sometimes be insufficient to ensure proper pollination and seed set. Insufficient pollination in cucumber and other cucurbits leads to higher number of underdeveloped seeds (Gupta et al. 2021). Therefore, introduction of the supplementary bee hives and application of pollinator attractants improve pollination and seed set. However, care must be taken during application of plant protection chemicals, as indiscriminate use may affect the pollinators' population and activity. Therefore, application of chemicals should be avoided at peak flowering stage and if required, should be undertaken in the evening or late afternoon.

For restricting the spread of crop-specific diseases, it is advisable to space out the seed crop from the commercial crop either by space or time. Isolating the celery stalk production fields from celery seed fields may reduce the incidence of late blight caused by *Septoria* in seed fields, whereas isolation of lettuce seed fields from leaf and head production fields reduced the incidence of lettuce mosaic virus which is seed-borne (McDonald and Copeland 1997).

4.4 Isolation Requirements

Isolation distance is the minimum separation (distance) between two or more cross-compatible species/varieties to prevent the natural outcrossing. Satisfactory isolation of seed crop helps to maintain purity in the following ways:

(a) Minimize the risk of cross-pollination between cross-compatible species/crops (helps in maintaining genetic purity).
(b) Prevents mixing of different varieties of the same crop during harvesting (helps in maintaining physical purity).
(c) Reduces the risk of transmission of pests and diseases from alternative host crops.

Isolation between crossable varieties can be achieved in three different ways.

4.4.1 Temporal Isolation

Isolation by time allows seed of different varieties of the same crop to be produced at the same location by planting at an interval. In tropical and sub-tropical regions where the growing seasons are long enough to allow two production cycles of the cross-compatible crops, these can also be isolated by time. Normally, seed crop should be sown early or late by a margin of 15–20 days than the adjacent field to prevent the outcrossing. For example, seed production of the early and mid-maturity group of cauliflower varieties.

4.4.2 Spatial Isolation

The isolation distances are determined by the flowering behaviour of the species and the mode of its pollination (vector and distance travelled). The isolation distance also depends on the direction of insect flight (in the case of insect-pollinated varieties) or the direction of winds (in the case of wind-pollinated varieties). Table 3 lists the important vegetables and their recommended isolation distances, as followed in India, which are close to those followed elsewhere.

4.4.3 Isolation by Physical Barrier

The seed crop is isolated by any physical barrier (buildings, net houses, etc.) or surrounded with densely planted plants having higher height than the seed crops to prevent the introgression of foreign pollen. In vegetable seed production, moringa, maize, pearl millet, etc. are used as barrier crops.

5 Roguing

Many vegetable varieties, which are in seed chain, occasionally may tend to show genetic alterations over several generations. It is, therefore, necessary to take care to keep the natural variation within the acceptable limits. For this purpose, the crops are inspected at different growth stages and individual plants that do not confirm to the plant type defined in the official variety description, are removed. This procedure, known as roguing, is an important routine activity in organized seed production endeavours to maintain the genetic purity of a variety.

Plant breeders use positive selection to increase the proportion of desirable plants thereby enhancing the frequency of desirable alleles in a population, whereas roguing is a negative selection that removes relatively small proportions of

Table 3 Minimum isolation requirements for vegetable seed crops in India (IMSCS 2013)

Crop	Minimum isolation distance (m)			
	Varieties or OPs		Hybrids	
	Foundation	Certified	Foundation	Certified
Amaranth	400	200	–	–
Asparagus, celery, parsley	500	300	–	–
Beetroot, radish, turnip, spinach	1600	1000	–	–
Brinjal	200	100	200	200
Carrot	1000	800	1000	800
Cauliflower, cabbage, knol khol, Chinese cabbage	1600	1000	1600	–
Cowpea, beans, dolichos	50	25	–	–
Cucumber, bitter gourd, muskmelon, bottle gourd pumpkin, sponge gourd, ridge gourd, snake gourd, snap melon, winter squash, summer squash, watermelon	1000	500	1500	1000
Fenugreek, garden pea	10	5	–	–
Indian squash, long melon	1000	500	–	–
Ivy gourd, pointed gourd	20	20	–	–
Lettuce	100	50	–	–
Okra, chilli, capsicum	400	200		
Onion	1000	500	1200	600
Tomato	50	25	200	100

off-type plants (Faulkner 1984). The off-types may present in a seed crop due to any of the following reasons:

- The presence of different morphological types within a vegetable crop may vary. This tendency is greater in predominantly cross-pollinated vegetables (e.g. cauliflower, cabbage, cucurbits and onion) than self-pollinated (e.g. peas, tomato, cowpea) crops. Hence, the varieties of self-pollinated vegetables are generally more uniform and stable than varieties of cross-pollinated crops.
- Seeds that result from cross-pollination between the vegetable seed crop and other compatible varieties or wild plants. These are difficult to identify in the first generation but show up in later generations.
- Deviations from the normal type due to mutation.
- Accidental mixture of seeds of other varieties in the seed stock during its production, processing, handling or storage.
- Presence of volunteer plants which may arise from dormant seed of the previous crop grown in the same field or leftover vegetative pieces.

It is relatively easier to conduct intensive rouging in breeder seed plots than in large commercial seed production plots. The crop should be grown in such a way so that individual plants can be observed. Normally, a paired row system of planting

Fig. 1 A seed crop of onion at flowering (Courtesy: Dr. Yogeesha HS, ICAR-IIHR, Bengaluru).

may be followed so that each plant can be observed by walking between rows. Such a planting pattern also facilitates the detection of dwarf off-type plants. Field inspections are to be undertaken at right stages, by walking systematically through the crop so that each plant is observed well (Fig. 1). Plants bearing fruits showing undesirable characteristics (NOT true to type) must always be removed, rather than the plucking such fruits, as the remaining flowers of the off-type plant may continue to contribute to the production of undesirable fruits bearing off-type seeds in the next generation. In cross-pollinated crops, the undesirable (off-type) plants should be removed before flowering. Furthermore, cross-compatible weeds and their wild relatives, diseased and infected plants must be eliminated. Inspection of the crop with the sun behind is recommended as it is difficult to examine plants against the light. Proper training and supervision of the field staff involved in roguing is required. The roguing personnel should have adequate knowledge of the morphological characters of the variety. Morphological characters like leaf shape, flower colour, stem colour/pigmentation, fruit shape and colour are usually good markers for rouging the off-types, while characters which are strongly affected by environment, e.g. leaf colour, plant height and earliness of flowering, are not considered very reliable basis for roguing.

5.1   Different Stages of Rouging

(a) Pre-flowering: Plants having different morphological characteristics like plant
    height, foliage morphology, colour, etc. should be removed from seed produc-
    tion fields before flowering.
(b) Flowering: To prevent mixing of varieties of the same crop, rouging is done
    based on curd maturity in cauliflower, sex expression in cucurbits and flower
    initiation time in solanaceous crops.
(c) At fruit development: Based on true-to-type characteristics of developing fruits
    like fruit shape, ripening colour, size, etc., the off-types are removed.
(d) At maturity: Late maturing plants in case of early maturing varieties (specially in
    fruit vegetables) and vice versa are to be discarded.

6   Harvesting, Threshing and Seed Extraction

The best time of harvesting any seed crops is at a stage when the highest yield of best
quality seed could be obtained. Premature or delayed harvest often adversely affects
seed quality. The appropriate stages of harvest, as determined by visual maturity/
harvest indices in some vegetable crops, are presented in Table 4.

In vegetables, seeds are either extracted from dry seed heads or fruits or from
mature wet fruits. Hence, vegetable crops are classified as dry-seeded, which include
the brassicas, legumes (Fig. 3), chilli and onion, and wet-seeded crops such as
cucurbits, brinjal and tomatoes. The right stage of harvesting the seed crop is
determined by morphological or physiological indices such as the colour of the
fruit, the colour of the calyx and firmness, and fruit cracking. Tomato fruits harvested
from pink to red-ripe stage and seed extracted through the fermentation method
showed higher seed quality in terms of germination, field emergence and vigour
index (Pandita et al. 1996). The methods of harvest and extraction depend on the
type of fruit, with threshing done manually or mechanically. Utmost care should be
taken during the shifting of harvested produce from the field to threshing/processing
floor. The trolley/vehicle, threshing floor, processing machines and jute bags should
be clean/free from the seed/plant parts of other varieties of the same crop or weeds to
avoid contamination at this stage. Threshing machines should be properly cleaned to
avoid admixture and run at a safe speed to avoid mechanical damage to the seed.

In some fleshy fruit/vegetables like cucumber, pumpkin, melons, etc., the post-
harvest ripening (PHR) of fruit is needed to maximize the seed yield and quality, as
the seeds continue to develop and mature in the fruit even after harvest until the seed
extraction (Gupta et al. 2021).

Fruit vegetables like tomato (Fig. 4), brinjal, cucumber, watermelon, muskmelon
and ash gourd (Fig. 5) bear ripe seed in the fleshy fruits with a gelatinous layer
around the fresh seed. The seed is separated from this gelatinous material by any of
the following methods.

Table 4 Harvest maturity indices for different vegetable seed crops

Crop	Harvest maturity stage	Harvesting time
Solanaceous vegetables		
Brinjal	Colour of fruit turn yellow/straw (Fig. 2c)	Day time
Chilli/ capsicum	Green colour of fruits changes to red or yellow (Fig. 2d)	Day time
Tomato	Skin colour changes from green to red or pink	Day time
Peas and beans		
Beans	Basal pods become dry like parchment and turn yellow	During the day
Peas	Majority of pods become parchment like	Morning hours
Other pod vegetables		
Okra	Pods become grey or brown (Fig. 2a)	Day time
Bulb crops		
Onion	Seeds become black on ripening in silvery capsules	Morning hours
Cole crops		
Cauliflower, cabbage	Plants start drying and pods become brown in colour	Morning hours
Leafy vegetables		
Fenugreek	Plants turn brown	Morning hours
Spinach	Plants start turning yellow	Morning hours
Root crops		
Carrot	Secondary and third-order umbels turn brown	Morning hours
Radish	Pods turn brown parchment like	Day time
Turnip	Haulms turn brown parchment like from green colour	Morning hours
Cucurbits		
Cucumber	Fruit stalk shows withering. Actual seed maturity can be ascertained by cutting several fruits longitudinally while mature seeds separate easily from the flesh	Day time
Muskmelon	On seen maturity, fruits tend to separate (full slip) from the stem easily. Skin coat becomes waxy and fruit aroma increases	Day time
Squashes/ pumpkin	The rind of fruits becomes hard and its colour changes from green to orange yellow or straw (Fig. 2b)	Day time
Watermelon	At maturity the tendrils have withered on shoot bearing the fruit. Skin colour of the fruits changes from green to yellow on the underside of fruit touching the soil	Day time

Fig. 2 Seed harvesting stages of vegetables: (**a**) Okra, (**b**) Pumpkin (**c**) Brinjal, (**d**) Chilli

Fig. 3 (**a**) Standing vegetable pea crop for seed production. (**b**) Harvested mature pea crop for seed extraction

6.1   Fermentation

In this method, the crushed fruit pulp containing seeds are kept in a container and allowed to ferment for 1 to 5 days depending upon the vegetable, ripeness and temperature of the surrounding. The pulp mixture is stirred daily to avoid discolouration of the seed and to maintain a uniform rate of fermentation. After completion of the fermentation process, the seeds are washed thoroughly by a displacement method. The good seeds settle at the bottom layer, while the pulp and other debris which float on the surface are removed. After the separation, the clean seeds are dried in shade or using low-temperature dryers.

6.2   Acid Treatment

The seed can also be separated from the gelatinous coating by acid treatment. In this method about one litre of commercial grade HCl (hydrochloric acid) is mixed with 100 kg of tomato pulp and kept for 30 min with occasional stirring, after which the seed is washed thoroughly. It is important not to allow the seed to remain for a longer duration in the acid as this will reduce the seed germination. Containers used for fermentation should be made of a safe alloy to avoid seed discolouration.

Fig. 4 Different stages of maturity, harvesting and seed extraction in tomatoes

Fig. 5 Ash gourd seed crop at harvest and manual cleaning

6.3   Alkali Treatment

Alkali extraction is also equally effective, where an equal volume of alkali (in a ratio of 84 g ordinary washing soda with one litre of boiling water) is added to the fruit pulp in an earthen pot. After thorough mixing, it is allowed to stand overnight for seeds to settle. After decanting the clear liquid, the seeds will be washed thoroughly.

7 Seed Drying

At the time of harvest or seed extraction, the moisture content of seed is usually higher than the optimum range for good germination and storability. The pulpy fruits like tomato, brinjal, cucumber, watermelon, muskmelon and ash gourd possess higher moisture content at harvest and absorb still more during wet extraction, whereas other vegetable seeds like peas, cowpea, French bean, onion, brassicas and fenugreek have relatively lower moisture at harvest. Sometimes the seed may also have high moisture content due to adverse climatic conditions. Hence, it is very important to bring the seed moisture to an optimum level before storage. It is reported (Harrington 1973) that most of the vegetable seeds attain equilibrium moisture content (EMC) of 9–12% at >75% relative humidity (RH) at 25 °C in storage, whereas these are around 6–8% at 45–60% RH. Thus, for storage of seed in unsealed containers at ambient conditions, seed moisture should be kept in the range of 9–12%, and for vacuum/sealed storage it should be less than 6–8%. Both natural and artificial/forced drying methods can be used (see Chap. 10) for drying vegetable seeds, but the temperature should not be more than 35 °C. However, seed drying methods are described in detail for field and vegetable crops in Chap.10.

8 Seed Processing

Raw vegetable seed that is received for processing after extraction and drying usually contains some percentage of undesirable materials. Seed processing operations remove undesirable components and plant debris such as empty pod, chaff, straw, flower heads as well as stone pieces, soil clods, coarse dust, etc., and damaged and broken seeds and seeds of other crops and weeds (Fig. 6). Thus,

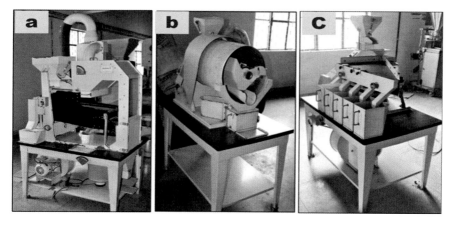

Fig. 6 Machine type used for vegetable seed processing. (**a**) Air screen cleaner, (**b**) Indented cylinder, (**c**) Specific gravity separator

Table 5 Screen aperture sizes for vegetable seed processing (Trivedi and Gunasekaran 2013)

Vegetable crops	Screen aperture size (mm)	
	Top*	Bottom*
Bitter gourd, bottle gourd pumpkin	11.00r	6.50r
Cucumber	8.00r	2.00r
Muskmelon	5.00r	1.00r
Ridge gourd, sponge gourd	9.50r	6.40r
Watermelon	6.00r	1.80r
Brinjal, chilli, tomato	4.00r	0.80s/2.10r
Okra	6.00r	4.30r
Methi	3.25r	1.20r
Spinach (round seeded)	5.00r	2.75r
Spinach (sharp seeded)	8.00r	2.50r
Cauliflower	2.75r	1.10r
Cabbage	2.75r	0.90r
Onion	3.80r	2.00r
Carrot	2.30r	1.00r
Radish	4.50r	2.00r
Turnip	1.80r	1.20r

Where *r – screen with round perforation, s – screen with slotted perforation

cleaning improves the physical purity and appearance of the seed lot. If the seed lot contains substantial amounts of inert materials, it is first passed through a scalping machine, whereas if it possesses more of plant appendages, clumping, etc., then a debearder is used to ensure a free flow of seed (see Chap. 10 for different techniques involved in seed processing). Usually, the air and screen cleaner machine with two vibrating perforated screens is used in the processing of vegetable seed. The upper screen eliminates impurities larger than the seed whereas, the lower screen separates out any seed or other impurities smaller than the optimum seed size of the crop. Table 5 presents the screen aperture sizes for seeds of different vegetable crops as recommended in India.

9   Seed Quality Control

Quality of seed represents the overall value of the seed for its intended purpose. Poor quality seed leads to loss of money and potential crop failure. Seed quality involves genetic and physical purity, germination potential and seed health.

Physical purity represents the pure seed fraction of seed lot on percentage by weight basis. Whereas, germination is one of the most important and widely accepted indices of seed quality (Chap. 13 describes the concept of seed quality in detail). Germination testing is carried out in accordance with the specifications for material, testing conditions and procedures specified in ISTA Rules, with some minor modifications, if needed. Normally, a minimum sample of 400 seeds, using

Table 6 Standard laboratory germination testing protocols for some vegetable seeds (ISTA 2019)

| Crop | Prescription for | | | | Remarks (additional treatments) |
	Substrate	Temperature (°C)	First count (days)	Final counts (days)	
Cauliflower, cabbage	TP; BP	20 ↔ 30; 20	5	10	Prechill/KNO$_3$
Okra	TP; BP; S	20 ↔ 30	4	21	–
Onion	TP; BP; S	20;15	6	12	Prechill
Chilli/capsicum	TP; BP; S	20 ↔ 30	7	14	KNO$_3$
Muskmelon	BP; S	20 ↔ 30; 25	4	8	–
Cucumber	TP; BP; S	20 ↔ 30; 25	4	8	–
Pumpkin	BP; S	20 ↔ 30; 25	4	8	–
Tomato	TP; BP; S	20 ↔ 30	5	14	KNO$_3$
Pea	BP; TPS; S	20	5	8	–
Brinjal	TP; BP; S	20 ↔ 30	7	14	–
Broad bean	BP; S	20	4	14	–
Cowpea	BP; S	20 ↔ 30; 25	5	8	–

TP – top of paper, BP –between paper, S – sand

four replicates of 100 seeds each, is recommended for a statistically dependable germination test. The time required for germination testing depends on the respective crop species. The conditions (media and temperature) and duration for germination tests of some vegetables are given in Table 6, while the minimum germination and purity limits as per the Indian system of seed certification are presented in Table 7.

10   Genetic Purity

Maintenance of the genetic constitution of an improved vegetable variety is achieved through the selection of true-to-type plants and seed collection at the nucleus seed stage and careful multiplication of seed in subsequent generations. Trueness to type can be assessed by post-control plot test or grow-out tests. Post-control plot test is recommended for hybrid seeds, as well as breeder seeds of open-pollinated varieties, specially of cross-pollinated species. Moreover, biochemical, molecular and cytological methods can also be used to access the variety/hybrid purity, if needed (see Chap. 15 for more details pertaining to assessment of variety/hybrid purity). Seed certification schemes, including the post-control grow-out plots, can effectively

Table 7 Minimum seed germination and purity standards of vegetable crops as per OECD seed scheme and IMSCS, 2013

S. No.	Crop	IMSCS Minimum (%)		OECD Minimum (%)	
		Germination	Purity	Germination	Purity
1.	**Cucurbits**	60	98	80	98
	Ridge gourd, bitter gourd, bottle	60	98	75	98
	gourd, sponge gourd, pumpkin	60	98	80	98
	Muskmelon, watermelon				
	Cucumber, squash				
2.	**Solanaceous**	70	98	75	97
	Brinjal, tomato	60	98		
	Chillies, capsicum				
3.	**Peas and beans**	75	98	75	98
	Dolichos	75	98	80	98
	Frenchbean	75	98		
	Peas				
4.	**Sundry vegetables**	65	99		
	Bhindi				
5.	**Bulb crops**	70	98	70	97
	Onion				
6.	**Cole crops**	65	98	70	97
	Cauliflower	70	98	75	97
	Cabbage, khol khol				
7.	**Leafy vegetables**	70	95	70	97
	Amaranthus	70	96	75	95
	Asparagus	70	98	70	97
	Methi	70	98		
	Lettuce	60	96		
	Spinach beet (palak)				
8.	**Root crops**	60	96	65	95
	Carrot	60	96	70	97
	Beetroot	70	98	70	97
	Radish	70	98	80	97
	Turnip				

control the genetic purity of the seed crop (see Chap. 8 for standard procedure for conduct of grow out test in seed crops).

11   Opportunities in Vegetable Seed Production

With increasing living status and consciousness of the masses towards health and nutritional food habits, the consumption of vegetables has increased over the years, increasing the demand for quality vegetable seeds. Major issues of concern in vegetable seed sector are high seed price and quality. Farmers face the problem of spurious seed, low germination percentage and timely availability of quality vegetable seeds (Roy et al. 2020).

Seed germination standards in India are rather low in most of the vegetables such as cucurbits, carrot, radish, tomato, etc. In crops like carrot where seed matures at different times on different order umbels, germination of seed lots remains low, whereas in case of cucurbits such as cucumber, higher number of unfilled seeds is the cause of low germination. In lesser known vegetables including chenopodium, basella and moringa, seed production, processing and testing protocols, and seed standards need to be developed, as these vegetables are nutritionally rich and the demand for the seed of such crops is in the rise.

Seed production under protected conditions of cultivation needs to be promoted in view of better quality, higher yield and low chances of pest attack. This offers a better scope for organic seed production and of seed export.

Acknowledgements Authors acknowledge their appreciation to Dr. H.S.Yogeesha, Principal Scientist, ICAR-IIHR, Bengaluru for Fig. 1, and Director, ICAR-IIVR, Varanasi for the rest of the photographs used in this chapter.

References

Barker RE, Berdahl JD, Jacobson ET (1984) Registration of Rodan western ryegrass. Crop Sci 24: 1215–1216

Coyne DP (1969) Breeding behavior and effect of temperature on expression of a variegated rogue on green beans. J Am Soc Hortic Sci 94:488–491

Faulkner GJ (1984) Maintenance, testing and seed production of vegetable stocks. National Vegetable Research Station, Wellesbourne, England

Gupta N, Kumar S, JainS K, TomarB S, Singh J, Sharma V (2021) Effects of stage of harvest and post-harvest ripening of fruits on seed yield and quality in cucumber grown under open field and protected environments. Int J Curr Microbiol App Sci 10(01):2119–2134

Harrington JF (1973) Seed storage and longevity. In: Kozlowski T (ed) Seed biology, vol III. Academic Press, New York, p 209

ISTA (2015) International rules for seed testing. International Seed Testing Association (ISTA), Bassersdorf, p 276

ISTA (2019) Seed testing rules. International Seed Testing Association, CH-Switzerland

Kim HH, Chun C, Kozai T, Fuse J (2000) The potential use of photoperiod during transplant production under artificial lighting condition on floral development and bolting using *Spinacia oleracea* L. as model. Hortic Sci 35:43–45

Mastenbroek C (1982) Maintenance breeding of potato varieties. In: Seeds, Proc, FAO/SIDA techn. conf. on improved seed Prod., 2–6 June1981, Nairobi. Kenya, FAO, Rome, pp 151–157

McDonald MB, Copeland LO (1997) Seed production—principles and practices. Springer Science +Business Media, Dordrecht, p 754

Pandita VK, Randhawa KS, Modi BS (1996) Seed quality in relation to fruit maturity stage and duration of pulp fermentation in tomato. Gartenbauwissenschaft 61(1):33–36

Pennisi G, Orsini F, Landolfo M, Pistillo A, Crepaldi A, Nicola S, Gianquinto G (2020) Optimal photoperiod for indoor cultivation of leafy vegetables and herbs. European J Hort Sci 85(5): 329–338

Rick CM (1950) Pollination relations of *Lycopersicon esculentum* in native and foreign regions. Evolution 4:110–122

Roy S, Gupta N, Manimurugan C, Singh PM, Singh J (2020) Research and development issues prioritization for vegetable seeds in India: a perception study. Indian J Agric Sci 9(1):34–38

Seed Act (1966). https://seednet.gov.in

Singh PM, Singh B, Pandey AK, Singh R (2010) Vegetable seed production—A ready reckoner. Technical Bulletin No 37. ICAR-IIVR, Varanasi, pp 8–13

Trivedi RK, Gunasekaran M (2013) Indian minimum seed certification standards. The Central Seed Certification Board, Department of Agriculture and Cooperation, Ministry of Agriculture., Govt. of India, p 569

Waycott W (1995) Photoperiodic response of genetically diverse lettuce accessions. J Am Soc Hortic Sci 120:460–467

Principles of Variety Maintenance for Quality Seed Production

Elmar A. Weissmann, R. N. Yadav, Rakesh Seth, and K. Udaya Bhaskar

Abstract

Development of new plant varieties is key to sustainable agriculture, specially given that climate change poses newer challenges with uncertainties of production ecosystems. However, to accrue the full genetic potential of a variety, it is essential to maintain the variety in true-to-type, as it was released for commercial use. The methodology adopted to maintain the genotypic constitution of a variety through the series of multiplication (generations) is known as variety maintenance or maintenance breeding. Its successful implementation needs a thorough understanding of the breeding methodology, varietal characteristics, and the influence of the environment on them. Though specific procedures of variety maintenance are followed for each crop group, which are based on their flowering, pollination behaviour and other essential traits, the basic principles are based on the mode of pollination (self- or cross-pollinated) and genetic constitutions.

Keywords

Variety maintenance · Nucleus seed production · Cross-pollinated spp. · Self-pollinated spp. · Often-cross pollinated spp. · Ear-to-row progeny · Plant-to-row progeny

E. A. Weissmann (✉)
HegeSaat GmbH & CoKG, Singen, Germany
e-mail: elmar.weissmann@eaw-online.com

R. N. Yadav · R. Seth
ICAR-Indian Agricultural Research Institute, Regional Station, Karnal (Haryana), India

K. Udaya Bhaskar
ICAR-Indian Institute of Seed Science, Regional Station, Bengaluru, India

1   Introduction

Productivity of a variety is the result of improved genetics (genetic gain) coupled with better crop management. Variety development and seed production are complementary activities and one without the other has little relevance in the context of agriculture. To realize the genetic potential of the new and improved varieties, a strong seed multiplication must be linked to their distribution, and marketing system (Seth et al. 2009). Significant strides in productivity in major cereals have been made in India since the 1950s, specially with the ushering of the Green Revolution, as the combined outcome of the enhanced genetic potential of improved varieties, availability of genetically pure seeds of such varieties, and improved crop management practices (Yadav et al. 2019).

Seed security is a prerequisite for food security, and this is amply demonstrated by the linear increase in grain production and availability of genetically pure seeds (Indian Seed Statistics: Perspectives 2019), thus, crop improvement is the key to a successful seed programme. Plant breeding per se is man-directed 'plant evolution' and one of the chief limitations of traditional breeding is that selection decisions are primarily based on phenotypes. The phenotype is the manifested expression of one or more traits, whereas genotype reflects the allelic composition of an individual at one or more loci (Singh and Singh 2017). The different traits of an individual can be categorized into two groups: (i) qualitative traits which are governed by one or few major genes/oligogenes, each of which produces a large effect on the characteristic phenotype, and (ii) quantitative traits which are controlled by many genes, each having a small effect on a characteristic phenotype, and are generally cumulative. From a breeding perspective, most of the traits of interest are metric or quantitative in nature, whose phenotypic expression is significantly impacted by environment and also genotype and environment interactions.

For a practising maintenance breeder, the following equation best expresses the phenotype and its constituents:

$$P = \mu + G + E + (G \times E)$$

where P is the phenotype of a quantitative trait (controlled by multiple genes), μ is the population means, G is the genotype effect of the individual concerned, E is the environmental effect on trait expression and $(G \times E)$ is the interaction component. An accurate assessment of G, E and $G \times E$ components of phenotypic variation for different quantitative traits is one of the perpetual pursuits for plant breeding (Singh 2012). This assessment and judgement about G, E and $G \times E$ also pose a dilemma in maintenance breeding, as the production of different classes of seeds, including the multiplication of nucleus and breeder seed, is commonly based on the phenotype, except in few cases where laboratory analysis is required to ascertain a trait.

1.1   Quality Control: An Essential Prerequisite of Varietal Maintenance and Seed Multiplication

The role of quality control in varietal maintenance and seed multiplication is of paramount importance for maintaining delineated traits and making them available to end-users in their exactitude. In seed multiplication, a generation system is followed to produce an adequate quantity of seed from a small quantity of the purest seed available, and a statutory system is devised for field inspection and approval of the scheme leading to seed certification. All certification systems viz., OECD schemes, AOSA and Indian seed certification systems are essentially based on the same philosophy of generation system, and similar procedures. Variety purity standards are maintained by using a set of harmonized procedures during seed production. The control scheme for varietal maintenance and seed multiplication per se follows a generation system with seed class denominations typical for the specific certification system. The case study in wheat depicts the systematic augmentation of seed quantities and multiplication area involved to increase the seed from the generation of 'nucleus' or 'breeder seed' (Table 1) from 0.4 t up to 64,000 t for the generation of the "truthfully labelled" or "commercial stage two" class seed (TL or C2) indicating the need of a robust system with varietal purity as central tenet at each multiplication seed class (Table 2).

2   Maintenance Breeding

The term maintenance breeding, often used interchangeably with varietal mainte-nance, is the foundation to a quality seed production programme. Without proper maintenance breeding, varieties deteriorate rapidly in terms of their genetic potential, and lose much of their value for cultivation, irrespective of their performance traits at the time of release. Maintenance breeding, therefore, is the key to the purification and stabilization of released varieties and varieties to be released, which help to disseminate and enhance the productive life of a variety. Combining the art and science of Plant Breeding with applied aspects of seed production is needed for the maintenance of plant varieties. It is based on the fundamental principles of plant breeding, but the real task during seed multiplication is to identify both obvious and cryptic variants and eliminate these in the initial generations of purification so that only the true-to-variety material is scaled up in subsequent seed multiplication stages. With wheat as a case study, screening for phenotypical off-types needs many visits through the fields within the growth stages (Fig. 1) of ear emergence and dough stage. The intense inspections have to be in the initial multiplication classes, irrespective of the national or international certification scheme. For exam-ple in wheat, many breeders and seed multiplication services recommend 8 inspec-tion visits through the fields of "nucleus seed" production, while reducing their number successively to two in fields of "certified seed" production (Weissmann 2022, unpublished data).

Table 1 Generation system in wheat followed for systematic seed multiplication in India *vis-a-vis* multiplication stages in OECD with emphasis on seed multiplication ratio

Seed type	Generation system	Seed class: Indian context	OECD/EU seed class	Seed quantity (t) @multiplication ratio of 1:20	Seed yield (t/ha)	Area required (ha)
Technical	G1	Ear-to-row	Ear-to-row			
Technical	G2	Nucleus seed	Breeder seed	0.4	3.0	0.1
Technical	G3	Breeder seed	Pre-basic seed	8.0	3.0	3
Technical	G4	Foundation seed	Basic seed	160	3.0	53
Commercial	R1/C1	Certified seed	Certified seed (C1)	3200	3.0	1067
Commercial	R2/C2	TL seed	Certified seed (C2)	64,000	3.0	21,333

Table 2 Analogy of field inspection in Indian seed certification and EU system of seed certification

Generation system	Seed class: Indian context	Indian Field inspection	Indian Seed certification	EU Field inspection	EU Seed certification
G1	Ear-to-row	Breeder	No	Breeder	No
G2	Nucleus seed (NS)	Breeder	No	Breeder	No
G3	Breeder seed (BS)	Breeder	No	Seed certification Agencies	Seed certification Lab
G4	Foundation seed (FS)	Concerned producer	Seed certification Agencies	Seed certification Agencies	Seed certification Lab
R1/C1	Certified seed (CS)	Concerned producer	Seed certification Agencies	Seed certification Agencies	Seed Certification Lab
R2/C2	TL seed (TL)	Concerned producer	No	Seed certification Agencies	Seed certification Lab

Minimum number of visits in the field

Nucleus seed: 8
Breeders seed: 6
Foundation seed: 4
Certified seed: 2

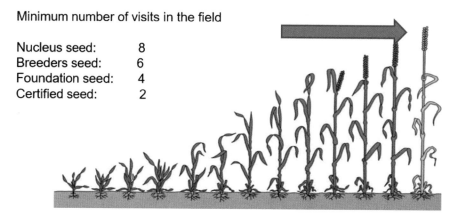

Fig. 1 Time span (blue arrow) and number of field inspections (visits) to check uniformity and purity during maintenance and multiplication in a wheat crop (Weissmann 2022)

2.1   Objectives of the Maintenance Breeding

The basic tenets of varietal maintenance, on which a robust seed multiplication programme relies, are as follows:

- Systematic multiplication of seeds (hastening up through generation system).
- Increase the homozygosity, which in turn can lead to an increase in the homogeneity.
- Elimination of off-types thereby maintaining the stability.

Maintaining the stability depends on contrivances for the occurrence of off-types viz. unintentional mixing during seed multiplication, genetic segregation at individual loci, pollination from other plants in the neighbourhood (1–3%) and natural mutation (with a frequency of 1×10^{-6}). Hence, breeders involved in varietal maintenance should have a thorough understanding of breeding behaviour and the impact of environmental conditions, varietal descriptors of the variety/parental lines (of hybrids) and specific requirements like isolation, land requirements together with the impact of the pressure of biotic factors.

During different stages of seed multiplication, there may be contamination and even complete loss of certain desirable trait (s). Hence, the prevention of contamination gets top priority in variety maintenance programmes (Priyadarshan 2019). This requires a comprehensive knowledge of the flowering and pollination behaviour of the respective species, typical morphological characteristics of the variety as per DUS testing, and the ability to identify possible variation(s) due to growing conditions. For example, there could be multiple GxE interactions during seed production influenced by the micro-environments, which may trigger the expression of various off-types. Differentiating the true variants from the temporal variants due to growing conditions poses a big challenge to the seed production professionals during the maintenance breeding. Therefore, it is desirable to take up the maintenance breeding programme of the variety in its area of adaptation following the GAP recommended for the crop.

When maintenance breeding is undertaken carefully, by combining field observation for morphological characters and some laboratory tests for quality traits viz. organoleptic and cooking quality; fatty acid profiles; protein profiles (gliadins for bread making purpose) and/or useful nutrient(s) for which the variety has been bred; it results in quality seeds, desired agronomic performance, and hence, a longer life in cultivation.

2.2 Methodology

The maintenance breeding would be a function of any of the four fundamental breeding schemes (Table 3), as described by Simmonds and Smartt (2014). Hence, the practices adopted for varietal maintenance of different crops primarily depend upon the mode of reproduction (asexual or sexual) and the mode of pollination (self or cross or often cross-pollination). These reproductive modes/mating systems are responsible for the kind of variants/off-types, which might be expected and observed in a seed production programme.

A sound understanding of the mating system of the crop is a prerequisite for undertaking an appropriate maintenance breeding programme (Yadava et al. 2022). In self-pollinated crops, like wheat and rice, nucleus seed is produced by growing plant-to-rows, i.e. evaluating selected true-to-the-type plants on the basis of performance of their progenies. The plant-to-row method is suitably modified, depending upon the growth habit of the crop plants, to panicle-to-rows in rice, ear-to-rows in wheat or cluster rows in cowpea (Yadav et al. 2003). Plant-to-row method is used as

Table 3 Fundamental populations in plant breeding (based on Simmonds and Smartt (2014))

Mating system	Life cycle, propagation	Population type	Characteristics
Inbreeding	Annual, seed-propagated	Inbred pure lines (IBL)	Homogeneous, homozygous; isolated by a selection of transgressive segregants in F_2-F_7 generation of crosses between parental IBL
Outbreeding	Annual/biennial/perennial, seed-propagated	Open-pollinated populations (OPP)	Heterogeneous, heterozygous; constructed by changing gene frequencies by selection (population improvement) or by making synthetics via parental lines or clones. This population is heterogeneous but not necessarily heterozygous
Outbreeding	Annual or biennial, seed propagated	Hybrids (HYB)	Homogenous, highly heterozygous; constructed by crossing inbred lines selected for combining ability; close to OPP
Outbreeding	Perennial or quasi-annual, vegetative propagation	Clones (CLO)	Homogeneous, heterozygous; isolated by a selection of transgressive clones in subsequent vegetative generations of F_1 between heterozygous parental CLO

such, in crops like green gram, chickpea, field pea and lentil, where single plants can be taken out easily. In cross-pollinated and often cross-pollinated crops like maize, pearl millet, pigeon pea and mustard, the plant-to-row methodology is not very effective, as it does not exclude genetic contamination through pollen of off-type plants. The method used in such crops is Reserve Seed Method. In this method, single plants typical of the variety are selected, individually harvested, threshed and screened for seed traits. In the next year, only a small part of the seed of each selected plant is planted as plant rows; the remaining seed is stored as reserve. The plant rows are carefully screened before and after flowering and until harvest. True-to-type uniform plant rows are identified. The reserve seed from the plants that produced true-to-type progenies is bulked and used as the nucleus seed.

2.2.1 Self-Pollinated Crops

Rice

Most of the seed production systems in the world follow 3 or 4 generations of seed multiplication starting from pre-basic or breeder seed. In India, there are three acknowledged classes of seed, i.e. breeder (BS), foundation (FS) and certified (CS). The seed chain follows a three or four-tier system of multiplication (BS $\rightarrow$ FS $\rightarrow$ FS/CS $\rightarrow$ CS). The initiation of seed multiplication chain is from breeder seed (a product of highest quality nucleus seed). If the breeder seed is not of high genetic purity, contaminants present get multiplied exponentially in the succeeding generations. This may result in complete erosion of the identity and loss of desirable attributes of a variety. Avoidance of contamination and prevention

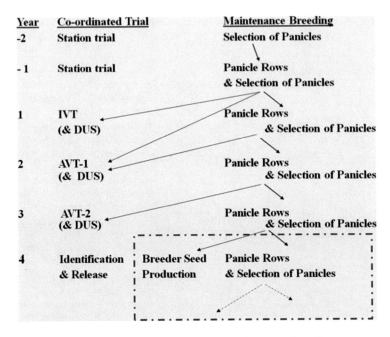

Fig. 2 Systematic procedure and ideal stage of initiating maintenance breeding in any rice varietal development and seed production programme (Atwal et al. 2009)

of genetic deterioration are therefore essential prerequisites of any effective seed programme. Varietal maintenance of basmati and non-basmati rice varieties for decades at the Regional Station of the Indian Agricultural Research Institute, Karnal is a leading example of the efficacy of this approach in enhancing the longevity of these varieties.

Rice varieties Pusa Basmati 1 and Pusa 44 were notified in 1989 and 1994, respectively, but the demand for the breeder seed of these varieties is still very high. This is made possible only because of the appropriate and effective maintenance breeding by the concerned institutions/researcher(s). The best approach for varietal purification, maintenance, and nucleus seed production of rice is the ***Panicle- to-row*** method (Figs. 2 and 3) and outlined below:

- Around 350 to 500 true-to-type single panicles are selected.
- Selected panicles are threshed individually and are thoroughly examined for seed characteristics (seed length, width and shape, etc.).
- In the case of basmati varieties, a portion of the seed is also tested for cooking analysis (kernel elongation, aroma, etc.).
- In the case of molecular marker-assisted backcross breeding (MABB) varieties, incorporated genes are also screened for their presence (e.g., Pusa Basmati 1718 possessing bacterial blight (BB) resistance genes namely *xa13* and *Xa21*) or

Fig. 3 Varietal maintenance plots (Panicle-to rows) of Pusa Basmati 1121 at ICAR IARI, Regional Station, Karnal

another basmati variety Pusa Basmati 1847 having two genes (*xa13* and *Xa21*) for BB resistance and two genes (*Pi54* and *Pi2)* for blast resistance).

- Seeds of panicles not matching to defined seed characteristics or not meeting the defined cooking quality benchmarks or any single panicle having an accidental plant without having the R-allele of the disease resistance gene are straightway rejected.
- Seeds of remaining (about 200–250) panicles are grown in panicle-to-rows or paired rows (from a single panicle).
- Thorough examination of panicle rows is done for their standard diagnostic traits at different crop growth stages.
- Panicle rows not matching to the typical plant type of the variety are completely removed.
- The remaining selected panicle rows are harvested and threshed individually and the seed of each panicle row is critically examined.
- Eventually, the seed of the selected true-to-the-type panicle rows is bulked to get genetically pure highest quality nucleus seed.

The incorporation of cooking tests (Fig. 4) and scrutiny for disease resistance genes using molecular markers in the varietal maintenance programme has significantly improved the market acceptability of these varieties. The emphasis is on stable diagnostic traits, cooking and quality characteristics (kernel length, elongation ratio, aroma) along with screening for disease resistance genes (MABB varieties), wherever applicable.

It is to be noted that roguing (in terms of taking out off types) is never undertaken in the varietal maintenance plots (nucleus seed plots). It is the straightway rejection of panicle rows expressing any sort of variants. Roguing operations are undertaken

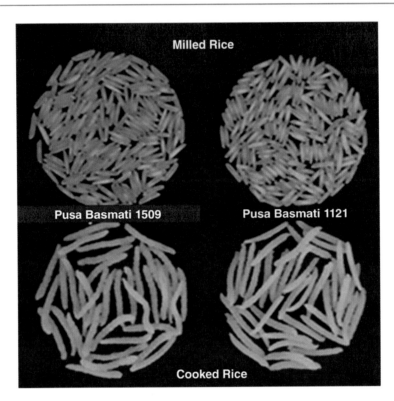

Fig. 4 Cooking Test showing kernel elongation: An integral part of varietal maintenance of Basmati rice

only in large-scale seed production plots (breeder, foundation, certified and truthfully labelled seed plots) (Seth et al. 2022).

Pusa Basmati 1121

A superb basmati rice variety with remarkable grain and cooking quality was notified for commercial cultivation in 2005. The superior linear cooked kernel elongation of this unique variety was derived from its parents Basmati 370 and Type 3. Amassing of favourable loci for extra-long grain and exceptionally high linear cooked kernel elongation was because of transgressive segregation due to selective inter-mating of the sister lines showing better linear kernel elongation in the segregating generations. A total of 13 rice varieties/enhanced germplasm including Basmati 370 and Type 3 were used to bring together the favourable alleles at multiple loci for grain and cooking quality characteristics and agronomic advantage in the development of Pusa Basmati 1121 (Singh et al. 2018). These novel varieties are the outcome of multiple crosses and intricate pedigree (Fig. 5), thereby making the varietal maintenance of these a highly specialized task. A range of off-types show up in the repeated cycles of seed multiplication of these varieties, and often it

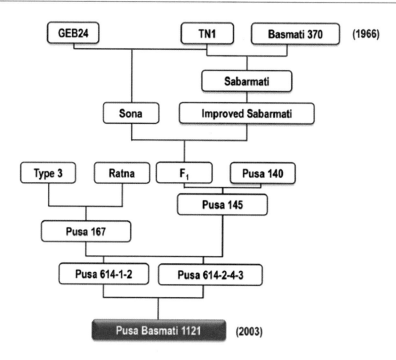

Fig. 5 Pusa Basmati 1121 pedigree showing the contribution of several varieties. Years in parentheses indicate the year in which crossing was initiated and release of the variety, after Singh et al. 2018

becomes very challenging to maintain the precise combination of favourable alleles of the specific variety during the repeated cycles of seed production.

Pusa Basmati 1121, a variety in great demand, is a distinctive example of the importance of variety maintenance, in a variety having a complex parentage, throwing different types of variants such as (i) Tall and dwarf off-types (Fig. 6); (ii) Grain size variants and (iii) Long awned off-types (Seth et al. 2022). Repeated cycles of varietal maintenance have enabled this variety to retain its predominance among Basmati rice varieties both in the domestic and international markets, making a significant contribution to farmers' prosperity.

Pusa 44

A non-basmati variety is popular in northern India (Joshi et al. 2018, Dwiwedi et al. 2021), due to its high yield (8–10 t/ha) and suitability for mechanical harvesting. It is a semi-dwarf *indica* rice variety, with a sturdy culm, with long slender grains and high head rice recovery, and has a significant share in domestic consumption as well as rice export (Indian Seed Statistics (2019). Unlike Pusa Basmati 1121, Pusa 44 is a very stable variety to maintain. Only a few variants, mainly grain size, are observed in the Nucleus/Breeder seed plots. The popularity of this variety has grown

Fig. 6 Pusa Basmati 1121 purification and maintenance. Distinct off- types in a paired row, raised from single true to type panicle (Seth et al. 2022)

manifolds since its release due to sustained varietal maintenance and availability of quality seed.

The above case studies amply showed the critical role played by varietal maintenance in any effective seed multiplication programme. With every step in the generation system of seed multiplication, the ratio of the volume of seed increases significantly from one generation to the next, depending upon the seed multiplication ratio of the crop. Maintaining varietal purity by examining every plant in large plots of Breeder, Foundation and Certified seed, is neither practicable nor cost-effective. Hence, utmost care should be taken at the initial nucleus seed production (Mandal et al. 2010).

Wheat
The variety maintenance in wheat presents a typical case of a self-pollinated species. A seed sample of a particular variety obtained from the concerned breeder is raised in isolation and around 500 spikes or ear-heads which are true-to-type (based on DUS characteristics), are selected and threshed separately. These single-ear seeds are examined for distinct varietal traits and those not conforming to the variety are rejected, and seeds from the true-to-type ear-heads, which did not show any variants upon table examination, will be stored separately for the subsequent stage of nucleus seed production (Table 4).

Table 4 Seed multiplication and maintenance breeding in wheat

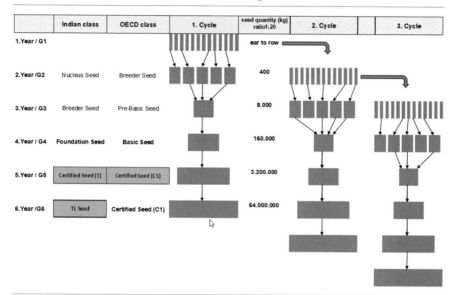

	Indian class	OECD class	1. Cycle	seed quantity (kg) ratio1:20	2. Cycle		3. Cycle
1.Year / G1				ear to row			
2.Year /G2	Nucleus Seed	Breeder Seed		400			
3.Year / G3	Breeder Seed	Pre-Basic Seed		8.000			
4.Year / G4	Foundation Seed	Basic Seed		160.000			
5.Year / G5	Certified Seed (1)	Certified Seed (C1)		3.200.000			
6.Year /G6	TL Seed	Certified Seed (C1)		64.000.000			

Fig. 7 Nucleus seed production of wheat at ICAR-IARI, Regional Station, Karnal. (Ear-rows of varieties HD 2428 or HD 2967 NSSII surrounded by breeder seed of the same varieties)

Nucleus Seed Stage I (NSS1)

Seeds from selected ear-heads (G1) are planted in rows of 3 m length with isolation distance of 5 m from other varieties. Preference should be given to planting of nucleus seed plot surrounded by breeder seed crop of the same variety on all the sides to prevent any chance of out-crossing (Fig. 7). NSS1 seed is usually sown in single rows or may be in paired rows for sake of ease in the inspection. With any sort of deviation from delineated varietal descriptors, the entire row is discarded. If a genetic variation is detected in the progeny at or after the flowering stage, reject the progenies on both sides of the deviant progeny to prevent variation due to natural out-crossing in subsequent progenies, if the variation appears obviously to be heritable in nature. Progeny from each ear-head is to be harvested and threshed separately for multiplication in the next generation.

Nucleus Seed Stage II (NSS2)

If the need for source seed for breeder seed production is high, NSS2 is undertaken. In this method, the seed produced from each ear row of NSS1 is grown separately in larger plots. The number of such plots may vary depending upon the requirement of Breeder seed (BS). The plots are examined for any variation and variant plots are rejected as and when observed. True-to-the-type plots are harvested and threshed in bulk. This constitutes NSS2 which is usually sufficient enough to produce the desired volume of BS.

Maintenance of Hybrids in a Self-pollinated spp., e.g., *Rice, Wheat(EU).*

Hybrids are developed using CMS system by involving three lines (Table 5) viz.
A or CMS line (male sterile);
B or maintainer line (male fertile) and
R or restorer line (male fertile).

The success of hybrid seed production programme depends on the purity of parental lines. Maintenance and multiplication of the A-line are done by crossing it with B-line (A×B). Genetic purity of parental lines is the most important prerequisite to ensure the purity of hybrids. Maintenance of the A-line can also be achieved

Table 5 Maintenance breeding and seed production of hybrids in self-pollinated crops

Year Generation	Indian class	OECD class	A-Line + B-Line	R-Line
1.Year / G1				
2.Year /G2	Nucleus Seed	Breeder Seed		
3.Year / G3	Breeder Seed	Pre-Basic Seed		
4.Year / G4	Foundation Seed	Basic Seed		
5.Year / G5	Certified Seed (C1)	Certified Seed (C1)		

Fig. 8 Line maintenance (ear-to-row) isolation with plastic walls, in hybrid wheat, Germany (Weissmann 2022, unpublished)

by using plastic barriers (Fig. 8). Since B and R lines are fertile, their maintenance and nucleus seed production are similar to that of self-pollinating inbred varieties.

2.2.2 Often Cross-Pollinated Crops, e.g., Pigeon Pea

Being an often cross-pollinated species, maintaining a proper isolation distance, field inspections just before and at the time of flowering, and roguing are essential aspects of varietal maintenance in pigeon pea. Pigeon pea varieties may be bred as OPVs or hybrids.

Nucleus Seed Production of Varieties and/or Restorer Lines of Hybrids

Initially, 500–1000 true-to-type plants are selected from the basic/foundation seed plot, which is in isolation of 250 m from the plots of other varieties. Selected plants are bagged for selfing before the onset of flowering. Selfed seeds are harvested separately and are examined for distinct varietal descriptors. Non-conforming individual plant progenies are rejected and selected individual plant progenies raised in plots of 4 m row length with a spacing of 60–75 cm (row to row) keeping an isolation distance of 5 m between the progeny rows. Plant progeny rows showing variation should be completely rejected. Uniform progeny rows are harvested separately and examined for seed size, shape, and other distinct characteristics. Uniform progenies are bulked to constitute nucleus seed. In case, there is a higher requirement of nucleus seed, one more cycle of multiplication may be repeated.

If the variety is newly released and still segregating or needs purification then this method may not be very effective. In such case, we should go for reserve-seed method for purification of the variety, as described below:

1. Select a large number of single typical plants of the variety from base population raised from seed supplied by the breeder or from nucleus seed plot.
2. Harvest and thresh the selected single plants individually and examine for seed traits. Variant seed packets are rejected. Seed of the so retained seed packets is divided into two parts maintaining their identity.
3. One part is used for sowing in plant rows for evaluation in the next year, while the remaining part is stored as reserve seed maintaining its identity.
4. The single plant progeny rows are screened critically at regular intervals. A row with variant plant(s) is rejected.
5. During flowering the nucleus seed crop is examined daily or at alternate days to avoid cross-contamination. The rejected plant rows should be uprooted and removed from the field immediately after detection.
6. The retained rows are harvested and threshed separately and each separate seed lot is again examined for seed characteristics.
7. The reserved seed of single plants which produced uniform progenies typical of the variety is bulked to constitute the nucleus seed. The pure nucleus seed so produced is used for further multiplication or breeder seed production.

Nucleus Seed Production of A Line of Pigeon Pea Hybrids

In case of the cytoplasmic-genic-male sterility (CGMS) system, A and B lines are planted in a ratio of 4:1 with an isolation distance of 250 m. Any off-type plant should be rogued out before flowering. Conforming male sterile plants are harvested and threshed separately and after examining the seed characteristics, these are bulked as the nucleus seed of A line, whereas regarding B line, plants with designated varietal descriptors are separately harvested.

2.2.3   Cross-Pollinated Crops, e.g. Sunflower

Nucleus Seed Production of Open-Pollinated Varieties

About 5000 plants are selected in the base population which are raised under isolation. Selected plants are evaluated for varietal descriptors, of which 25% plants are advanced for progeny testing, these plants are raised in single progeny rows and for every 10 to 20 progeny rows a check cultivar is included for evaluation purpose. From each plant-progeny-row, a portion of selfed seed is retained as reserve seed. After harvest of progeny trial, laboratory evaluation is conducted to earmark superior progenies and reserve seed of superior progenies which was saved from individual plants is bulked to form nucleus seed (Fig. 9).

A and B lines obtained from the original populations are used for sowing in ratio 1B: 2A:1B of 4 m length with about 300 rows of each line. Plants conforming to all the varietal descriptors are tagged and the pollen is transferred manually from B line to A line (paired crossing), while in B line selfing is done. Adept bagging and

Fig. 9 Nucleus seed production of sunflower at the University of Agricultural Sciences, Bengaluru. *Maintenance of A & B Lines in the CGMS System*

tagging are enabled for selected capitula in about 400–500 A × B crosses. Selected B lines are harvested first and evaluated for seed varietal descriptors. A lines are harvested later than the corresponding B lines from which they received pollen and threshed separately. After laboratory examination, conformed seeds from selected plants are taken in progeny rows. Both A and B line progeny rows not conforming to varietal descriptors are rejected. Individual A line plants are observed for pollen shedders. If any pollen shedder is found the whole line and its corresponding B line are rejected. Reserve seeds from conformed progenies of both the lines, after bulking separately form the nucleus seed respectively (Nucleus and breeder seed production manual, DSR, Mau 2010).

Nucleus Seed Production of R Line

Seeds from original stock of the breeder are sown and about 1000 plants conforming to varietal descriptors are delineated. Of these, about 200 plants are selected, from which seeds are harvested and threshed separately and examined for seed descriptors. Plant-to-row progenies from selected plants conforming to varietal descriptors are taken up retaining a portion of seed as reserve. The progeny rows are examined and those with confirmed characteristics are identified and reserve seed from the respective plants is bulked. Depending on the quantity needed for breeder seed production, one more season may be taken up and bulked seed forms nucleus seed.

3   Measures to Evaluate Varietal Purity to Increase Homogeneity and Stability

Maintaining trueness to type, so that genetic purity does not get affected during varietal maintenance and cycles of seed multiplication due to out-crossing, mechanical admixtures, residual segregation and mutations, has to be tackled by scientific means of planning like isolation, field inspection, checking for varietal descriptors during different stages. Examination of seeds, seedlings, control plot testing and varied biochemical and molecular mechanisms can be used for ascertaining the presence of delineated traits contributing to varietal purity.

Morphological characters (colour, pigmentation, appendages, etc.) of the seeds, seedling characters (coleoptile anthocyanin colour, plant growth habit, etc.), application of biochemical tests (phenol colour reaction, peroxidase test and electrophoresis techniques) and molecular markers can be deployed if necessary, to ensure and evaluate purity and to draw inferences in varietal maintenance programme. As the Indian DUS testing system is widely adapted from the UPOV list of characteristics, these are equally relevant to the Indian system of characterization of varieties.

Inferences on varietal purity can be drawn by resorting to some tests mentioned in Table 6 as per International Union for the Protection of New Varieties of Plants (UPOV 2017) and International Seed Testing Association (ISTA 2022) validated methods for species and variety testing. Some of these methods can be deployed even before sowing and can have an estimate of varietal purity. Advent of biochemical and molecular means of varietal testing widened the possibilities of application, as environment-dependent expression associated with some of the morphological parameters can be surpassed and can be used effectively in variety maintenance programmes.

Isolation is a powerful technique often practised by breeders during maintenance breeding, so that cross-contamination is eliminated. Isolation can be done via sowing time, local or spatial, or via artificial isolation, e.g. bagging, plastic walls or pollination nets. Insulation should be stringently followed, particularly in maintenance breeding programme of the A line, by use of separate equipment for each genotype, and wearing protective gear.

4   Conclusion

Variety development, maintenance and seed production represent a continuum. For the success of any variety development and seed production programme, maintenance breeding plays a critical role to get the most buck from every penny spent on crop breeding and seed production.

Table 6 Examples of validated methods for species and variety purity testing (UPOV 2017, ISTA 2022)

Testing method	Crops	Remarks
Examination of seeds	Avena & Hordeum	Colour of seed under UV light
Examination of seedlings	Beta spp.	Seedling colour—White, yellow, pale red or red
	Brassica spp.	Turnip grown in dark—White fleshed & yellow fleshed cotyledons
	Lolium spp.	Root traces with fluorescence under ultraviolet light
Biochemical tests	Triticum spp.	Phenol test (1%)
	Lupinus spp.	Presence or absence of alkaloids (Lugol's solution)
	Sweet clover	Copper sulphate-ammonia test
	Oats	HCl test
	Rice and sorghum	KOH test
	Wheat	NaOH test
	Soybean	Peroxidase test
	Oats & Barley	PAGE—Alcohol-soluble proteins are extracted and separated
	Pisum & Lolium	SDS-PAGE
	Zea spp. & Helianthus spp.	Ultra-thin layer Iso-electric focusing
	Triticum spp.	Acetic acid urea polyacrylamide gel electrophoresis (A-PAGE)
DNA based methods	Triticum spp.	Analysis of minimum of eight microsatellite markers (Table 8B & 8C of ISTA Rules)
	Zea mays	Analysis of minimum of eight microsatellite markers (Table 8f of ISTA Rules)

Acknowledgements The authors express their gratitude to Dr. S.P. Sharma and Dr. S.S. Atwal, Former Heads of IARI Regional Station, Karnal, and Dr. V.P. Singh, Former Principal Scientist, IARI, New Delhi for initiating the pioneering work on maintenance breeding at IARI, RS, Karnal and New Delhi.

References

Atwal SS, Singh AK, Sinha SN (2009) Maintenance breeding in basmati rice. Indian Farming 59(1):37–41

Dwiwedi P, Ramawat N, Dhawan G, Gopala Krishnan S, Vinod KK, Singh MP, Nagarajan M, Bhowmick PK, Mandal NP, Perraju P, Bollinedi H, Ellur RK, Singh AK (2021) Drought tolerant near isogenic lines (NILs) of Pusa 44 developed through marker assisted introgression of qDTY2.1and qDTY3.1 enhances yield under reproductive stage drought stress. Agriculture 11:64

Indian Seed Statistics: Perspectives (2019) Technical bulletin. ICAR-IISS/2019/5, pp 1–16

International Rules for Seed Testing (2022) International Seed Testing Association, Zurich, Full Issue i–19-8 (300)

Joshi K, Joshi PK, Khan MT and Kishore A (2018). Sticky rice: variety inertia and ground water crisis in a technologically progressive state of India. IFPRI discussion paper 01766: 1–26

Mandal AB, Sinha AK, Natarajan S (2010) Nucleus and breeder seed production manual. Directorate of Seed Research, Mau, pp 1–142

Priyadarshan PM (2019) Maintenance breeding and variety release. In: Plant breeding: classical to modern, pp 561–570

Seth R, Atwal SS, Sinha SN (2009) Role of seed in spread of basmati varieties. Indian Farming 59(1):34–36

Seth R, Singh AK, Gopala Krishnan S (2022) Maintenance breeding of pusa basmati varieties. In: Yadava DK, Dikshit HK, Mishra GP, Tripathi S (eds) Fundamentals of crop breeding. Springer Nature, Singapore. ISBN 978-981-16-9256-7. Pp 677-701

Simmonds NW, Smartt J (2014) Principles of crop improvement (second edition) (first Indian reprint). Wiley India Pvt Ltd., New Delhi

Singh BD (2012) Plant breeding, principles and methods., 9th edn. Kalyani Publishers, New Delhi

Singh BD, Singh AK (2017) Marker assisted plant breeding: principles and practices. Springer, New Delhi

Singh VP, Singh AK, Mohapatra T, Gopala Krishnan S, Ellur RK (2018) Pusa basmati 1121—a rice variety with exceptional kernel elongation and volume expansion after cooking. Rice 11:19

UPOV (2017): TG/3/15 wheat. Guidelines for the conduct of tests for distinctness, uniformity and stability, UPOV TG3/15, International Union for the Protection of new varieties of plants, Geneva, 2017/05/04

Weissmann EA (2022) Studies and practical experiences in setting up maintenance breeding schemes for self pollinating varieties (line varieties and hybrid varieties). HegeSaat GmbH & CoKG, unpublished data

Yadav OP, Singh DV, Dhillion BS, Mohapatra T (2019) India's evergreen revolution in cereals. Curr Sci 116(11):1805–1808

Yadav RN, Seth R, Atwal SS (2003) A modified method (cluster-row) of nucleus seed production in cowpea. Seed Tech News 33(3–4):1–2

Yadava DK, Dikshit HK, Mishra GP, Tripathi S (2022) Fundamentals of field crop breeding. ISBN 978-981-16-9256-7, pp 1–1323

Hybrid Seed Production Technology

Shyamal K. Chakrabarty, Sudipta Basu, and W. Schipprach

Abstract

Hybrid technology, harnessing the advantage of heterosis between two diverse genotypes to achieve maximum hybrid vigour, is widely recognized and commercially used for crop variety improvement both in field and vegetable crops. Hybrids can be developed using appropriate technology, irrespective of the mating and pollination system in the plant species. Production of hybrid seed depends on plant, pollinator and environmental factors, which influence it individually or in interactive ways. Hence, an understanding of these components is important to undertake hybrid seed production of a given crop species. The basic requirements for hybrid seed production at a commercial scale are (a) a unisexual flower or a bisexual flower with sterile pollen in anther or self-incompatible flower/plant; or pistillateness; or large conspicuous bisexual flowers for easy emasculation of flowers in plants to be used as the female parent and (b) abundant pollen production, dispersal and its easy transfer from the male parent to the female parent for satisfactory seed setting. These are dependent on floral biology, flower features, mode of pollination and reproduction of the crop species. Agronomic crop management with scientific insights is equally important for successful hybrid seed production. These are discussed in this chapter with appropriate examples.

S. K. Chakrabarty (✉) · S. Basu
Division of Seed Science and Technology, ICAR-Indian Agricultural Research Institute, New Delhi, India

W. Schipprach
University of Hohenheim, Stuttgart, Germany

Keywords

F_1 hybrid seed · Male sterility systems · Row ratios · Emasculation pollination · Interspersed staminate flowers (ISF) · CMS · GMS · CGMS · Synchronization of flowering

1   Introduction

Meeting the food demand of an ever-increasing population is the primary objective of agriculture. Hence, it is the constant endeavour of plant breeders to breed varieties that yield high under diverse agro-climatic conditions. A large number of hybrid varieties of the field crops, vegetables and flowers with higher productivity and other desirable traits are developed adopting appropriate selection and crossing methods of required genotypes. The F_1 hybrids, thus developed, are superior to their parental lines for productivity, quality and/or adaptability in diverse situations. Hence, there is a growing demand for hybrid seeds by farmers globally.

In simplest terms, the 'F_1 hybrid' is defined as the first filial generation of offspring of distinctly different parental types and refers to a plant cultivar derived by crossing two diverse parental lines/cultivars, each of which is an inbred line and is near homozygous. Crossing between two such genetically divergent but compatible parental lines produces hybrid seeds by employing controlled pollination. Hybrid seeds are heterozygotes in their genetic constitution and highly uniform in morphological features. The divergence between the parental lines results in better heterosis, whereas the homozygosity of the parental lines ensures a phenotypically uniform F_1 population. The basic principles of hybrid seed production are similar to those of an open-pollinated (OP) variety in terms of selection of the site and growing season, source of seed, cultivation methods, etc. However, special care is needed in terms of isolation, pollination techniques and manipulation of growing conditions for better seed yield and maintenance of the parental lines.

The primary factors that control hybrid seed production are the plant system, mainly the floral biology, mating type, pollination system and its underlying mechanism, and agronomic conditions like soil, season, irrigation, fertilizer, chemical, planting system, harvesting, etc. (Virmani 1994). All these factors determine the proper requirements of various inputs at an appropriate time for successful hybrid seed production. These principal factors or conditions can be considered as principles in hybrid seed production. In this chapter, we discuss those principles with associated practices.

Hybrid seed production depends on the method/system used for development of hybrid and pollination control mechanisms that govern the various activities in determining the isolation distance, planting ratio, synchronization of flowering, rouging and supplementary pollination. These are discussed below:

A hybrid is produced by crossing two genetically diverse parents. Pollen from male parent (pollen parent) pollinates and sets seed in female (seed parent) parent to produce hybrid seed. The development of a hybrid in a cross-pollinated crop is easier and more economical than that in a self-pollinated crop due to the higher outcrossing percentage in the former.

In nature, to create genetic variability and wider adaptation to different environmental conditions, the flowering plants have adopted various mechanisms for cross-pollination. Hermaphrodite flowers have both male and female reproductive organ in a single flower whereas the presence of unisexual/imperfect flowers favours out-crossing, which results in genetic heterogeneity and show wider adaptations (Frankel and Galun 1977). Flowering plants have evolved various mechanisms to favour cross-pollination. These are as follows:

1. Dicliny: The plants are unisexual.
   (a) Monoecious: The male and female flowers/inflorescence are borne on different nodes of the same plant e.g., cucurbits, maize, castor etc.
   (b) Dioecious: The male and female flower/inflorescence is borne on different plants. Field crops belonging to this class with hybrids are limited (e.g., spinach).
2. Dichogamy: The anther dehiscence and stigma receptivity occur at a different time that favours cross-pollination. Non-synchronization of male and female flowers may vary from one to a few days.
   (a) Protoandry: The anthers dehisce before the stigma becomes receptive, e.g. maize, castor, sunflower.
   (b) Protogyny: The stigma becomes receptive before the dehiscence of anther, e.g. pearl millet, Indian mustard, onion, cauliflower, etc.
3. Herkogamy: The stigma is covered with a waxy layer which does not become receptive until the waxy membrane is removed by honeybees resulting in cross-pollination, e.g. lucerne and alfa-alfa.
4. Heterostyly: The flowers have styles of different lengths (long, medium, short, pseudo-short), wherein the stylar length determines the outcrossing percentage, e.g. brinjal (Fig. 1).
5. Male sterility: Absence/atrophy/misformed/malformed male sex organ (stamen) or absence of functional pollen grains in a complete flower that does not allow self-fertilization.
6. Self-incompatibility: Failure of pollens to fertilize the ovule of the same flower, or that of other flowers on the same plant, e.g. *Nicotiana* and *Brassica*.

2   Genetic Principles in Hybrid Seed Production

Hybrid seed production requires a female plant in which viable male gametes are absent naturally or removed by artificial means. Hand emasculation is done to make a plant devoid of pollen so that it can be used as a female parent. Another simple way to use a female line for hybrid seed production is to identify or create a line that is incapable to produce viable pollen. This is called a male sterile line.

Fig. 1 Flower types in brinjal
(long, medium and short style;
from bottom to top)

2.1   Male Sterility

Male sterility prevents self-pollination, facilitates cross-pollination and promotes heterozygosity. Male sterility is exploited in agricultural crop plants for hybrid seed production. Male sterility is of three types: genetic, cytoplasm and cytoplasmic-genetic male sterility.

2.1.1   Genetic Male Sterility (GMS)

Male sterility is controlled by mutations in nuclear genes that affect stamen and pollen development. It can be controlled either by dominant/recessive genes. A male sterile line is maintained by crossing a male sterile line with a heterozygous male fertile line. Genic male sterility-based hybrids are available in safflower and pigeon pea. A GMS line (A-line) is maintained by backcrossing with the heterozygous B-lines (maintainer line). The A-line (seed parent) has 50% fertile and 50% male sterile plants. In hybrid seed production plot using GMS system therefore, it is required to rouge out 50% male-fertile plants. Seed and seedling markers that are closely linked to male sterility genes in the A-line can help to identify off-types and remove those male fertile plants from the field before flowering. In vegetable crops, GMS has been commercially used for hybrid production in muskmelon.

2.1.2   Cytoplasmic Male Sterility (CMS)

Cytoplasmic male sterility (CMS) is governed by extra-nuclear genes mainly present in the mitochondrial genome. These show non-Mendelian inheritance pattern and are under the regulation of cytoplasmic factors. In the majority of the cole crops in

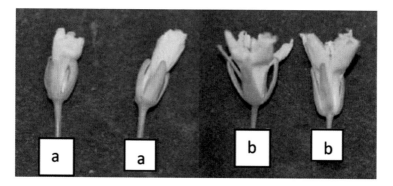

Fig. 2 Flowers in CMS (**a**) and male fertile (**b**) parent in cauliflower

Brassicace family, cytoplasmic male sterility has been used commercially for hybrid seed production (Fig. 2).

2.1.3 Cytoplasmic-Genetic Male Sterility (CGMS)

Male sterility is controlled by an extra-nuclear genome and often nuclear genes restore fertility in the hybrid plants. The male sterility is controlled by both the nuclear and cytoplasmic genes. In cytoplasmic-genetic male sterility, restoration of female fertility is undertaken using restorer lines carrying nuclear restorer genes in the crop. The male sterile line is maintained by crossing with a maintainer line that has the same genome as that of the MS line but carries normal (N) fertile cytoplasm. The fertility restoration is done by fertility restorer genes (Rf). The Rf genes do not have any expression of their own unless the sterile cytoplasm is present. The Rf genes are required to restore fertility in S cytoplasm which causes sterility. Thus N cytoplasm is always fertile and sterile cytoplasm with Rf-- gene produces male fertile plants; while S cytoplasm with rfrf genes produces only male-sterile plants. Another feature of these systems is that Rf mutations (i.e. the mutations to rf or no fertility restoration) are frequent, so N cytoplasm with Rfrf is the best for stable fertility. Cytoplasmic-genetic male sterility systems are widely exploited in both field and vegetable crop plants for hybrid development due to the convenience to control the sterility expression by manipulating the nuclear gene–cytoplasm combinations in any selected genotype. Incorporation of these systems for induction of sterility evades the need of emasculation thus facilitating the production of hybrid seed under natural conditions.

2.2 Self-Incompatibility

Self-incompatibility (SI) is a mechanism that prevents self-fertilization through recognition of self (own) pollen on stigma on the flower in the same plant or that of other plants of the same genotype. But pollen from other plants carried by wind, insects and other vectors deposit on stigma of such flowers and set seeds. Therefore, self-incompatibility prevents self-fertilization and facilitate cross-fertilization. SI is observed in both hermaphrodite and homomorphic flowers. The self-incompatibility response is genetically controlled by one or more multi-allelic loci and relies on a series of complex cellular interactions between the self-incompatible pollen and pistil.

Besides the use of male sterility and self-incompatibility systems, the following methods/systems are also followed for effective hybrid seed production in both field and vegetable crops.

2.3   Emasculation and Pollination

The male flowers are pinched off in the female lines a day before anthesis, or the stamens are manually removed from a bisexual flower in a female plant before flower opening (anthesis). This system is feasible when the male and female parts of a single flower or plants are separate. This is practised in bisexual perfect flowers where the androecium could be removed easily. In the female parent, the anther column is removed from a bisexual flower, the process called emasculation, and pollen of desired male line is dusted manually on the stigma of the emasculated flower in female parent to facilitate pollination and fertilization. This technique is commercially feasible in crops which have large, conspicuous flowers, easy removal of stamens/anthers, high seed set rate per pollination, low seed rate/ha and higher cost of hybrid seed, e.g. cotton, tomato, brinjal, chilli, melon, etc. This technique requires trained labour for emasculation, pinching, bagging, pollen collection and pollination adding to the cost of hybrid seed. It is, therefore, vital to know the floral biology, flowering time, crop morphology, and synchronization of flowering in the parental lines to plan emasculation and pollination in seed parent. The male and female rows are grown in recommended row ratios (male:female) or blocks. The fruit set on female lines is harvested for hybrid seed extraction.

2.4   Use of Gynoecious Sex Form

The gynoecious sex form has been commercially exploited for hybrid seed production in cucurbits. For hybrid seed production of cucumber, sponge gourd, bitter gourd and musk melon, the female and male rows are planted in a specific row ratio of 4:1 in the northern states of India under favourable climatic conditions to achieve high seed yield. The female parent bears only pistillate flowers and pollination is accomplished by insects (honeybee and wasp). To ensure good fruit, seed set and seed recovery, a sufficient population of the honeybee is maintained at the boundary of seed production plots. The male parent line is maintained by selfing (mixed pollination) and rouging out undesirable plants before contamination may take place. The female lines, i.e. gynoecious lines are maintained by inducing the staminate flowers with the application of silver nitrate (200 ppm) at two to four true leaf stages and followed by selfing. The weather conditions at seed production

Fig. 3 Pistillate flowers in a gynoecious line of cucumber

location play an important role, as the gynoecious lines are unstable under high temperature and long photoperiod conditions (Hormuzdi and More 1989). For this reason, the gynoecious cucumber did not become popular in tropical countries. However, a few true-breeding tropical gynoecious lines in cucumber and musk-melon have been developed (Fig. 3). These homozygous gynoecious lines are maintained by applying GA_3 at 1500 ppm or silver nitrate at 200-300 ppm or sodium thiosulphate at 400 ppm to induce staminate flowers at two and four true leaf stage. Homozygous lines are planted in strict field isolation. The gynoecious lines are crossed with a monoecious male parent to produce the F_1 hybrid.

2.5   Use of Chemicals and Growth Regulators

The hybrid seed can be produced by inducing femaleness and maleness with the application of chemicals and growth regulators. Spraying of etherel (2-choloro-ethyl-phosphonic acid) at 200–300 ppm at two and four true leaf stages and flowering is effective in inducing the pistillate flowers successively in the first few nodes on the female parent in bottle gourd, pumpkin, and squash, which are employed in hybrid seed production.

2.5.1   Sex Modification through Hormones and Chemicals

Though the sex expression in dioecious and monoecious plants is genetically determined, it can be modified to a considerable extent by environmental and introduced factors such as mineral nutrition, photoperiod, temperature and phytohormones. Amongst these, phytohormones are the most effective agents for sex modification and their role in the regulation of sex expression in flowering plants has been documented. The morphological differences in various sex types and their specific metabolic characteristics result from the possession of specific patterns of proteins, enzymes and other molecules. Modification of sex expression in cucurbits has been induced both by changing the environmental conditions and by applying treatments with growth regulators. Auxin treatment increases femaleness while gibberellins cause a shift towards maleness. Application of plant growth regulators is reported to alter sex expression and flower sequence in cucurbits when applied at the two to four true leaf stages, the critical stage at which a particular sex type can be suppressed or encouraged (Hossain et al. 2006).

Chemicals inducing femaleness:

- Auxins – Naphthaleneacetic acid (NAA), Etherel, Ethephon, Maleic hydrazide. Cytokinins, Brassinosteroids.

Chemicals inducing maleness:

- Silver nitrate ($AgNO_3$), Abscisic acid (ABA), Gibberellic acid (GA_3), Thioporpinic acid, Phthalimide, Paclobutrazol, etc.

In cucumber, $AgNO_3$ was found to be a potent inhibitor of ethylene action leading to femaleness. It should be sprayed when two to three true leaves are fully expanded. Gibberellic acid spray leads to excessive elongation and weakening of plants and there will be an increased number of malformed male flowers with less pollen. In gynoecious cucumber, there will be an increased number of male flowers on the vine when sprayed with silver nitrate or gibberellic acid, which made it possible for the multiplication of gynoecious lines in hybrid seed production. The sex ratio could be increased by the application of plant growth regulators like etherel or ethephon, gibberellic acid, naphthalene acetic acid and maleic hydrazide (Shailendra et al. 2015; Shiva et al. 2019).

2.6 Manipulation of Environment for Sex Modification in Hybrid Seed Production

The environment has a strong influence on sex modification. The role of environmental conditions in hybrid seed production and maintenance of the parental lines are described below.

2.6.1 Rice

The expression of male sterility and its restoration in rice is influenced by environment-sensitive genic male sterility (EGMS). It has been further classified into photosensitive and thermosensitive genic male sterility genes. The hybrids developed using these systems are called two-line hybrids, as no maintainer is required for the multiplication of the female line. The EGMS lines are multiplied by growing these in a season or a location in which the flowering period coincides with the required sterility/fertility change. For example, temperature-sensitive genic male sterility (TGMS) lines change to fertility at lower temperatures and the most ideal regime to induce a higher level of fertility is 27/21 °C. Therefore, in such cases, the TGMS lines need to be planted in such a way that the crop is at a fertility-inducing stage (say 5–20 days after planting) when favourable temperatures are prevailing. Similarly, the sterility-inducing stage coincides with a photoperiod of more than 13.75 h and temperatures above 32/34 °C. The hybrid seed production following this system depends on the critical fertility-inducing factors and their duration.

2.6.2 Castor

Castor is a monoecious plant species with staminate flowers and pistillate flowers located at different positions in a raceme (Fig. 4a). There are genotypes with a predominantly higher proportion of pistillate flowers governed by both genetic and environmental conditions (Fig. 4b). Low temperature promotes pistillate plants in particular genotype that reverts to monoecism with an increase in temperature and higher order of branches. Maximum female sex expression is seen when the daily temperature during raceme formation and development is less than 30 °C. It is also very high at the early growth stages of the female line in a higher soil nutrient condition. Female lines are multiplied at higher temperature condition that induces temperature-sensitive interspersed staminate flowers (ISF) and hybrid seed production is taken up at relatively low-temperature condition in which the ISF is not formed in a pistillate line. Therefore, two-line hybrid seed production has been possible in castor.

Fig. 4 Sex types in castor (**a** monoecious; **b** pistillate)

Male sex expression of several plant species is favoured at high temperatures and female sex expression at low temperatures. In tomatoes, male-sterile mutants develop male-sterile flowers at a temperature of above 30 °C and normal flowers at lower temperatures. In Brussels sprouts, low temperature affects the development of the androecium. In onion, the male-sterile plant produces viable pollen above 20 ° C. In cucumber, high temperature and long day length (>14 h) favour male flowers.

3   Agronomic Principles of Hybrid Seed Production

3.1   Environmental Requirements

Optimum growing season, conditions, and location are critical in obtaining good yield and quality of hybrid seeds. The regions with abundant sunshine are preferred for seed production. Unless the parental lines are specifically sensitive to a particular temperature and photoperiod for flowering and male sterility expressions, such as rice and castor, the hybrid seed crops can be raised in conditions favourable for the species. Sunshine hours are kept in planning in case of photoperiod-sensitive crops like lettuce and spinach, which require long-day conditions for flowering and seed set. Some species, on the other hand, require a low temperature to promote flowering (vernalization). Many temperate vegetables like cabbage, cauliflower, beetroot, European type radish and carrot need vernalization. High rainfall areas are not suitable for hybrid seed production due to the adverse effect not only on pollination but also on seed viability and vigour, whereas excessive wind speed may hamper the activity of pollinators, carry wind-borne foreign pollen from long distances resulting in contamination and cause seed shattering.

3.2   Land Requirement

The field selected for raising a hybrid seed crop should be free from 'volunteer' plants. Volunteers mean the plants originating from the seed/plant material of the previous commercial or seed crop. In vegetables, volunteer plants are seen in spinach, tomato, etc. The land should be levelled with proper drainage and should have sufficient organic matter in it. The cultural operations of hybrid seed production are similar to the production of open-pollinated varieties.

3.3   Isolation Distance

The spatial separation of the plots of hybrid seed production from any kind of contaminant is a critical requirement. Appropriate distances are to be maintained from other hybrids of the same crop, same hybrid seed production field not conforming to genetic purity standard, other related species, which may cross-pollinate, and fields affected by designated diseases to prevent genetic and disease contamination. Proper isolation standards, as required for a given species, are followed at all stages of maintenance and seed production to maintain the genetic purity of the hybrid. Isolations can be maintained in terms of distance or time, or a combination of both. For time isolation between seed production plots, information on the number of days to flowering of the parental lines should be known, and the planting schedule must adhere to this information. Two fields of hybrids may be planted without maintaining safe spatial isolation, provided an appropriate time interval has been kept between the two sowings, so that the pollination in the first planted field would be over before flowering/pollination starts in the second field or vice versa. The off-types, i.e. very late or early flowering plants should be removed/discarded from the field to ensure genetic purity.

The isolation distances for hybrid seed production are recommended based on the flowering behaviour of the crop, the movement pattern of its pollen and the medium of its dispersal viz., wind, insect, etc. Some examples of isolation distances in the field and vegetable crops as per the Indian seed regulation (Annonymous 2013) are given in Table 1.

The mode of pollination determines the isolation distance to be maintained. In the case of self-pollinated species the isolation distance is relatively less, but in the case of cross-pollinated species relatively longer distances are maintained from other varieties. The isolation distance also depends on the direction of insect flight or of winds, which have a direct influence on the pollen movement.

Isolation by time allows seeds of different varieties of the same crop to be produced in nearby fields. If the season is long enough to allow two production cycles, then two cross-compatible hybrids can be isolated by time. In certain cases, a barrier crop is raised between the fields of two cross-compatible fields to minimize contamination.

---

4   Stigma Receptivity

The stigma of the flower is the organ on which pollen lands to facilitate fertilization and consequently seed formation. Variations in the stigma morphology support pollen germination and tube growth in the compatible pollen. Longer stigma receptivity in cytoplasmic male sterile (CMS) lines is a desirable trait that favours higher hybrid seed yield. It is important both for manual or natural pollination by vectors like wind, insects, etc. Initiation of stigma receptivity is reported to be highly variable across the plant species and genotypes. Generally, stigma becomes receptive at the time of flower opening (anthesis). In the case of CMS, the protogynous and protandrous type stigma becomes receptive one or few days before anthesis and is extended by a few more days afterwards (Lloyd and Webb 1986). The duration of stigma receptivity is usually longer in CMS lines than in their fertile counterparts. The stigma receptivity in CMS lines of rice, for instance, lasts for up to 4 days (Gupta et al. 2015). In *Brassica* spp. CMS lines with *Moricandia* cytoplasm, the stigma receptivity was recorded up to 6 days after anthesis (Chakrabarty et al. 2007). In another study, stigma of CMS lines of *B. juncea* was receptive for 6–8 days after

Table 1 Minimum isolation requirements for foundation seed of the parental line and certified seed of hybrids of some field crops and vegetables

S. No.	Crop	Isolation distance (m) FS@	CS#	Remarks
1.	Rice	200	100	Barrier crops may be grown to further minimize the pollen flow
2.	Maize	400* 600** 400$	200* 300** 5***	*Same kernel colour and texture; same or different hybrid not conforming to varietal purity **Different kernel colour and texture ***with common male parent $Not conforming to varietal purity requirements
3.	Sorghum	300* 400**	200* 400** 5***	*Other variety; same and other hybrid not conforming to varietal purity **Johnson grass, forage sorghum ***with common male parent
4.	Pearl millet	1000*	200* 5**	*Other varieties; same and other hybrid not conforming to varietal purity requirements for certification **with common male parent
5.	Rapeseed mustard	200* 100**	50* 50**	*Self-compatible **Self-incompatible
6.	Castor	600	300	–
7.	Cotton	50* 5**	30* 5**	*Other varieties of same species; same variety not conforming to varietal purity **Other varieties of different species; blocks of parental lines of same hybrids
8.	Tomato	50	25	–
9.	Brinjal	300	150	–
10.	Cauliflower, cabbage, beetroot, radish, turnip	1600	1000	–
11.	Bottle gourd, muskmelon, watermelon, sponge gourd, bitter gourd, pumpkin	1000	500	–

@*FS* foundation seed of parental lines, #*CS* certified seed of hybrids

anthesis (Mankar et al. 2007). In protogynous lines of *B. juncea*, the duration of maximum stigma receptivity was reported to be up to 3 days after stigma protrusion (Chakrabarty et al. 2011), though the seed set was observed up to 10 days after stigma protrusion. In comparison to the CMS and protogyny systems, the open-pollinated varieties showed stigma receptivity up to 4 days after anthesis(Maity et al. 2019).

In rice, the stigma remained receptive for six to seven days without pollination, with its maximum receptivity up to three days after the opening of the spikelet. In pearl millet CMS line 'Tift 23A' higher seed set was observed when pollinated 2–3 days before anthesis than the open-pollination after anthesis (Burton 1966). In sorghum, stigma becomes receptive one to two days before blooming and unpollinated inflorescence remains receptive up to a week or more (Ayyangar and Rao 1931; Maundar and Sharp 1963). In pigeon pea, stigma becomes receptive 68 h before anthesis and continues for 20 h (Prasad et al. 1977). In castor, pistillate flower retains receptivity for five to six days. In safflower, stigma remains receptive for 72 h. However, the duration of stigma receptivity is much influenced by the climatic conditions during flowering.

Duration of stigma receptivity showed variation in Cucurbitaceae family. Stigma remained receptive in *Luffa* species from 6 h before to 12 h after flower opening (Singh 1957). Stigma was more receptive at 3–4 h after anthesis for pollination in melons (Tarbaeva 1960). Nandpuri and Brar (1966) observed that the maximum stigma receptivity prevailed 2 h before anthesis and 2–3 h after anthesis in the case of muskmelon. Stigma is receptive 36 h before anthesis and remained so until 60 h after anthesis (Nandpuri and Singh 1967) in bottle gourd. Nepi and Pacini (1993) reported that the stigma became receptive 1 day before anthesis and remained receptive for 2 days in *Cucurbita pepo*. Stigma remained receptive for 24 h before and after anthesis in bitter gourd (Miniraj et al. 1993).

5   Pollen Viability

Pollen maturation to its release and dispersal, germination on a receptive stigma and fertilization are essential steps to plan the timing for hybridization for successful hybrid seed production. The response of pollen during its mobile phase, i.e. after its release and detachment from the anther of the pollen parent, is also important as pollen passes through different environmental conditions. The duration of pollen viability varies greatly among crop species and varies from few minutes to hours (Stanley and Linskens 1974; Shivanna and Johri 1985; Barnabas and Kovacs 1997). In cereals (*Poaceae*) pollen grains lose viability within 20–30 min from dehiscence. In rice, for example, the pollen grain loses its viability within 10 min due to prevailing high temperature that leads to pollen desiccation. In some species of Solanaceae and Fabaceae, pollen viability is reported for up to several weeks. A favourable environment needs to be identified for hybrid seed production. Pollen viability/longevity depends on several climatic factors like temperature, humidity, etc. as well as pollen vigour that might be influenced by the nutrition and disease status of the plant. Low atmospheric humidity (0 ± 40%) is favourable for longer pollen viability. In maize, pollen grain remains viable for a few minutes after its dehiscence. Pollen in sorghum remains viable for three to six hours in anther and the viability gets over within 20 min after its detachment from anther. In cotton, pollen remains viable for up to 24 h. However, pollen collected from younger buds showed viability for 44 h or more. Similarly, castor pollen remains viable for one day after anthesis.

Knowledge of both stigma receptivity and duration of pollen viability is important for successful hybrid seed production both in open fields and under protected conditions. If pollen viability can be prolonged with proper drying under vacuum and storing in a sealed container at 4–5 °C, the pollen parent may not be grown each year and stored pollen under controlled conditions could be used for hybrid seed production.

6 Pollination Control

Pollen dispersal and its transfer to the seed parent are very important steps in hybrid seed production. Usually, the two parental lines used for hybrid seed production are different in their morphological traits, including differential flowering time. The parental lines should flower simultaneously and follow similar flowering patterns so that there is effective pollination, followed by fertilization and seed set. The pollination/crossing is achieved either by manual operation or by allowing natural pollination by the wind and/or insects depending on the flower type and pollen characteristics.

7 Synchronization of Flowering

Failure to achieve proper synchronization of flowering between the parental lines is the most commonly encountered problem in hybrid seed production, resulting in very poor or no seed set. Hence, synchronization of flowering of the parental lines is a prerequisite for successful hybrid seed production. This is because the seed set on the female parent depends on the availability of viable pollen supplied by the male parent during the flowering period, while the stigma is receptive. Synchronization of flowering means that the seed parent and pollen parent flowers simultaneously. In general, parental lines of most of the hybrids across the crop species differ in their growth duration and consequently flowering. Therefore, it is essential to determine the flowering behaviour in terms of its initiation, peak flowering, termination and thereby flowering duration to take up an appropriate seeding/sowing plan. This helps achieve synchronized flowering between the parental lines involved in hybrid seed production. It is observed that the seed parent should flower a day or two earlier than the male parent but not vice versa as the seed parent, particularly with the cytoplasmic male sterility, protogyny, and protandry systems, shows longer stigma receptivity duration. The pollen parent usually flowers profusely but terminates quickly, particularly in plants with non-branching and determinate growth habits. Knowledge about the flowering pattern of the parental lines/combinations in particular agroecological and seasonal conditions help in planning the sowing of the parental lines for hybrid seed production. Achieving synchronized flowering through proper seeding intervals, staggered sowing, seed treatment, spraying of chemicals, fertilizers and other agronomic interventions are possible only with prior knowledge of the flowering behaviour of the parental lines for hybrid seed production. The seed parent with CMS, protogyny, and protandry systems having an extended flowering period, with insufficient pollen availability will result in a low seed set. To supply an adequate quantity of pollen during the flowering of the seed parent, staggered sowing may be followed in the pollen parent at an interval of 3–5 days.

Following are some measures practised for achieving synchronization of flowering during hybrid seed production of different crops:

7.1   Rice

1. Staggered sowing: In general, the parental lines involved in hybrid development differ in respect of flowering time and duration due to their growth and development. Therefore, it is necessary to determine the seeding intervals between the parental lines and adopt staggered sowing to ensure synchronized flowering.
2. Fertilizer application: Young panicle development is compared by observing the primordium development of the parental lines 30 days after transplanting. The primordium development stages are indicators of the flowering of parental lines (Yuan 1985). Following this principle, it has been suggested that during the first three stages of panicle differentiation, the early parent is applied with a quick release of nitrogenous fertilizer (spray of urea at 2% delays flowering of the early parent) or spray the later developing parent with potassium di-hydrogen phosphate at 1.5%. This adjusts developmental differences up to 4 to 5 days.
3. Water management: In case a difference in the development of panicles is observed during later stages draining water from the field will delay early parent panicle development while higher standing water will speed panicle development to the late parent (Feng 1984; Xu and Li 1988). Restorer lines are more sensitive to water management.

7.2   Sorghum

1. Staggered sowings depending on plant growth, development, and flowering time and duration in the parental lines need to adopted (Murthy et al. 1994).
2. The advancing parent needs to be sprayed with 500 mg of Maleic hydrazide in 1 L of water, 45 days after sowing.
3. Urea (1%) solution can be sprayed on late parent (House 1985; Kannababu et al. 2002).
4. One irrigation may be skipped for the advancing flowering.
5. Spraying CCC (Chloro Chlorine Chloride) at 300 ppm for delaying flowering.

7.3   Pearl Millet

1. Staggered sowing of parental line.
2. Urea solution (1%) can be sprayed on late parental line.
3. Jerking of early parent delays the flowering.
4. Additional nitrogen fertilization or foliar spray to the late parent (Govila and Singhal 2003).

7.4   Sunflower

1. There should not be more than 3–4 days difference in the flowering to avoid staggered sowing problems.
2. Seed treatment with simple hydration of the female parent and with GA_3 at 50 ppm to the male parent for 18 h before sowing and spraying urea (1%) thrice on alternate days at the button formation stage to the male parent resulted in a good degree of synchrony in flowering between the parental lines in the hybrid, namely, KBSH-1 and TCSH-1 under North Indian conditions (Chakrabarty et al. 2005; Chakrabarty 2008).

7.5   Cauliflower

1. Staggered sowing of parental line.
   (a) Application of GA_3 at 100 ppm and IAA at 50 ppm thrice (at the initiation of bolting, 7 days after the first spray, and at the bud initiation stage) was found effective in achieving synchronization of flowering between the parental lines of cauliflower hybrids (Personal communication, S. K. Chakrabarty, unpublished).

8   Planting Ratio

Hybrid seed production depends on the seed set on the female plants out of the total area planted for hybrid seed production. Therefore, a higher hybrid seed yield is expected with a higher proportion of female (seed parent) population, in terms of rows/blocks. But it, in turn, depends on the pollen produced and supplied by the male parent population/rows. Assuming a good pollen producer, it is also critical to have information about its pollen dispersal ability and pollen viability till it lands on the stigma for effective pollination and fertilization. Therefore, the planting ratio/row ratio is one of the important factors that determine the hybrid seed yield. To estimate the optimum row/planting ratio systematic experiments are conducted on each crop and with different parental combinations. The role of pollinators and other vectors like wind play important roles in determining the planting ratio. The distance between the seed parent and pollen parent also becomes an important factor in

Table 2 Recommended planting/row ratio for hybrid seed production

Crop	Pollen parent:seed parent	References
Rice	2:8;2:10	Viraktathamath et al. 2003
Maize	1:2; 1:3; 1:4 or 2:4; 2:6; 2:8	Mac Robert et al. 2014; Sharma et al. 2020
Sorghum and pearl millet	2:6	Singhal and Rana 2003; Govila and Singhal 2003
Cotton	1:4; 1:5	Lather and Singh 1997
Sunflower	1:3; 2:6 or block planting of the male and female parent	KempeGowda and Kallappa 1992; Ranganatha et al. 2003
Pigeon pea	1:3	Tikle et al. 2014
Rapeseed and mustard	2:8	Maity et al. 2012
Castor	1:3;1:4	Chakrabarty 2003
Cucumber, sponge gourd, pumpkin	1:3; 1:4	Sharma et al. 2004
Tomato, brinjal, Chilli	1:3; block planting of the male parent	–

hybrid seed production plots, particularly in a predominantly self-pollinated crop like rice as natural wind is helpful to a great extent in increasing seed set. Planting geometry is also important in wind-pollinated hybrid seed production plots. Planting the seed/pollen parent in rows perpendicular to the wind direction is the principle in case of wind as pollen vector.

Lal and Singh (1990) observed that 2:1 (female:male) ratio was optimum for hybrid seed production in muskmelon. Soto et al. (1995) compared manual pollination and open pollination for the production of the parental lines with a female:male ratio of 3:1, 6:1 and 9:1 in cucumber and they reported that 6:1 ratio was economic for hybrid seed production. Kushwaha and Pandey (1998) recommended a planting ratio of 4:1 for hybrid seed production in bottle gourd. Sharma et al. (2004) reported a 3:1 ratio for hybrid seed production of cucumber. Satish Kumar (2005) reported that there was no significant difference among planting ratios compared for seed yield and its attributes in bitter gourd hybrid seed production. Recommended planting/row ratio for hybrid seed production is given in Table 2. Hybrid seed production plots in maize and cauliflower are shown in Figs. 5 and 6, respectively.

9   Supplementary Pollination

Supplementary pollination is a method to ensure adequate pollination of the seed parent by physical or mechanical methods or by maintaining an abundant vector population. It is achieved by different practices as per the flowering behaviour and flower morphology of the crop. For instance, it can be performed manually by rubbing the capitulum of the sunflower with desired pollens; shaking or jerking the rice inflorescence (panicle) with bamboo sticks or pulling a rope through the

Fig. 5 Field layout in hybrid maize seed production

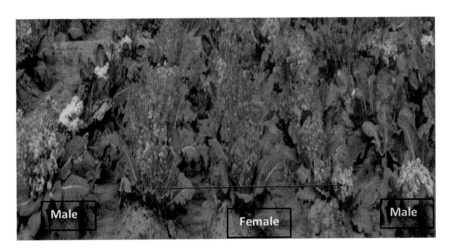

Fig. 6 Seed production of CMS-based cauliflower hybrid with a planting ratio 1:3 (M: F)

rows of the male parent. Supplementary pollination has to be done for 7–10 days during flowering. Time, duration and frequency of supplementary pollinations are important for higher seed settings in the seed parent. In rice, for example, the first supplementary pollination is performed in the morning during the initiation of spikelet opening. This process is then repeated 4–5 times in a day at an interval of

30 min till the completion of flowering in both the parental lines. In the case of cross-pollinated crops, insects are the primary pollinators. In such cases, supplementary pollination by honey bees or bumblebees results in higher hybrid seed set and yield. Keeping beehives in hybrid seed production plots coinciding with flowering is recommended for realizing higher quantity and better quality of hybrid seed in sunflower, rapeseed, etc.

Vegetable crops viz., tomato, garden pea, fenugreek and cowpea are primarily self-pollinated, whereas, other vegetable crops like onion, carrot, cole crops, cucurbits and *Brassica* are cross-pollinated in nature. The natural insect population (honey bees/solitary/bumble bees, wasp, butterflies, moths, beetles, flies) is normally sufficient under open conditions to ensure satisfactory pollination, but with high plant populations maintained in seed production plots, there is a possibility that the natural insect population may be insufficient to ensure proper seed set. Therefore, the introduction of supplementary bee hives improves pollination and seed set. However, care must be taken to ensure that pest protection chemicals are not used in a way to harm useful pollinating insects. The spray of chemicals should be avoided at peak pollination period to support insect activity and should be done only in the late afternoon.

10   Roguing

A quality seed must be free from any genetic or physical admixtures. To ensure high genetic and physical purity, it is important to have proper isolation and regular removal of off-types and volunteer plants from the female and male/restorer parental lines at all growth stages. Rouging is the removal of undesirable plants from seed production plots, which helps prevent further genetic contamination from off-types (plants of the same species but different genotypes) by cross-pollination. The undesirable plants may be volunteer plants from the previous crop, off-types produced by the out-crossing with contaminants and admixtures during the process of harvesting, threshing, packing and handling, or the presence of 'pollen shedders' a term used for male fertile plants in the seed parent population.

Rogues or off-types may occur in a crop due to any of the following reasons.

- The diversity of the morphological types within a crop may be wide. This tendency is greater in predominantly cross-pollinated (e.g. cauliflower, cabbage, cucurbits and onion) than self-pollinated (e.g. peas, tomato, fenugreek) crops.
- Some plants may display deviation from the normal type due to developmental variation.
- Volunteer plants may arise from vegetative pieces or dormant seed of the previous crop grown in the same field.

The timely and careful removal of all off-type plants in both female and male lines ensures the high purity of hybrid seed. Plants having seed-transmittable diseases, in either the female or male parent population, are also to be removed at the early stage to maintain good seed health.

Vegetable varieties often show genetic variability in morphological traits after growing over several generations. It is, therefore, necessary to exert control and keep the natural variation within acceptable limits by inspecting the crops at different growth stages and removing individual plants which do not conform to the defined limits of that variety.

It is always easier to conduct intensive rouging in breeder seed plots than in large commercial seed production plots. Hence, maintaining the highest genetic purity of the parental line seed during their multiplication at the pre-basic, breeder (equivalent to basic of OECD) and foundation (equivalent to Certified 1 of OECD) seed stage is as important as the following roguing in the hybrid seed production plots.

Variety description based on morphological characteristics like leaf shape, flower colour, fruit shape and colour generally form the basis for rouging, though some of these like leaf colour, plant height, and earliness of flower initiation are known to be influenced by environmental conditions.

11   Harvesting, Threshing and Seed Extraction

The best time for harvesting seed crops is at a stage when the highest yield of the best quality seed can be obtained. The seed of various field and vegetable crops are extracted from dry seed heads, dry fruits or from fruits (as in many vegetables) in which the seeds are wet at the time of extraction. Harvesting of the male parent should be done before the female (seed) parent. Threshing can be done by hand or by machines. Care is taken while transporting material from the field to the threshing floor. Both the trolley and the threshing floor are cleaned of any seed/ plant parts to avoid genetic admixture or contamination by weed seeds at this stage. Threshing machines should also be properly cleaned to avoid any admixture.

12   Seed Drying

At the time of maturity, the seed contains higher moisture content than the optimum for better germination and storability. Seeds from pulpy fruits like tomato, watermelon, muskmelon, cucumber and brinjal have very high moisture content at harvest and absorb more during wet extraction. Seeds of vegetables like onion, carrot, *Brassica*, etc. have relatively low moisture at harvest, but in drier climates may be prone to shattering loss. Sometimes, the seed may also have high moisture content due to adverse climatic conditions such as pre-harvest showers. Therefore, to reduce seed moisture to the optimum level, threshed and cleaned seeds are dried before storage. For ambient storage, seed moisture should be kept under 9–12% and for sealed packaging and storage it should be 6–8%. Natural and artificial methods of forced drying may be used for this purpose.

The hybrid seed production technologies in maize, castor and cotton under open field conditions and that of tomato and bitter gourd under protected conditions are described here.

13   Hybrid Seed Production in Maize (*Zea mays* L.)

Maize hybrids are derived from inbreds that are resulted from repeated inbreeding of specific maize population. Hybrids have early plant maturity, better disease resistance, food processing and nutritional quality. Based on the number and type of parents involved in a cross, maize hybrids can be classified as (i) single-cross, (ii) three-way, (iii) double-cross, and (iv) top-cross. Crossing between two inbred lines results in a single-cross hybrid, whereas crossing a single-cross hybrid with inbred line results in a three-way hybrid. Double-cross hybrid is derived by crossing two single-cross hybrids. To produce a top-cross hybrid, an inbred is crossed with an open-pollinated variety. Among all types of hybrids, single-cross hybrids are inherently more heterotic and uniform than any other form of hybrid.

For the production of the quality seed of a single cross maize hybrid, the respective parental lines are multiplied in isolation in the breeder and foundation seed stage and the single cross is produced in the certified seed production stage. In a hybrid seed production plot, male and female parents of foundation seed are planted at consecutive rows in the seed production plot. Depending on the pollen dispersal capacity of the pollen parent and the nature of the seed parent, the ratio of male and female rows can vary considerably. The tassel of the female plant is removed before initiation of pollen shedding so that the pollen of the male parent is available in the seed production plot. Detasseling of the female is necessary to avoid 'female-selfing'. Mechanical removal of tassels from the female parent before dehiscence is called detasseling. Tassels can be removed manually or mechanically. In a CMS-based hybrid system, detasseling is not required as female lines possess male-sterile cytoplasm.

The management of both the parents is important and requires adequate attention during seed production. The key factors that determine a successful quality hybrid seed production are as follows:

- Purity of the female and male parents.
- The ratio of female to male parents in the seed production field.
- Timely removal of off-types, diseased and rogues to prevent contamination.
- Detasseling of the female parent at an appropriate time.
- Synchrony of silking in female and pollen shedding in male parents.
- Careful and separate harvesting from the female and male parent rows to avoid mechanical seed mixtures.

13.1   Selection of Area

The area for seed production should have a similar climatic condition (temperature and photoperiod) as that of the area where the variety is intended to be recommended for cultivation. The area should have mild and dry weather with abundant sunshine. During seed production especially flowering, temperature, R.H., and wind velocity should range between 25 and 30 °C, 60–70%, and 2–4 km/h, respectively. Rain-free season (post rainy) with good irrigation facility is recommended for seed production as it favours better seed set and lesser incidence of diseases and insect problems. Areas prone to excess rains, high humidity, extreme temperature, strong winds and hail storm should be avoided.

13.2   Field Selection

The field selected for seed production should be well-drained, levelled with fertile soil preferably sandy loam to loamy soils as maize is very sensitive to water logging and drought. In addition, should be away from commercial crop and related crop species, free from pest and disease incidence, volunteer plants, weed seeds, off-types, diseased plants, soil-borne pathogens. The field should not have maize in the previous season.

13.3   Isolation

Maize is a cross-pollinated crop and proper isolation distance from other maize fields is required to produce genetically pure seed. Following measures are generally taken to maintain purity during seed production. Seed plots must be temporaly or spatially isolated to avoid contamination during the flowering by wind-borne pollen from neighbouring fields. Temporal isolation is achieved when pollen shedding of the male parent and silking of the female parent is either early or later (15 days) from the flowering of other maize fields in the vicinity. The minimum standard for isolating maize seed production fields is often established by national seed regulatory agencies taking account of factors like pollen count, wind speed and direction, pollen dispersal, and insect activity. Minimum isolation distance ranges from 200 m to 600 m.

13.3.1   Spatial Isolation

Pollen from other sources is excluded by maintaining physical distance of the seed production plot from the contaminating fields. Isolation distance of 400 m is the minimum seed certification standards for maize seed production. Component isolation required for different types of hybrids may vary from 5 to 7 depending on the number of parent involved in the hybrid (Table 3).

Table 3 Number of isolations required for hybrid maize seed production

S. No.	Type	Combinations	Parents	Breeder seed	Foundation seed	Certified seed	Total
1.	Single-cross	Inbred x inbred	2	2	2	1	5
2.	Three-way	F_1 × inbred	3	3	2	1	6
3.	Double-cross	F_1 × F_1	4	4	2	1	7
4.	Top-cross	Inbred × Var.	2	2	2	1	5
5.	Double top cross	F_1 × Var.	3	3	2	1	6

13.3.2 Temporal Isolation

In circumstances where spatial isolation is not possible, the same can be achieved through adjusting sowing time between two plots differing by at least 40 days difference.

13.3.3 Border Rows

Distances less than 200 m can be modified by planting border rows. Border rows must be planted with seed used for planting male rows in the seed field. Seeds saved from male rows of the previous production of the same cross cannot be used for planting border rows, or for planting within the isolation distance.

The isolation distance is maintained at 300 m if the kernel colour or texture of the contaminating maize is different from that of the seed parent, or if the contaminating field is planted with sweet or popcorn. In this case, modification of isolation distance by planting border rows will not be permitted.

Mandatory distances in the case of spatial isolation can be reduced to not less than 100 m in the case of:

- Planting of border rows of the pollen parent adjacent to the production plot (max. 10 rows).
- Presence of natural obstacles (forests, dense hedges, dams) of sufficient height, width and vegetation to shield from pollen flight.
- Isolation strips of the pollen parent or pollen-sterile corn planted between production and commercial fields.

Differential blooming dates are permitted for modifying isolation distances, provided five per cent or more of the plants in the seed parent do not have receptive silks when more than 0.50 per cent of plants in the field, within the prescribed isolation distance, are shedding pollen.

13.4 Use of Border Rows

Isolation is facilitated through the use of border rows, which are the rows of male parents planted around the borders of the seed production plots. The higher the number of border rows, the lesser the chance of contamination from undesirable pollen and better pollination efficiency due to the availability of desirable pollen of the male parent.

(i) The minimum number of border rows to be planted for modifying isolation distances less than 200 m is determined by the size of the seed field and its distances from the contaminant.

(ii) Border rows are sown in the seed field adjacent to it, but in no case separated by more than 5 m from the seed field.

(iii) Border rows must be sown at the same time with that of the male parent in the seed production field.

(iv) Border rows should be planted on all the sides of seed production field.

(v) Seed fields having diagonal exposure to contaminating fields should be planted with border rows in both directions of exposure.

(vi) If two hybrid seed fields, with different pollinator parents, are within the isolation distance of one another, border rows are necessary for each of them to avoid contamination of the respective seed parent.

13.5 Planting Pattern

Hybrids are produced by controlled mating among genetically distinct parents. Achieving high seed yield requires manipulation of the physical location of the male and female parents. Planting patterns for hybrid depend principally on the pollen-shedding ability of the male parent which determines the amount of pollen that is available and nicking or synchronization of flowering of the male and female parents (which determines the likelihood that pollen will be available when the silks have emerged from the female parents and are receptive). For a single cross hybrid, the ratio of female to male rows in the field is usually 3:1, whereas it may extend up to 8:2. The recommended planting patterns include 4:1 (four rows of the female parent to one row of the male parent), 4:2, 6:1, and solid female with inter-planted male.

13.6 Pollen Control

Pollen control is extremely important as it is vital for the production of a genetically pure seed crop. The popular pollen control methods are detasselling and the use of male sterility. Detasseling involves physically removing tassels from the female parents before it sheds pollen. Manual detasseling is achieved by grasping the stalk just below the tassel and removing the tassel with an upward jerk. In mechanical detasseling, cutting or pulling of the tassel is undertaken with machines which improves the efficiency as compared to the manual approach but a loss of 1–3 uppermost leaves in the former leads to a significant loss of seed yield. Detasseling should be initiated when the top 3–4 cm of the tassel are visible above the leaf whorl and continue every day until no more tassels are left. Regular monitoring before and after detasseling phase is essential for ensuring genetic purity.

13.6.1 Precautions During Detasseling

1. Grasp the complete tassel so that all pollen-bearing parts are fully removed.
2. Immature detasselling should be avoided. It may cause a few spikelets to be left in the leaf whorls, which may emerge and shed pollen. Also, the top leaves are likely to be pulled out, leading to a reduction in yield or attack of the disease.
3. Do not hold the tassel too low on the stalk to prevent pulling out of plant tops.
4. Once detasselling starts in a field it must be repeated daily in all weather. A fixed time should be observed every day. Be particular to start detasselling from the same side every day, in the case of a large field.
5. Mark all the male rows at both ends by driving long wooden markers in the ground, or by some other suitable means. The markers should be painted white.
6. Look out for suckers (tillers) on female plants and also for lodged or damaged plants in female rows, as they are likely to pass unnoticed during detasselling.
7. The detasseller should drop the tassels on the ground after removing them and not to carry them in hand, as this may involve the danger of contaminating receptive silks.
8. Put an experienced detasseller in charge of this operation. He should walk behind the other detassellers and check that no tassels are left in the female lines.

13.7 Flowering Manipulation

Achieving synchronization of flowering among parental lines, i.e. reaching the flowering stage simultaneously (a phenomenon known as nicking) is most important in hybrid seed production. For optimum seed setting the male parent should shed pollen when the first silks become receptive silks. The most common technique for achieving synchronization is split date planting of the male and female parents. Split date planting is made based on some combination of days to flowering, growth stages, and heat units accumulated from the date when the first parent was planted. As a precautionary measure, male parents may be planted on two or more dates to extend the pollen-shedding period. In case the gap of flowering of male and female parents is high (>5 days), an adjustment in the sowing dates will be required to ensure synchronization of flowering whereas where the gap is less than 5 days, foliar application of growth regulators, fertilizers or mechanical measures could be used.

13.8 Rogueing

Ideally, seed production plots should not have been planted with maize in the preceding season to prevent contamination by volunteer plants. In addition, all undesirable off-type plants showing variability in morphological characteristics (e.g. plant height; leaf shape and size; flowering habit; silk and ear characteristics; kernel shape size and colour) should be removed from the field. Start removing distinctly tall and extra vigorous plants at the knee-high stage. At the pre-flowering stage, rogue out off-type plants which are easily distinguishable based on plant characteristics such as leaf shape, size, plant height, etc. Continue rouging during the flowering stage to remove plants differing in tassel or silk character. Roguing for off-types and malformed plants should be completed before pollen shedding. Diseased plants affected by stalk rot should be rogued from time to time. At harvesting, off-textured or off-coloured ears are to be discarded. During hybrid seed production, four field inspections are undertaken, one before flowering and three during flowering.

14 Hybrid Seed Production in Castor (*Ricinus communis* L.)

Castor is an important non-edible oilseed crop possessing a unique fatty acid composition that is used for various industrial purposes, mainly as lubricants and innumerable derivatives for pharmaceutical use. India is a major castor-producing country with a phenomenal increase in area and productivity due to the commercial use of hybrids. The development of hybrids in castor is based on the pistillate mechanism. Some critical points in hybrid seed production in castor are given below:

14.1 Isolation

Castor is a cross-pollinating species with the wind as a primary source of pollen dispersal and transfer. The extent of cross-pollination mainly depends on the direction and velocity of the wind, and the proportion of female to male flowers on the raceme. Genotypes that produce mostly female or 100% female racemes easily get pollinated by foreign pollen from sources located as far as 1000 m distance. Besides the wind, insects like honey bees, butterflies, moths, etc. are also known to play a role in pollen dispersal that results in variable levels of cross-pollination leading to contamination of varieties and parental lines. Based on systematic research under AICRP on Oilseeds in ICAR, India the following isolation distances for different seed categories are recommended (Table 4).

Considering the requirement of long isolation distances, to produce genetically pure seed, it is ideal to take up hybrid seed production of castor in non-traditional areas and off-seasons to avoid contamination.

Table 4 Recommended isolation distances for different seed categories of castor hybrids/varieties

S. No.	Seed class	Isolation distance (meter)
Varieties and male parents of hybrids		
1.	Nucleus (equivalent to pre-basic of OECD) and breeder equivalent to basic of OECD)	1500
2.	Foundation (equivalent to Certified 1 of OECD)	1000
3.	Certified (equivalent to Certified 2 of OECD)	600
Female parents of commercial hybrids		
1.	Nucleus and breeder	2000
2.	Foundation	1500
3.	Certified hybrid seed	600

14.2 Season and Planting Condition

The time of planting and production season has a profound influence on sex expression in castor. While warm and rainy seasons with an average daily temperature of about 30 °C provide an ideal environment promoting male flowers for undertaking seed production of varieties, and multiplication of the male parents of hybrids, the mild winter season with an average temperature ranging between 15 and 25 °C during flowering is the most conducive for taking up hybrid/certified seed production as it favours the production of female flowers. In the case of varieties and male parents, such exposure to male promoting environment, i.e. rainy-summer season encourages good expression of plants bearing mostly male spikes which could be easily eliminated through timely rouging. Similarly, the female parents, which are environmentally sensitive, when raised in male-promoting environments preferably in summer season with an average daily temperature above 32 °C produce interspersed staminate flowers (ISF) which are crucial for the multiplication of the female parents.

14.3 Breeder/Foundation Seed Production of Female Parents

It is one of the most critical steps in hybrid castor seed production. The seed to be used to produce the breeder seed of female parents is the nucleus seed. The ideal season to undertake seed production of pistillate lines is summer (planting at about 25 °C during January's second fortnight). The following two methods have been adopted in India for producing breeder seed of female parents based on their pistillate nature (Ramchandram and Ranga Rao 1990).

14.3.1 Conventional Method
- As per the prevailing standards of Indian seed certification (Annonymous 2013), 20 to 25% of monoecious plants are allowed in seed production plots to ensure adequate pollen supply to pistillate plants.

- Prior to flower opening in primary racemes (at least 2–3 days before), all deviants that did not conform to the diagnostic characters, especially with respect to the node numbers (up to primary spike), nature of internodes, bloom and leaf characteristics are discarded.
- At the stage of flower opening in primary racemes, pistillate plants are identified conforming to the diagnostic morphological characteristics of the female parent and tagged distinctly at the base of primary racemes.
- All monoecists which exclusively bear male flowers beyond three whorls at the base of the spike are removed.
- Plants with interspersed staminate flowers, if any, should be retained subject to the condition that the retained plants fulfil all other prescribed standards.
- The numbers of female and male plants are counted in each row and the monoecious plants are removed if these are above the stipulated percentage.
- Tagged female plants are examined regularly for possible reversion to monoecism in secondary, tertiary and quaternary order racemes. Remove the tags as and when the female plant reverts to monoecism up to the fourth sequential order of the branches.
- On maturity, female plants bearing the tags are harvested and seeds from each picking are kept separately after proper drying, packing and labelling.

14.3.2 Modified Method
- Unlike in the conventional method, all monoecious plants are rogued out at least 2 to 3 days before the flowering begins in the primary raceme.
- Female plants are examined critically for various morphological characteristics, particularly the number of nodes up to primary raceme (most of the female flowers on primary raceme fail to set fruits due to the non-availability of pollen).
- A large number of interspersed late male flowers may appear on primary as well as subsequent order racemes in about 35 to 50% of the female populations which provide sufficient pollen for the late developed female flowers on the same raceme, or on later order racemes.
- Plants are examined regularly for any reversion to monoecism up to fourth order raceme and the off-types are rogued out. However, the pistillate plants reverting to monoecism in fifth sequential order onwards can be allowed in the population as supplementary source of pollen.
- Seed are collected from all female plants and kept as per the picking order after proper drying and labelling.

14.3.3 Other Precautions
- The number of ISF may vary from 1–2 to >10–15 male flowers per spike. These plants should be retained as pollen sources.
- However, the primary spikes with a highly ISF nature, i.e. 5–6 male flowers per whorl tend to revert to monoecious in the later orders which are to be closely observed.

- The majority of the primary spikes may not set seed due to the non-availability of pollen. However, the later orders or on the matured primary spikes itself interspersed male flowers appear and fertilize the female flowers.
- Observe the female plants for any revertants at any stage (secondary to pentenary). Remove such revertants.
- In case the number of revertant is high (>30%), those in the third or fourth order may only be removed and the seed may be harvested from the earlier orders only.
- In case the proportion of late revertant female plants with interspersed male and occasional bisexual flowers increases in the population, those may be allowed to remain.
- However, seeds should be collected only from all-female plants and picking-wise seed lots be kept separately.
- Seed from late-order revertant female plants should not be allowed to be mixed with those from all the female plants.

14.4 Certified Hybrid Seed Production

The cross-pollinated nature of castor with differential sex expression due to its high sensitivity to environmental factors (climate, nutrition, etc.) makes hybrid seed production complicated. In castor, pistillateness, a polygenically controlled character, is highly variable but can be managed to a large extent by various agronomic manipulations like sowing time, nutrition and irrigation. Hence, knowledge of crop's adaptation and following location-specific agronomic recommendations are prerequisites for obtaining higher seed yield.

The practices for successful hybrid seed production of castor in different situations are detailed below:

Genetic contamination should be avoided by strictly following the stipulated isolation distance (1000 m) and timely rouging. Sowing should be done towards the end of the monsoon season with an average temperature of about 25–30 °C depending on the location so that the emergence of the primary and secondary spikes coincide with the cool season. If delayed, the flowering period might experience higher temperatures resulting in ISF in female parents. A planting ratio of 3:1 or 4:1 (female:male) lines with two male rows as border rows all around the hybrid seed production plot is suggested.

Rouging is an essential activity in the certified hybrid seed production plots to keep the female parent plants completely pistillate and get these fertilized by the desirable pollen parent. The possible rouge plants in the female line are monoecious, pistillate with ISF, sex revertant and plants with hermaphrodite or bisexual flowers. The stages of rouging are as below:

First: Within 30 days before primary spike initiation, off-types are removed both in female and male lines based on morphological characteristics.

Second: At the time of primary spike initiation, monoecious and pistillate plants with ISF or hermaphrodite flowers in the female parent should be removed. In the male parent population, all morphological deviants should be removed. The male plants with male flowers in more than 2–3 whorls in the primary spike should be removed.

Third: At the time of secondary spike initiation, in addition to the above, revertant in the second order should also be removed. Deviants based on the spiny or non-spiny nature of capsules should be removed. Plants with ISF may increase with increasing temperatures. The removal of plants or spikes depends on the population size and the extent of ISF plants in the plot.

Fourth: At the time of tertiary spike initiation, early revertant, plants with ISF and hermaphrodites should be removed. This depends on the population size and extent of the revertant. If the number of revertant is high and the population size is low, only reverted spikes should be cut off and removed after harvesting the primary spike.

A seed yield of 10–12 q/ha and 8–10 q/ha in the male and female lines, respectively could be achieved depending upon the soil and growing conditions. An average yield of 12–15 q/ha of hybrid seed is achieved in a hybrid seed production plot with good management.

15   Hybrid Seed Production in Cotton

Hybrids are commercially cultivated in cotton on a large scale in India. Both intra- and interspecific cotton hybrids have been developed in India since 1970. The cultivation of conventional F_1 hybrids significantly increased the lint yield over the best open-pollinated varieties available in upland and Asiatic cotton (Tuteja et al. 2011). Despite adopting the emasculation and pollination technology for hybrid seed production, and raising its cost, hybrids are popular among farmers. Cotton hybrids cover more than 90% of a total of 12.2 mha area (Singh 2016).

Hybrid seed in cotton, including GM varieties, is produced either by manual emasculation and pollination technique in case of inter-varietal and inter-specific hybrids or only by pollination in case of male sterility-based hybrids.

Cotton is an often cross-pollinated crop. The average natural outcrossing is about 6%. Cotton pollen being heavy and sticky cross-pollination occurs only by insects, i.e. honey bees and bumblebees. The majority of the hybrids released in India are based on the inter-varietal crossing. Development of such hybrids involves three steps viz.: (i) growing of female and male parents in separate blocks nearby, (ii) emasculation of female parent plants and (iii) pollination of the female parent with the pollen from the male parent.

Hybrid seed production using male sterility eliminates emasculation since the pollens are sterile in female parents. However, pollination has to be done manually. In cotton, mainly two types of male sterility such as genetic male sterility and cytoplasmic genetic male sterility are used for seed production. Thus the cost of hybrid seed production can be reduced.

For hybrid cotton seed production, the soil should be well drained and medium to heavy deep. Maintenance of genetic purity for certified seed production of conventional or male sterility-based hybrids depends upon the use of safe isolation distance.

For hybrid seed production in cotton female and male parents are planted in the same field with 5 m isolation between parents and keeping a minimum distance of 30 m from other cotton crop in the area (Meshram 2002). The sowing dates of parental lines are adjusted in such a way that there is the synchronization of flowering between the female and male parents with continuous supply of pollen till the crossing season is over. Staggered planting of male parent is suggested depending on the date of flowering in male and female and pollen production in the male parent (Doddagondar 2006; Doddagondar et al. 2008).

15.1 Emasculation of the Female Parent

Flower buds that are likely to open the next day are chosen for emasculation in the early afternoon. Different emasculation methods can be adopted depending upon the flower types.

15.1.1 Doak Method or Thumb Nail Method

This is the most successful method used in hybrid seed production of tetraploid cotton with more than 40% seed set. The method involves the removal of the corolla along with anther sheath by giving a shallow cut at the base of the bud using the thumbnail and applying a jerk/twisting action (Doak 1934). Care should be taken so that the white membrane covering the ovary is not damaged that affects the boll setting. It should also be ensured at the time of emasculation that no anther sac remains attached at the base of the ovary causing selfing and genetic impurity in the hybrid seed lot. Emasculated flower buds are generally covered with tissue paper bags (9 cm × 7 cm) to prevent contamination from foreign pollen and marking the emasculated flowers.

However, this method is not suitable for hybrid seed production in diploid cotton in which the flower buds are small and the style is short and fragile rendering the method unsuitable.

15.1.2 Pinching off of the Top of Corolla

Also known as Surat method, this is useful for the emasculation of diploid flowers where the top portion of the flower bud is pinched off using the thumb and first fingernails so that the tip of the stigma gets slightly exposed, and the bud is lightly covered with mud. As the buds mature the stigmatic head extends sufficiently for pollination. In genotypes where stigma protrusion/exsertion is relatively low, the entire corolla is removed and mud is applied on the unopened anther sacs. Pollination is done on the following morning (Mehta et al. 1983).

15.1.3 Straw Tube/Copper Straw Method

In this method, the top of the corolla is pinched off and a piece of straw tube is inserted into the style to separate all anthers in anther column and leaving the tube in the same position till pollination is done on the next morning. However, during the process of emasculation more time is required.

15.1.4 Removal of Petals and Dusting off Anthers

Petals are removed by the thumbnail method, and the pollens are brushed off from the anthers by lightly touching and moving the thumb and first finger down and up along the staminal column. A light tapping at the flower pedicel dislodges any excess anthers sticking to the bracts. This method is useful in G. *Herbaceum* species in which anthers are of granule type, which can be removed easily and dropped to the ground.

15.2   Pollination

The emasculated buds are covered with red-coloured tissue bags for easy identification. The emasculated buds are pollinated the following day between 8 and 11 am (or longer in some cases) when the stigma receptivity is maximum. When the crossing is done during October-November (temperature ranges from 31 to 35 °C and RH ranges from 63 to 70%) the male flowers are collected and dried under the sun for a few hours for effective pollen dehiscence and dusting. The androecium with pollen is shaken gently or rubbed on the stigmatic surface. Sufficient pollen is to be used on the stigmatic lobes for proper development of the locules. The crossed flowers are again covered with a white-coloured tissue bag to distinguish them from emasculated buds awaiting cross-pollination. A thread is tied to the pedicel for the identification of crossed bolls. Fertilization occurs after 12–30 h of pollination and hence the crossed buds are kept covered for 3–4 days (Deshmukh et al. 1995).

For an effective seed setting, the crossing programme should be continued for about 10–14 weeks after the initiation of flowering. Light and frequent irrigation during this period facilitates good boll development.

Besides a strict roguing both in the female and male parent plots, all un-emasculated buds and flowers in the female parents need to be removed to avoid any admixture in the resultant hybrid seeds.

Fully matured and completely opened bolls are to be picked and collected for seed. These are cleaned of any lint, washed, dried, cleaned and stored after tagging. Damaged/undeveloped/underdeveloped bolls should be discarded before sorting.

16   Hybrid Wheat Seed Production

Though, not commercially very popular, hybrid wheat varieties are in use in some parts of the world, specially in Europe and the UK. In India due to the low level of heterosis reported in wheat, hybrids are not yet in use. Conventionally, three-line hybrid wheat seed production is taken up using specific female-male line in a specified proportion and following recommended methods for proper growth of the parental lines, to synchronize flowering and effective pollination. These tend to increase the cost of production of hybrid seeds. In order to make hybrid wheat seed production easy and cost-effective, the blend hybrid seed production method has been proposed which maximizes hybrid seed production with cost reduction. A recent study has reported that a blending of 6 to 8% restorer line seeds with the seed parent seed increased the yield of mixed hybrid seed production with higher purity (Nie et al. 2021).

---

17   Hybrid Seed Production in Tomato under Polyhouse Conditions

Tomato seed production under open field conditions is often affected by various environmental stresses resulting in poor seed quality and quantity. Thus, seed production of tomato is preferred to be undertaken under protected conditions, i.e. low-cost poly house or climate-controlled poly house, which give higher seed yield and of better quality with lesser use of chemical pesticides.

17.1   Growing Seedling

Sowing of the parental line seeds is done at least 25–30 days before the transplanting. About one week before transplanting, the irrigation is withheld in the nursery to harden the seedlings.

17.2   Transplanting

The male and female lines are planted with 25–30 days old seedlings on ridges (in raised beds) 90 cm apart with the plant-to-plant spacing of 60 cm. A planting ratio of 4:1 for female:male plants are recommended.

17.3   Intercultural Operations

The parental lines should be stacked before flowering not only to protect the fruits from rotting but also to make more space to move between rows for emasculation and pollination operations. The seed plot is regularly weeded to keep it free from weeds. Irrigations at 4–5 days interval are sufficient. Flowering and fruiting are two critical stages of irrigation. The lower branches should be continuously pruned to encourage the proliferation of upper branches. Off-types and virus-infected plants must be removed before hybridization. As the hybrid fruits are developed on the female plants, sturdy stacking to withstand the fruit load should be given.

In the absence of a functional male sterile system, the hybrid tomato seed is produced by manual emasculation and pollination. For this, the crossing should be initiated at 40–45 days after transplanting.

17.4   Emasculation

The female flower must be pollinated by the pollen from the male line. To prevent self-pollination, the stamens from the flower buds of the female line are removed before they shed their pollen. Emasculation is initiated at about 50–60 days after sowing in the first cluster of 3–4 flowers, selected for crossing. Buds are forced-open with sharp-pointed forceps and anthers, which are fused to form a cone-like structure around the stigma, are removed by a vertical splitting of the cone. Emasculation in the bud stage is done in the evening hour.

17.5   Pollen Collection

The best time for pollen collection is during the late evenings before the pollen has been shed. The anther cones are removed from the flowers and are placed in suitable containers viz. Petri dish or cups. The anther cones are dried by sun-drying or by placing under a 100-W lamp overnight. The dried anther cones are taken in a cup with a lid of fine mesh and shaken to extract pollen. Pollen is collected in cups or rings and used for pollination. These pollens remain viable for at least one day.

17.6   Pollination

Emasculated flowers are generally pollinated the next day in the morning hours. The stigma is dipped into the pollen container or pollinated by touching the stigma with the tip of the index finger dipped in the pollen pool. Successful pollination is easily detected within one week by the enlargement of the pollinated fruit. After crossing operations are completed, any non-crossed flowers on the female plants are removed to avoid any chance of contamination from selfed fruits before harvest.

17.7   Harvesting

On an average 50 or more fruits are retained on medium fruited female parent plant. Tomato fruits ripen in about 50–60 days after pollination. Fruits are harvested after full maturity and collected in non-metallic containers, such as polythene bags, plastic buckets or crates.

17.8   Seed Extraction

The ripened fruits are crushed manually or by machines. The bags of crushed fruits are kept in big plastic containers for fermentation and for separating the gel mass embedded with the seeds. Seeds can be extracted from the fruit mass following fermentation, acid or alkali methods, washed thoroughly, cleaned of any debris, sieved and dried.

An average yield of 3 to 3.5 kg hybrid seed can be expected from an area of 100 sq m. polyhouse.

18 Hybrid Seed Production in Bitter Gourd under Insect-Proof Net House

Hybrid seed production of bitter gourd under the insect-proof net-house is a lucrative and environment-friendly technology, wherein the crop is vigorous, insect-free, and exhibits higher fruit and seed yield with better seed quality as compared to open field conditions. It also helps in reducing the cost of seed production and indiscriminate use of pesticides for insect and pest control. The essential steps in hybrid seed production technology of bitter gourd under net house conditions are given below:

Seedlings of the male and female parent lines are raised from genetically pure seed stocks and transplanted, maintaining a planting ratio of 3 female rows to 1 male row requiring a total of 30 female and 10 male plants for a 100 sq. m area. Intercultural practices are followed as recommended for the commercial crop at a given location and season.

Bitter gourd requires trailing to support the rapid growth of vine after planting. Staking reduces the fruit rot and diseases, ensures better pollination, facilitates easy harvesting, and gives higher seed yield. Bamboo poles, wooden stakes, PVC pipes, or similar materials could be used as trellis to support vine and keep the fruit and foliage off the ground. The trellis are either placed as an erect pole with horizontal support or in a dome-like structure. For seed production, vertical stakes are better than dome-shaped structures. The trellis should be 2.0–2.5 m high above the seed bed. The staking supports the climbing vines and lateral stems. Strings/ropes should be used to secure adjoining stakes. When the primary branch of the vine reaches the top of the trellis, all the lateral branches of the vine should be cut to promote pistillate flowers.

18.1 Flowering Behaviour

Bitter gourd is monoecious and bears staminate and pistillate flowers separately where the proportion of staminate flowers is very high as compared to pistillate flowers. The average ratio of staminate to pistillate flowers in monoecious lines varies from 12:1 to 9:1. To achieve higher yield, high sex ratio (higher pistillate flowers per vine) is desirable. Application of growth regulators like GA_3 at 50 ppm sprayed thrice at three leaf, tendril and bud initiation stage was found promising for higher induction of pistillate flowers, fruit set, seed yield and quality (Nagamani et al. 2015).

18.2   Pollination

Bitter gourd is a highly cross-pollinated crop. Female and male flowers are borne on different nodes of a vine. Hybrid seed production is done by manual hand pollination. For successful hybrid seed production, pollination is to be initiated one week after first female flower opening in a vine. All opened flowers and fruits are removed from female parent before pollination. Unopened female and male buds are covered with butter paper bag and cotton, respectively (to prevent contamination). Pollination is done next morning between 7 am and 12 noon. Pollen from male flower of male parent is manually dusted on stigma of female flower. Pollinated flowers are covered with butter paper bag to prevent contamination. Although stigma remains receptive for one day in spring-summer and more than one day in rainy season but pollination is undertaken between 7 am and 12noon on the day of anthesis.

Generally, 12–14 fruits/vines are retained for higher seed yield and quality. Ripe fruits are harvested when they turn orange in 25–30 days and 20–25 days after pollination in the spring-summer and monsoon seasons, respectively. Seeds should not be harvested from partially orange or burst fruits as it affects seed quality.

18.3   Seed Extraction

Seed extraction is done manually by opening the ripe fruits and removing the seeds from the fruits, followed by macerating to separate the red mucilaginous seed coat. Extracted seeds should be shade-dried to a moisture content of 6–8% before packaging.

An average seed yield of 2.5 kg is expected from an area of 100 sq. m. insect-proof net house.

19   Cauliflower

For very long, cauliflower hybrids used in commercial cultivation were based on self-incompatibility (SI) system. However, instability of SI system and problems associated with the multiplication of the parental lines using bud-pollination are its major drawback. Hence, CMS systems were identified and a large number of CMS-based hybrids are now available for commercial cultivation. The methods for hybrid seed production are specific to the system of male sterility and location of adaptation. ICAR-Indian Agricultural Research Institute has developed *Ogura* cytoplasm-based CMS lines for hybrid development in early, mid and late maturity groups of cauliflower and standardized the seed production technology of CMS based mid maturity cauliflower hybrid, Pusa Cauliflower Hybrid-3 which has recently been introduced for commercial seed production (Anonymous 2021).

Thus, it is evident from the above deliberations that hybrid seed production is a specialized activity, which requires a thorough understanding of the flowering behaviour, sex expression under varying environments and pollination dynamics of each crop before undertaking it.

References

Anonymous (2013). Indian minimum seed certification standards. The Central Seed Certification Board, Department of Agriculture & Co-operation, Ministry of Agriculture, Govt. of India, New Delhi, p. 569

Anonymous (2021) Cauliflower hybrid seed production. IARI News 37:5

Ayyangar GNR, Rao UP (1931) Studies in sorghum anthesis and pollination. Indian J Agric Sci 1: 445–454

Barnabas B, Kovacs G (1997) Storage of pollen. In: Shivanna KR, Sawhney VK (eds) Pollen biotechnology for crop production and improvement. Cambridge University Press, New York, pp 293–314

Burton GW (1966) Photoperiodism in pearlmillet, *Pennisetum typhoides*, its inheritance and use in forage improvement. In: Proceeding of the 10th international grassland conference. Finnish Grassland Association, Helsianki, pp 720–723

Chakrabarty SK (2003) In: Singhal NC (ed) In: Hybrid seed production in field crops (principles and practices). Kalyani Publishers, New Delhi, pp 123–136

Chakrabarty SK (2008) Synchronization of flowering in parental lines for sunflower hybrid seed production under north-Indian conditions. Curr Sci 95:1077–1079

Chakrabarty SK, Chandrashekar US, Prasad M, Yadav JB, Singh JN, Dadlani M (2011) Protogyny and self-incompatibility in Indian mustard (*Brassica juncea* (L.) Czern and Coss)—a new tool for hybrid development. Indian J Genet Plant Breed 71(Special Issue):170–173

Chakrabarty SK, Yadav SK, Yadav JB (2007) Evaluation of stigma receptivity in cytoplasmic male sterile lines of *Brassica juncea* L. Czern & Coss. Full-length paper. In: Proceedings of the 12th International Rapeseed Congress held at Wuhan, China during March 25–30, 2007. Genetics and Breeding: Heterosis Utilization, ISBN 1-933100-20-6, pp 5–7

Chakrabarty SK, Vari AK, Sharma SP (2005) Manipulation of synchronization of flowering of parental lines of sunflower (*Helianthus annuus*) hybrid. Indian J Agric Sci 75:786–790

Deshmukh RK, Rao MRK, Rajendran TP, Meshram MK, Bhat MG, Pundarikaskshudu R (1995) Crop management techniques for increasing hybrid seed production in genetic male sterile cotton.Tropical. Agriculture 72:105–109

Doak CC (1934) A new technique in cotton hybridization: suggested changes in existing methods of emasculating and bagging cotton flowers. J Hered 25:201–204

Doddagondar SR (2006) Influence of planting ratio and staggered sowing of male parents for synchronisation of flowering in DHH 543 cotton hybrid seed production. Crop Res 32:250–254

Doddagondar SR, Sekhargouda M, Khadi BM, Eshana MR, Biradarpatil NK, Vyakaranahal BS (2008) Studies on planting techniques for synchronization of flowering in DHB-290 cotton hybrid seed production. Indian J Agric Res 42:195–200

Feng YI (1984) Regulation of flowering date in seed production of hybrid rice. Guangding Agric Science

Frankel R, Galun E (1977) Pollination mechanism, reproduction and plant breeding (monograph on theoretical and applied genetics). Springer-Verlag, Berlin Heidelberg, p 281

Govila OP, Singhal NC (2003) Pearl millet. In: Singhal NC (ed) Hybrid seed production in field crops (principles and practices). Kalyani Publisers, New Delhi, pp 253–268

Gupta R, Sutradhar H, Chakrabarty SK, Ansari MW, Singh Y (2015) Stigmatic receptivity determines the seed set in Indian mustard, rice and wheat crops. Commun Integr Biol 8(5): 042630. https://doi.org/10.1080/19420889.2015.1042630

Hormuzdi SG, More TA (1989) Studies on combining ability in cucumber (*Cucumis sativus* L.). Indian J Genet Plant Breed 49:161–165

Hossain D, Karin MA, Pramani MHR, Rahman AAS (2006) Effect of gibberellic acid (GA$_3$) on flowering and fruit development of bitter gourd. Int J Bot 2:329–332

House LR (1985) A guide to sorghum breeding. ICRISAT, Patancheru., India, p 206

Kannababu N, Tonapi VA, Rana BS, Rao SS (2002) Influence of different synchronization treatments on floral behaviour of parental lines and hybrid seed set in sorghum. Indian J Plant Physiol 7:362–366

KempeGowda H, Kallappa VP (1992) Effect of plant design and staggered sowing of parental lines on seed yield in KBSH-1hybrid sunflower. Abst Seed Tech News 22:27

Kushwaha ML, Pandey ID (1998) Hybrid seed production in bottle gourd. Punjab Veg Grower 33: 27–29

Lal T, Singh S (1990) Effect of plant density on hybrid seed production in muskmelon. Seed Res 18: 11–14

Lather BPS, Singh DP (1997) Hybrid seed production technology of cotton. In: Dahiya BS, Rai KN (eds) Seed technology. Kalyani Publications, New Delhi, pp 104–112

Lloyd DG, Webb CJ (1986) The avoidance of interference between the presentation of pollen and stigma in angiosperms. I. Dichogamy. New Zealand J Bot 24:135–162

Mac Robert JF, Setimela PS, Gethi J, Worku M (2014) Maize hybrid seed production manual. D.F., CIMMYT, Mexico

Maity A, Chakrabarty SK, Yadav JB (2012) Standardization of planting ratio of parental lines for seed production of mustard (*Brassica juncea* (L.) Cozern and Coss.) hybrid NRCHB 506. Bioinfolet 9:299–302

Maity A, Chakrabarty SK, Pramanik P, Gupta R, Singh PS, Sharma DK (2019) Response of stigma receptivity in CMS and male fertile line of Indian mustard (*B. juncea*) under variable thermal conditions. Int J Biometeorol 63:142–152

Mankar KS, Yadav JB, Singhal NC, Prakash S, Gaur A (2007) Studies on stigma receptivity in Indian mustard. Seed Res 35:148–150

Maundar AB, Sharp GL (1963) Localization of out-crosses within the panicle and fertile sorghum. Crop Sci 3:449

Mehta NP, Badaya SN, Patel. (1983) A modified hybrid seed production technique for Asiatic cottons. J Indian Soc Cotton Impro:36–37

Meshram LD (2002) Determination of isolation distance for quality seed production using cytoplasmic male sterility in cotton. J Cotton Res Dev 16:17–18

Miniraj N, Prasanna KP, Peter KV (1993) Bitter gourd Momordica spp. In: Kalloo G, Bergh BO (eds) Genetic improvement of vegetable plants. Pergamon Press, Oxford, UK, pp 239–246

Murthy DS, Tabo R, Ajayi O (1994) Sorghum hybrid seed productionand management. Information bulletin no. 41. ICRISAT Patancheru, India, p 67

Nagamani S, Basu S, Singh S, Lal SK, Behera TK, Chakrabarty SK, Talukdar A (2015) Effect of plant growth regulators on sex expression, fruit setting, seed yield and quality in the parental lines for hybrid seed production in bitter gourd (*Momordica charantia* L.). Indian J Agric Sci 85:1185–1191

Nandpuri KS, Brar JS (1966) Studies on floral biology in muskmelon (*Cucumis melo* L.). J Res Ludhiana 3:395–399

Nandpuri KS, Singh J (1967) Studies on floral biology of bottle gourd (*Lagenaria siceraria*). Punjab J Res 4:53–54

Nie K, Cui S, Mu X, Tian. (2021) Blend wheat AL-type hybrid and using SSRs to determine the purity of hybrid seeds. Seed Sci Technol 49:275–285. https://doi.org/10.15258/sst.2021.49.3.08

Nepi M, Pacini E (1993) Pollination, pollen viability and pistil receptivity in *Cucurbita pepo*. Ann Bot 72:527–536

Prasad S, Prakash R, Haque MF (1977) Floral biology of pigeonpea. Trop Grain Legume Bull 7:12–13

Ramchandram M, Ranga Rao V (1990) Seed production in castor. Directorate of Oilseeds Research, Hyderabad, India, p 62

Ranganatha ARG, Pradeep Kumar C, Hanumantha Rao C (2003) Sunflower. In: Singhal NC (ed) Hybrid seed production in field crops (principles and practices). Kalyani Publishers, New Delhi, pp 93–108

Kumar S (2005) Studies on pollination requirements for hybrid seed production in bottle gourd (*Lagenaria siceraria* M. standl.). Ph.D. thesis. P. G. School, Indian Agriculture Research Institute, New Delhi

Shailendra VM, Valia RZ, Sitapara HH (2015) Growth, yield and sex-expression as influenced by plant growth regulators in sponge gourd cv. Pusachikni. Asian J Hort 10:122–125

Sharma SP (1993) Studies on determination of isolation distances for seed production of pearl millet. Seed Res (Spl. issue):103–110

Sharma V, Bhushan A, Kumar D, Mishra S, Kumar R (2020) Single cross hybrid of qpm maize seed production as influenced by row ratio and spacing in North Western Himalayan region. Int J Curr Microbiol App Sci 9:1491–1496

Sharma RK, Pathania N, Joshi AK (2004) Studies on hybrid seed production in cucumber. Seed Res 32:103–104

Shiva D, Mahesh K, Pritee S, Aarati GC (2019) Effect of ethephon doses on vegetative characters, sex expression and yield of cucumber (*Cucumis sativus* cv. Bhaktapur local) in Resunga municipality, Gulmi, Nepal. Int J Appl Sci Biotechnol 7:370–377

Shivanna KR, Johri BM (1985) The angiosperm pollen: structure and function. Wiley Eastern, New Delhi

Singhal NC, Rana BS (2003) Sorghum. In: Singhal NC (ed) Hybrid seed production in field crops (principles and practices). Kalyani Publishers, New Delhi, pp 215–232

Singh RP (2016) Yield increase in major oil seeds in India through replacement by new climate resilient varieties. Clim Chang Environ Sustain 4:211–223

Singh SN (1957) Studies in floral biology of luffa spp. Hort Adv 1:775–776

Soto W, Alexandrova M, Kostov D, Ivanov L (1995) Economic efficiency from seed production intensification of pickling cucumber heterosis cultivars and their parent components. Bulgarian J Agric Sci 1(4):455–458

Stanley RJ, Linskens HF (1974) Pollen: biology biochemistry and management. Springer-verlag, Berlin

Tarbaeva LP (1960) The biology of flowering and pollination in the watermelon. Bjull Glay Bot Sada 38:76–78

Tikle AN, Saxena KB, Yadava HS (2014) Pigeonpea hybrids and their production: A manual for researchers, p 68

Tuteja OP, Verma SK, Banga M (2011) Heterosis for seed cotton yield and other traits in GMS (genetic male sterility) based hybrids of cotton (*Gossypium hirsutum* L.). J Cotton Res Develop 25:14–18

Viraktathamath BC, Ilyas Ahmed M, Ramesha MS, Singh S (2003) Rice. In: Singhal NC (ed) Hybrid seed production in field crops (principles and practices). Kalyani Publishers, New Delhi, pp 183–214

Virmani SS (1994) Heterosis and hybrid rice breeding. Monograph on theoretical and applied genetics 22. International Rice Research Institute. Springer-Verlag, Berlin Heidelberg

Xu S, Li B (1988) Managing hybrid rice seed production. In: Hybrid rice. IRRI, Manila, Phillipines, pp 157–163

Yuan LP (1985) A concise course in hybrid rice. Hunan Technol Press, China, p 168

Seed Processing for Quality Upgradation

J. P. Sinha, Ashwani Kumar, and Elmar Weissmann

Abstract

Pre-harvest and post-harvest operations are integral parts of the quality seed production system. If due attention is not paid during the first phase, probably therewill not be any means left to obtain quality seed. Similarly, good quality seed produced with utmost care may lose much of its value if proper management is not followed during the post-harvest period. The seed quality attributes are genetic purity, appearance and physical purity, germination potential, vigour and seed health. Proper attention should be given during both the phases of seed production, regarding all the attributes of seed quality. Genetic purity is primarily linked with production or pre-harvest phase. Whereas, physical purity is primarily achieved in post-harvest phase. The harvested raw seed mass consists of various materials other than seed (MOS). MOS may be trash, plant parts, inert matter, weeds, other crop seeds and non-viable seeds, e.g. cut grains, insect pest damaged, off sized or physically damaged kernels, etc. Such contaminants reduce seed quality and increase the volume of harvested seed mass that requires additional space during storage. Moreover, these contaminants often attract pest infestation, which can further cause seed loss both in terms of quality and quantity. Proper storage of processed and clean seed is also an essential component of seed programmes as the sowing season normally falls at least six to nine months after the production season.

J. P. Sinha (✉)
Water Technology Center, ICAR-Indian Agricultural Research Institute, New Delhi, India

A. Kumar
ICAR-Indian Agricultural Research Institute, Regional Station, Karnal, India

E. Weissmann
HegeSaat GmbH & CoKG, Singen OT Bohlingen, Germany
e-mail: elmar.weissmann@eaw-online.com

© The Author(s) 2023
M. Dadlani, D. K. Yadava (eds.), *Seed Science and Technology*,
https://doi.org/10.1007/978-981-19-5888-5_10

Keywords

Conditioning · Physical attributes · Separation · Cleaning · Grading · Seed treater · Seed drying

1   Introduction

Soundness, well-filled size and uniformity of physical attributes of seed (e.g. size, shape, density, colour, appendages) are significantly linked with its quality, as well as the ease of mechanical planting and for better crop establishment. Seed processing is an effective and efficient process to upgrade the quality of harvested seed and prepare it for safe storage until used. The seed processing has to be carried out with a goal to enhance seed lot quality by exploiting the physical attribute of seed as well as constituents of harvested seed mass. The three basic processes, conditioning, grading and application of protectants or enriching with nutrients, are accomplished during seed processing, following standard principles and procedures (McDonald and Copeland 1997; Agrawal 1996; Jorgensen and Stevens 2004). Operational modifications and machine adjustments are made as per the specificities of the crops and prevailing conditions.

2   Conditioning

The conditioning enables the raw seed lots to be handled safely in subsequent operations in seed processing and harnessing maximum efficiency of implied machines. It may comprise of moisture conditioning, removal of appendages, removal of major dockage or a combination of these in order to make the lot free flowing and without any mechanical abuses in further processing operations. In general, moisture content ~10% is considered safe for mechanical processing. Too low or too high moisture content may induce mechanical injury especially in pulses. Thus, drying or humidification of seed lots are recommended seed conditioning operations for maintaining safe moisture content before further processing operations.

2.1   Moisture Conditioning

It may require drying to reduce seed moisture content (if too high) or humidification to increase moisture content (if too low), with the ultimate aim to prepare the raw seed lots suitable for safe mechanical handling.

3   Seed Drying

The process of decreasing the seed moisture content to safe moisture limits is called 'seed drying', which is an important operation, as.

- it permits early harvest and reduces losses due to bad weather
- it is an aid for efficient and economical labour management
- it reduces the mechanical damage during handling and processing
- it reduces loss in seed quality and quantity during storage

The moisture in seed is distributed in two ways. The first is surface moisture and second one is internal moisture. Hence, during seed drying, movement of moisture is also in two ways, i.e. transfer of moisture from seed surface to air around the seed, and movement of moisture from the inner parts of the seed to the seed surface. The drying is controlled by: (i) air flow rate, (ii) drying temperature, (iii) drying rate and (iv) drying time.

The transfer of moisture from the seed surface to surrounding air is a function of the gradient in vapour pressure between the seed surface and the surrounding air. Free air movement is essential to progress the drying operation. Drying rate is a function of air movement. As the air flow increases, the drying rate increases up to a point at which the air absorbs all the moisture that is available to it. The required minimum air flow rates for efficient drying are given below:

Drying with non-heated air		Drying with heated air	
Moisture content (%)	Air flow rate m^3/min per m^3 seed	Seed type	Air flow rate m^3/min per m^3 seed
25	7.5	Light	2.8
22	6.0	Heavy	4.6
18	4.2		
15	2.8		

The vapour pressure gradient between the seed surface moisture and the air can be increased by increasing the temperature of the air. The higher air temperature can dry seeds more rapidly to low moisture levels. But the high temperature can cause heat injury to the seed. In case high moisture content seed is dried with high temperature air, the seed viability deteriorates significantly. Hence, the seed cannot be dried safely with high temperature air, especially when seed moisture is high. In general, seed drying temperature should not be higher than 43 °C. The critical temperature for drying varies with the moisture content of the seed. Hence, it is advisable to dry the seed depending on its moisture level.

Seed moisture content (%)	Drying temperature (°C)
Over 18	32
10–18	37
Under 10	43

(Sinha et al. 2010)

Sometimes the situation may be such that the seed cannot be dried to the required level at safe drying temperature condition if the relative humidity (RH) of the air is high. In such situations, the drying air should be dehumidified chemically or by refrigeration. Certain chemicals, e.g. silica gel, calcium chloride, activated alumina, activated charcoal and anhydrous calcium sulphate, that have a strong affinity for moisture are used to dehumidify the air in dryers. These chemicals are reactivated in dehumidification system to sustain the drying process. An alternate method of dehumidifying air is to employ refrigeration to drop the air temperature below its dew point. The dehumidification process either raises or lowers the air temperature abruptly. In the case of chemical dehumidification, the temperature of air rises, while in refrigeration dehumidification, air temperature declines. In order to attain optimum drying rate, cooling or heating arrangements are used in combination with chemical and refrigeration methods of dehumidification, respectively.

Vacuum drying is another way to remove moisture from the seed. It is somewhat like drying with dehumidified air. When the total pressure in the vacuum drying chamber is lowered, the component of this pressure due to vapour is reduced proportionately and drying of the seed is attained. In this case, the heat for evaporation is supplied mainly by conduction or radiation.

Dehumidified air or vacuum drying is costlier than heated air drying, but these methods are useful to dry heat-sensitive seed as well as for safe drying of seed to very low moisture contents.

The removal of moisture from the seed surface to the surrounding, i.e. the first phase of drying, can be regulated by considering the above facts. But the movement of internal moisture to the surface, i.e. the second phase of drying, is more critical. The rate of moisture movement in the second phase should be at least equal to the first phase of drying. It is important to note that the second phase of drying greatly varies with the species and cultivar. If this phase is slow, the cultivar should be dried in multi-stages. Otherwise, cracking or splitting of the seed coat may occur, which will reduce the seed quality or destroy it completely.

The drying rate is an important aspect in seed drying. Though fast drying helps in controlling moulds, but if the rate is too fast, seed coats of some species (such as many legumes) tend to shrink or split and/or they may become impermeable to moisture (developing into hard seeds), especially in large-seeded legumes, even though the inner parts of the seed might still remain moist. Rapid-dried seeds with cracks and splits are more prone to disease infection and insect pest attacks. Slow drying, on the other hand, increases the chance of mould growth which lowers the seed quality significantly. Hence, an optimum drying rate is adjusted by applying the rule of thumb, so that about 0.3% of the moisture can be removed per hour with an air flow rate of 4 m^3/min/m^3 of seed at 43.5 °C. However, some adjustments are needed considering the kind of seed (structure, composition), temperature and initial moisture content. The rate will be less if the initial moisture content is low, and if the temperature of drying air is below 43.5 °C. The time required for drying a seed lot depends on its initial moisture content, final moisture content (MC) to be attained, drying rate, air flow rate, relative humidity (RH) and temperature of the drying air.

The seed drying systems are mainly of three types:

1. Batch drying system
2. Continuous drying system
3. Heated air-drying system

3.1   Batch Drying System

This is a system in which products are kept in a bin and the heated air or drying air is ventilated through them. The different types of batch drying are bin dryer, tray dryer and tunnel dryer.

3.1.1   Bin Dryer

This type of dryer is useful when a small quantity of seed is to be dried. The air may be pushed or pulled through the seed bed and may go from top to bottom or vice versa. The depth of the seed bin is limited because the power required to force air through the seed is proportional to the depth. In deep bins, the entering air picks moisture from the seed, cools and may deposit moisture on the last layer of seed before it is released out. This causes mould growth and results in deterioration of seed in this layer.

3.1.2   Tray Dryer

In a tray dryer, many small trays are kept one above the other with a gap in between the drying chamber. The trays normally have perforated bottoms for better air flow. Seed is kept in thin layers in the trays and these may be manually shifted to allow uniform drying in all trays. The gap between the stack of trays allows ventilation.

3.1.3   Tunnel Dryer

It is similar to tray dryer. The only difference is that the group of trays is kept moving in a tunnel. The flow of heated air in a tunnel dryer may be co-current or counter-current.

3.2   Continuous Drying System

This is a system in which the products are kept moving either by gravity or by some mechanical means and the drying air is ventilated through the moving column/layer of seed.

3.2.1   Rotary Dryer

Rotary dryers are fitted with a drum normally of 1–3 m diameter and 3–6 m length, which rotates on its axis. The product flows downward through the rotating drum and is periodically lifted by inclined flights, then dropped, ensuring good air contact.

Such dryers are indirectly heated and are suitable for seeds with relatively low moisture content.

3.2.2   Column Dryer

Column dryers are usually for continuous flow of large lots of seed with two columns 20–35 cm thick. The seed flows through the air chamber in a solid column and is turned by baffles as it descends the column to allow uniform drying.

3.2.3   Belt Dryer

Seeds are spread in a thin layer over a perforated belt and the heated air is allowed to pass through the perforations. Such dryers are suitable for seeds with poor flowability.

3.2.4   Fluidized Bed Dryer

In this method of drying, products are dried under fluidized condition in a dryer. The seed is fluidized by the use of drying air with high velocity to cause suspension. In this process, higher rates of migration of moisture take place. Since every surface of the product is in contact with the drying air, uniform drying takes place. This method is normally used for seeds which have high moisture content and need to be dried quickly, such as vegetable seeds.

3.3   Heated Air Dryers

The drying air can be heated mainly in two ways.

3.3.1   Direct Fired Type

The combustion gas goes directly from burner into the drying air stream and then to the drying bin. Such driers are less safe because air burner may release particles of hot soot if it is not properly adjusted, and may cause carbon deposits on seeds. The thermal efficiency of this type of drier is higher than the indirect fired drier.

3.3.2   Indirect Fired Type

This type of dryer consists of the heat exchanger and the fire chamber. In the fire chamber, fuel is burned and heat is generated which heats the heat exchanger where drying air takes the heat. The heated air is forced into drying chamber by a blower (Fig. 1). The thermal efficiency is lower than the direct fired type, but is safe for seed drying.

4   Tempering

Besides drying, seeds of certain crops, e.g. soybean and peas, need tempering under very dry conditions to raise the moisture content to about 10%. Tempering is needed for crops which are sensitive to impact damage at a lower level of moisture content.

Fig. 1 Batch type hot-air dryer. (Photograph courtesy, Seed Processing Unit, IARI Regional Station, Karnal, India)

With the help of tempering, the moisture level can be increased to a safe level of ~10–12%.

5   Removal of Appendages

Seeds containing awns, beards, hairs, glumes and other appendages tend to interlock and cause clustering. These should be removed for improving the flow properties, cleaning characteristics and quality of the seeds. The general machines used for this operation are hammer mill, debearder and pebble mill.

5.1   Hammer Mill

It is a thresher with hammer type beaters in closed cylinder casing, concave, a set of oscillating sieves and an aspirating blower. Seeds that can be processed successfully in hammer mill are grasses, e.g. bluebunch wheat-grass, blue wildrye, and species like tall oat grass, bulbous barley, squirrel tail (Sinha and Srivastava 2003) and species with similar kind of seeds. Mechanical processing of Pusa Basmati 1, a popular long-awned paddy variety, is problematic due to the entangling of its awns. In order to make the seed mass suitable for mechanical processing, de-awing is practised with the hammer mill at optimized operational parameters by adjusting the cylinder speed, screen opening and feed rate (Sinha et al. 2010).

5.2   Debearder

It has a horizontal beater with arms rotating inside a steel drum. The arms are pitched to move the seeds through the drum. Stationary posts, adjustable for clearance with the arms, protrude inward from the drum. The seeds are rubbed against the arms and also against each other. By regulating the discharge gate, the processing time can be regulated. The degree of action is determined by the processing time, beater clearance and beater speed. Larger capacity, simple operation and lesser damage to seeds are its advantages over the hammer mill.

This machine can be used to remove cotton webbings from Merion bluegrass, remove the clip seed from oat seed, debeard barley seed, thresh white caps in wheat, split grass seed clusters, remove awns, beards and hull from some grass seeds as well as polish the same to upgrade the quality.

5.3   Pebble Mill

It has a drum rotating about a shaft inserted off-centre at opposite ends. The mill is loaded with seeds and smooth, half-inch pebbles and turned at a slow speed until the rubbing action of the pebbles rolls the fuzz from the seeds into small, round balls. The mixture of pebbles, seeds and matted fuzz is then run over a scalper to remove the pebbles. These types of machines are much effective in removing seed hairs and fuzz, e.g. for removing the cobwebby hairs from bluegrass and similar seeds.

5.4   Scalper

Seeds from a thresher or combine brought to a cleaning plant may contain a larger amount of trash, leaves, weed seeds, cut grains and infected seeds. Because of these materials, seed mass cannot be handled efficiently in cleaning or separation machines. Its removal is generally accomplished by scalpers. In general, two types of scalpers are in use: (a) a reel of perforated metal screen, which is inclined slightly and turns on a central shaft and (b) seeds fed into the higher end tumble inside the reel until they drop through the perforations. However, longer trashes remain in the reel and are discharged separately. In small capacity plants, the scalper is combined with the pre-cleaner air screen machine.

5.5   Huller and Scarifier

The hull or seed coat of some seeds are too hard to permeate water during germination. Such types of seeds require removal of the hull/husk, or scarification of the seed coat to absorb water and sprout properly. Hullers or scarifiers usually abrade the seeds between two rough surfaces, such as sandpaper. The severity of the abrasion or impact must be controlled carefully to prevent damage.

Commercially available scarifiers scarify the seed by forcing them against an abrasive material such as carborundum. Either an air stream or a centrifugal force may be used to bring the seeds in contact with the abrasive surface. Some scarifiers use a rubber abrading surface instead of carborundum to prevent damage, especially for fragile seeds. Seeds with high moisture content are hard to undergo hulling or scarification, due to the damaging effects induced by a huller or scarifier upon such seeds.

Seeds with long viability after hulling and scarifying can be processed immediately after harvest and stored until the following season. Others that lose viability rapidly may be stored and be hulled and scarified only before the planting season. Hulling and scarification may be performed separately or jointly, depending on the presence of unhulled seed, hard seed or both. Seeds such as that of Bermuda grass, bahia grass, buffalo grass require only hulling, whereas seeds of wild winter peas, hairy indigo, alfalfa, crotalaria, subclover and suckling clover may require scarification only. On the other hand, some seeds may require both scarification and hulling such as sweet clover, *Sericea lespedeza*, crown vetch, black medic and sour clover (Sinha and Srivastava 2003).

6   Removal of Dockage

Occasionally the harvested seed mass possesses more than 20% dockages by volume. Such lots cannot be subjected to mechanical processing as it will reduce the processing machine efficiency with increased seed loss through the reject port. In this condition, such lots are subjected to winnowing first, which exploits the differential aerodynamic behaviour of dockages to isolate them from the seed lot. An adjustable air flow with human safety winnower is advisable to deploy for the purpose. Various types of winnowers are available commercially.

7   Cleaning and Grading

The harvested seed mass may contain considerable amounts of moderate impurities like plant parts, broken kernels, soil clods, dust, stone etc., which have to be removed. Apart from these impurities, the seed mass comprises seeds with variation in shape, size, density, colour, texture etc., which reduce the seed quality with respect to biological value as well as commercial value. Cleaning and grading are required to upgrade both the physical and biological quality of seeds using the following units.

7.1   Air Screen Cleaner Cum Grader

It removes fine impurities and undersized seeds from the seed lot, which improves seed lot's physical purity as well as the flow of material in subsequent operations by the combination of air stream (aspiration) and screens (sieves). Basically, the difference in aerodynamic behaviour between seed and seed lot impurities is employed for cleaning and that of thickness of seed as a grade factor. This machine is also called Air Screen Thickness Grader cum Cleaner. It is important to note that it grades only on the basis of thickness, not the length of the seed.

An aspirating system is also installed in the machine providing air streams at two locations: (a) pre-suction channel—to remove lighter impurities before seed mass reaches to screens and (b) after suction channel—to remove impurities as well as lighter seed components that remain after the grading of seed mass.

There are two sets of screens for scalping and grading used in the Air Screen Cleaner cum Grader. By increasing the number of sets of screens the cleaning efficiency increases, however the throughput capacity decreases. The performance of the machine is a function of operational parameters, i.e. selection screen (size and type.), volume of air blast, oscillation speed, pitch of screen and feed rate. The size of the scalping (top) screen should be large enough to pass all the seed components and scalp off the coarse impurities, while the size of the grading (bottom) screen should be smaller than the optimum thickness/diameter of good seed so as to pass materials other than good seeds and ride over all good seeds. The selection of the type of screen should be according to the shape and size of the seed to be processed.

Shape of seed	Screen	Opening of screen
For round seeds	Scalping	Round hole
	Grading	Slotted hole
For oblong seeds	Scalping	Oblong hole
	Grading	Oblong hole
For lens-shaped seeds	Scalping	Oblong hole
	Grading	Round hole

However, the selection of size of opening of scalping and grading screens shall be done on the basis of sieve analysis for the specific seed lot with reference to available standards for the crop seed in respect of scalping and grading screens. Screen specifications for the processing of some important crop species as followed in India (Trivedi and Gunasekaran 2013) are presented in Table 1.

Oscillation speed, pitch and slope of screen cradle should be optimized so as the seed turns and tumbles at its own axis vertically and is subjected to grading by screen opening with cleaning and grading efficiency as well as a good throughput of processing. In general, the scalping screen is set at steep slope for fast removal of trashes and weed seeds, while the grading screen is set to flat slope so as to hold the seed longer for close separation. The pitch should be steeped for chaffy seeds and flattened for round seed in order to prevent bouncing over the screen, which affects the grading efficiency.

The feed rate of the seeds over the screens should be properly adjusted so as to get better cleaning and grading. It should be adjusted according to seed mass dockage level and shape of seed. Lesser feed rate should be maintained for higher dockage

Table 1 Screen aperture sizes for processing seeds of major field crops

Screen aperture size in millimeter		
Crop	Top screen	Bottom screen
Paddy		
Coarse grain/bold type	2.8s, 9.0r **(3.2s)**[*]	1.85s **(2.1s)**[*]
Medium slender	2.8s, 9.0r	1.80s
Fine/superfine	2.8s, 9.0r **(3.2s)**[*]	1.70s **(1.8s or 2.1s)**[*]
Wheat		
T. aestivum	6.00 r **(5.5r)**[*]	1.80s, 2.10 s, 2.30s **(2.3 or 2.4s)**[*]
Rapeseed & Mustard		
Mustard	2.75r, 3.00r, 3.25r **(3.2r)**[*]	0.90s, 1.00s, 1.10s, 1.40r **(1.6 or 1.8r)**[*]
Maize		
Maize except popcorn	10.50r, 11.00r **(10.50r)**[*]	6.40r, 7.00r **(6.0s or 6.40s+8.00r)**[*]

r round, *s* slotted apertures

seed mass. Feed rate should be optimized as maximum the three-layer seed remain on the screen.

The screen cleaning is also an important mechanism of the machine to harness maximum efficacy. There are three types of screen cleaning used in air screen machines, namely, (a) knockers or beater, (b) rubber balls and (c) brushes. Amongst all, the rubber ball system is the best as it induces least tear and wear to screen as well as minimizes the chance of mechanical mixing of seed (Fig. 2).

Screening efficiency increases as screen length increases, but the structural strength reduces with increasing the length of the screen. Hence, two Air Screen Cleaner cum Grader are deployed in seed processing in order to maximize the service life of the machine and cleaning and grading efficiency. The first machine primarily works for cleaning and the second one mostly for grading. It is important to note that the air screen machines are also effective in removing dockages and dead seeds (Sinha et al. 2001; Sinha et al. 2002).

7.2 Indented Cylinder Separator

It is also called length separator, as it utilizes the difference in length between the seed and impurities (especially lengthwise broken seeds) of seed mass which are either longer or shorter than the crop seed. The indented cylinder separator consists of rotating cylinder almost horizontal with adjustable horizontal separating trough mounted inside. The inside surface of the cylinder has small closely spaced semi-spherical indents (cells or pockets), hence the name.

The fed seed mass moves inside the bottom of the cylinder from the feed end to the discharge end. The cylinder revolves, turning the seed mass to be assessed by the recessed indents of the cylinder. Short seeds/impurities are lifted out of the seed mass and are dropped into the lifting trough. Long seeds remain in the cylinder and are discharged out to a separate spout at the end of the cylinder.

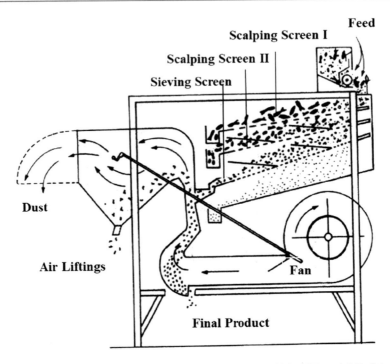

Fig. 2 Line diagram of air screen cleaner cum grader (Source: Sinha, J.P. and A.P. Srivastava (2003)

The rotation motion and recessed indents of the cylinder lift the shorter impurities to a certain height until the gravitational force exceeds the centrifugal force due to rotation and falls down in the separating trough. The length, centre of gravity, surface texture and size of the seed determine how they fit into the indent so that it can be lifted out of the seed mass. The indented cylinder separator can be used in two different ways.

1. Right Grading: Short seed/impurities are lifted and dropped into a separating trough and larger seed lot in the cylinder and discharged separately.
2. Reverse Grading: The crop seed is lifted and larger impurities are left in the cylinder. In this case, the capacity of the cylinder should be larger because all crop seed has to be lifted out. There is a higher risk of losing good seed with this application. To keep up with the capacity of the material passing through the air screen machine and right grading, multi-reverse grading indents are required.

The efficiency of separation depends on the operational parameters, cylinder speed, size of indent, feed rate, pitch of cylinder and retarder position. An increase in the cylinder speed increases the centrifugal force which holds the seed in the indents; hence, it will lift the seed longer or higher helping its separation.

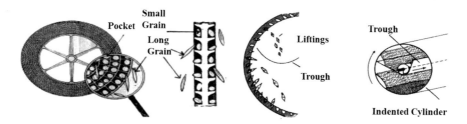

Fig. 3 Exploded view of indented cylinder. (Source: Sinha, J.P. and A.P. Srivastava (2003)

Conversely, a decrease in cylinder speed will decrease the centrifugal force and the seed will fall out of the indents at the lower level. The speed of the cylinder is so adjusted that the liftings should fall in the separating tray (Fig. 3).

The size and shape of the indent has to be selected as per requirement of the specific seed lot. The position of the trough also should be adjusted so that the trapped material in the indents falls in the trough. The longitudinal tilt or pitch of the cylinder and retarder position should be adjusted for fine separation.

7.3   Specific Gravity Separator

The differences in the density of good and poor quality seeds and trash are exploited by a gravity separator in separating the unwanted seed components or impurities from the seed mass subjected for processing. If the seed mass cleaned and graded by Air Screen and Indented Cylinder still possesses high-density material such as clods, stones or lower density grain components (immature, insect damaged, diseased seed) which are not separated by earlier operations, it has to be processed for density grading.

The machine employs the principle of floatation, in which seeds are vertically stratified in layers on the deck according to their density by vibration and inclination of deck the different density grain components or impurities of lot takes separate paths.

Seeds are fed onto the specific gravity separator in layers of three to five seeds thickness. The shaking deck and the differential air stream from the deck push the heavier seed uphill; while the lighter seeds follow the path to downhill. The machine consists of different adjustable grates to separate materials according to their density. Stones or clods are pushed to transverse edge where adjustable grates direct the materials to different outlets.

The performance of the machine is closely linked with the operational parameters: volume of air, end slope, side slope, oscillating speed and feed rate. Stratification of seed mass on the deck is achieved by differential air volume adjustment from feed to discharge end. It should be adjusted so that heavy seeds

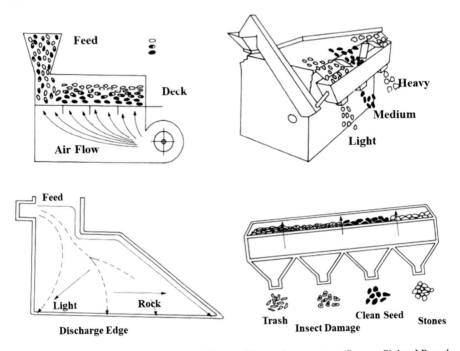

Fig. 4 Line diagram of the functioning of the specific gravity separator (Source: Sinha, J.P. and A.P. Srivastava (2003)

lie on the deck and the lighter seeds are lifted into the upper layer of the seed mass. The air volume on the feed side should be more than the outlet side.

The *slope* of the deck from the feed hopper to the discharge end controls the speed at which seed mass moves across the deck. The flat end slope is required when the difference between the seed and contaminants or seed to be separated is slight to hold the seed mass longer time on the check. On the other hand, the end slope can be increased if the differences in seed and contaminants are high in respect of specific gravity. *Side slope* is the tilt or inclination of the deck from low side to high side of the discharge end. It facilitates the light seed layers riding on a cushion of air slide downhill to low side of the cheek, while deck oscillation moves heavy seed uphill to the opposite side of the deck. *Feed rate* is critical and should be adjusted. Seed loss in reject port is inevitable during processing, but it can be minimized considerably by operating the machine at optimized parameters for the particular seed kind and lot subjected to processing. About 1 to 3% seed loss has been reported with specific gravity separation, and higher seed loss for poorer seed lots (Sinha et al. 2002) (Fig. 4).

7.4   Pneumatic Separator

It exploits the difference of terminal velocity in air stream of seed and impurities. Density, shape and surface texture are physical parameters which affect the resistance of particle to air flow and exhibit difference in terms of its terminal velocity. The seed mass is fed into the confined rising air stream where all particles with different terminal velocity are lifted to different levels. Particles of higher to lower terminal velocities are separated in different fractions.

In the pneumatic separator, the air system forces the air through the machine by creating pressure greater than atmospheric. On the other hand, in the aspirator, the fan is at the discharge end and induces a vacuum, which allows the atmospheric pressure to force the air through the separator. It consists of a control unit for variation in pneumatic pressure as per the requirement of seed mass which has to be processed (Fig. 5).

Fig. 5 Schematic diagram of pneumatic separator. (Source: https://www.usgr.com/seed-processing/wall-mount-air-seed-separators/)

7.5   Electric Separator

It separates seed mass constituents on the basis of differences in their electrical properties. The degree of separation depends on the relative ability of seeds in the mixture to conduct electricity or to hold a surface charge. A thin layer of seed is conveyed on the belt through a high-voltage electrical field, where it is given a surface charge. This charge is like the one that is picked up by a comb passed through the hair. As the belt rounds a pulley, seeds that have quickly lost their charge fall in the normal manner from the belt. Seeds that are poor conductors and slow to lose their charge will adhere to the belt and fall off gradually. Dividers in the drop path can then be positioned to collect any fraction of the distribution desired.

The operational parameters of the machine are feed rate, belt speed, electrode position, voltage and divider position. As the moisture content of the seed affects the conductance of electricity, all adjustments are linked with it.

Seeds of watercress from rice, ergot from bent grass, and Johnson grass from sesame can be separated by the electronic separators efficiently.

7.6   Spiral Separator

It classifies seed mass according to particle shape which is linked with rolling resistance. It consists of sheet metal strips fitted around a central axis in the form of spiral. The unit resembles open screw conveyor standing in a vertical position. The seed is fed at the top of the inner spiral. The round seeds roll faster as their rolling resistance is less than the flat or irregularly shaped seeds. The orbit of the round seed increases with speed on its flight around the axis, until it rolls over the edge of the inner flight into the outer flight where it is separately collected. On the other hand, the flat or irregularly shaped seed slide or tumble and move slowly and does not build up enough speed to escape from the inner flight. Hence, they are collected from inner flight outlet at the bottom separately.

The spiral separator is very useful for the separation of damaged seed in brassicas, vetch, lentil, peas and soyabean. In general, the spiral separator is useful to remove broken grains or texturally different seeds or impurity mixed in seed mass. It is also suitable to separate round seeds or damaged seeds (in respect of shape) from a mixture of round and flat seeds or damaged seeds (Sinha et al. 2008) (Fig. 6).

7.7   Inclined Draper

It separates the materials on the basis of differences in rolling or sliding characteristic, specific gravity and surface texture. The seed mass is fed from a hopper which distributes the mixture uniformly across the middle area of an adjustable tilted velvet, canvass or plastic-covered flat surface. The belt moves upwards, resulting the smooth or round seeds move to lower the end, while rough-surfaced or oblong or flat materials remain on the belt.

Fig. 6 Line diagram of a
spiral separator (Source:
Sinha, J.P. and
A.P. Srivastava 2003)

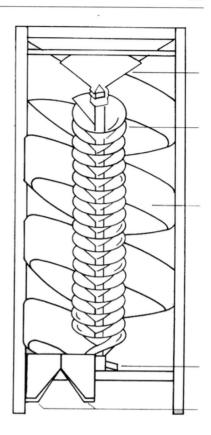

The rate of seed flow and the belt angle and speed can be adjusted according to the properties of the seed surface harnessing maximum separation efficiency. Multiple inclined drapers can be used for higher capacity processing.

The machine is useful to separate the fruits or seed clusters or plant debris and to remove buckhorn plantain (*Plantago lanceolata*) from red clover (*Trifolium pratense*) and to clean flower seed (Fig. 7).

7.8 Magnetic Separator

It also exploits the differences of surface texture of the seed. The seed lot is treated with iron fillings; rough seeds pick the fillings, but smooth do not. The treated seed is passed over a revolving magnetic drum. The seeds containing the magnetic materials are attracted to and fall near the drum or remain adhered to the drum. The adhering materials are brushed off at the end of a batch. The other materials fall farther to the drum and thus are separated.

The magnetic separator is used to remove *Stellaria media* (chickweed) from clover and alfalfa, dodder (*Cuscuta pentagona*) from clover, Lucerne (alfalfa), and

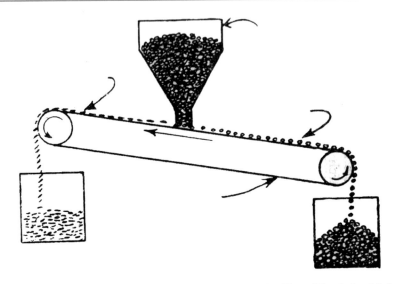

Fig. 7 Line diagram of an inclined draper. (Source: 'Guide to Handling of Tropical and Subtropical Forest Seed' by Lars Schmidt, Danida Forest Seed Centre 2000)

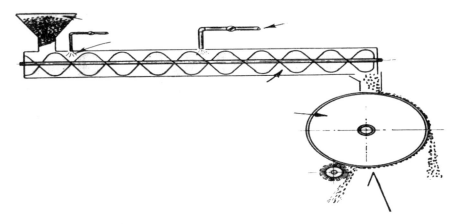

Fig. 8 Line diagram of a magnetic separator (Source: Sinha, J.P. and A.P. Srivastava (2003)

red clover, *Sinapis arvensis* (wild mustard) from brassicas (Sinha and Srivastava 2003) (Fig. 8).

7.9   Horizontal Disk Separator

It utilizes the difference in shape and surface texture with rollability characteristics and centrifugal force to separate different fractions. Seeds, confined to the centre of a flat rotating disk by a stationary circular plastic fence, are metered to the outer part of the disk through adjustable outlets. Centrifugal force causes round or smooth seeds to roll or slide off the disk. Irregular or rough seeds remain on the disk and are moved into a different hopper.

The feed rate through the outlets has to be adjusted so that each seed moves independently. The horizontal disk is similar to the spiral separator, but it is more selective because it has disk speed control that can change the proportion of seed retained or thrown off.

An added capacity can be obtained by mounting many disks on a single vertical shaft. The horizontal disk separator can separate dodder seeds from lucerne (*alfalfa*), curly dock from red clover, and other mixture in which one seed has a greater tendency to roll or slide than another.

7.10   Colour Sorter

The colour sorter is mainly used to separate discoloured seeds from the good seed lot when the other physical parameters like thickness, length, density and surface texture cannot differentiate them. The discolouration of seed occurs due to mechanical abuse of seed coat or pest infestation or adverse weather conditions. The machine uses photoelectric cells/sensors to compare the seed colour with the background filter. The background filter is selected to reflect the light of similar wavelength as that of good seeds. The seed mass to be colour sorted is metered by vibratory feeder so as single seed file is fed to the sorting chamber. Seed which differs in colour is detected by the photoelectric sensors by assessment of the reflected wavelength of light from the discoloured seed, which generates an electric impulse and activates an air jet to blow out the discoloured seeds from its normal path. As the discoloured seed has deviated its normal feed path, it is separated from the good seed which follows the normal path.

The colour sorter should be used for a seed lot only after the basic cleaning and grading. The efficiency of the machine depends upon the operational parameters, such as vibration of the feeder, speed of feeding belt, feed rate, position of discharge point, background colour filter, colour range, ejector timings and lag time of ejector (Fig. 9).

The electronic colour sorting significantly improves the purity, germination (%), vigour index and true density of the lot. However, colour separation process is effective only in the seed lots of colour purity level of around 70% or more (Sinha and Modi 2000; Sinha and Vishawkarma 2001).

7.11   X-Ray Sorter

A combination of radiography and image processing is utilized to process seed lots on the basis of physical parameters, i.e. size, shape, density, colour, texture as well as the anatomy of individual seeds. X-ray-based imaging provides a method for the non-invasive analysis of the internal structures even of treated, coated or pelleted seed.

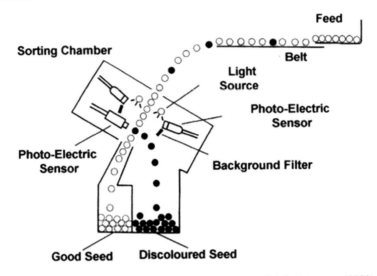

Fig. 9 Schematic view of a colour sorter (Source: Sinha, J.P. and A.P. Srivastava (2003)

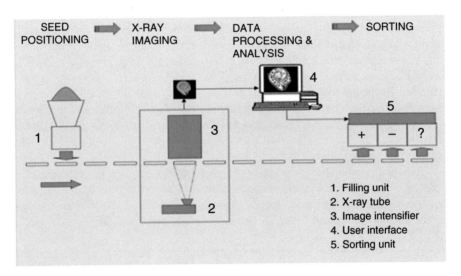

Fig. 10 Schematic view of an X-ray sorter. (Source: https://www.seedquest.com/technology/from/Incotec/upgrading/xray.htm)

Seeds in batches are placed in a special holder and exposed to long-wave X-rays, from where digital radiograph is generated of individual marked seeds. The radiograph is further converted into a 3D image, which subsequently is subjected to digital image processing (Fig. 10).

The processed image data are digitally analysed and compared with reference to the digital data of quality seed database. It is important to note that the X-ray imaging and sorting is based on internal morphology and not biological value.

8 Seed Treatment

In general, seed treatment refers to the application of a single pesticide, or a combination of two or more, to disinfect/disinfest seed from seed-borne pathogenic organisms or to protect the seed from storage pests. It also provides a safe microenvironment to planted seeds for developing into healthy seedlings. In some cases, coating or pelleting is carried out also with the objective of delivering essential nutrients to seeds for better nourishment and production of vigorous young plants or increasing the size and shape of seeds for enhancing the ease of mechanical planting (see chapter "Seed Quality Enhancement").

The equipment used to apply chemicals to seed in any form or method are classified as seed treater and these are of three types, namely, (a) dry seed treater, (b) slurry seed treater and (c) Mist-o-matic seed treater.

8.1 Dry Seed Treater

The seed is simply mixed by any means, e.g. shovels, scoops, rotating drum with treatment chemicals which are in the powder/dust form. The dusty condition that prevails during treatment and handling of the seed subsequent to treatment is the main disadvantage of this method. The low chemical use efficacy and non-uniformity of treatment are other limitations of this method.

8.2 Slurry Seed Treater

The chemical is mixed with water to form a thick suspension. The machines are equipped with an adjustable hopper to control the flow of seed into the machine, a slurry tank with mechanical agitator to stir the mixture constantly, a positive seed slurry metering device and a short mixing auger that mixes the fungicide slurry and seed and also moves the treated seed to the discharge spout of the machine. It overcomes some limitations of the dry seed treater, e.g. dusty environment and non-uniformity in coating of chemical. The average efficiency of this machine is about 60–70%.

8.3 Mist-O-Matic Seed Treater

This type of treater is designed to apply low doses of liquid/water soluble chemicals. It has an adjustable hopper that regulates the flow of seed into the machine; a positive metering device; a seed dispersing cone; and a rapidly spinning disk, which breaks up the liquid/water solution of chemicals into droplets. The dispersion cone causes the seeds to fall in a layer through the droplets. Further, the material is moved through a mixing chamber where the material is mixed through rotating auger or brushes and is moved to the discharge spout. The efficiency provided by the treater is

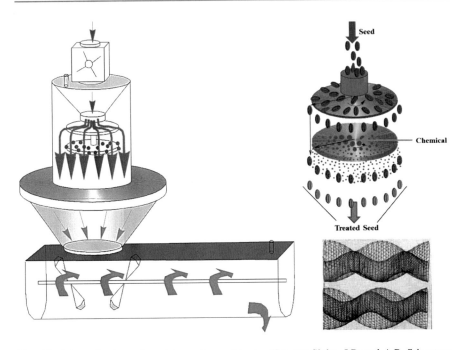

Fig. 11 Schematic view of a mist-o-matic seed treater (Source: Sinha, J.P. and A.P. Srivastava (2003)

in the range of 80–95%, as well as more uniformity in coating of chemical is associated with this method.

It is important to note that the solution of chemicals (in case of slurry seed treatment and mist-o-matic seed treatment) should be prepared so as the moisture content of treated seed does not increase by more than 1%. To attain it, chemical solution should be prepared according to the dose specified by pathological test and treatment method (Fig. 11).

8.4   Thermoseed Treatment

Developed by Lantmännen, Sweden, this is an innovative patented technology of thermal seed treatment for effective and economically competitive alternative to chemical seed treatment. Hot and humid air is used as heating medium and fluidized bed technology for seed mass movement in the system to ensure even exposure of thick seed layers. This system is capable to treat 30 tonnes of seed per hour effectively. The treatment plant consists of heating and cooling systems, sensing devices, system control software and transportation facilities. The exposure of seed for precisely conditioned hot humid air renders harmful pathogens present with seed without affecting seed quality https://www.lantmannen.com/.../thermoseed).

9   Seed Coating and Pelleting

Primary functions of coating /pelleting are to act as a binder and carrier for actives (pesticides or fungicides, nutrients, growth regulator, repellent), reducing the hazardous dust formation during handling, enhancer flowability, booster efficacy of actives and cosmetic improver. It improves seed-soil microenvironment (see chapter "Seed Quality Enhancement"). There are two components to a seed pellet: bulking (or coating) material and binder. The bulking material can be either a mixture of several different mineral and/or organic substances or a single component. The coating material changes the size, shape and weight of the seed. Desirable characteristics of a good coating material include: uniformity of particle size distribution, availability of material and lack of phytotoxicity. The second component, the binder, holds the coating material together. Binder concentration is critical because too much binder will delay germination. Too little binder will cause chipping and cracking of pellets in the planter box, which can cause skips and/or wide gaps in the plant rows. Many different compounds have been used as binders, including various starches, sugars, gum arabic, clay, cellulose and vinyl polymers.

The critical parameters for efficient coating and pelleting are: physico-chemical properties of the material, pan rotation speed, curvature, surface, angle of rotation, drying mechanism, temperature and operational parameters of machine with respect to seed to be given the treatment (Fig. 12).

Fig. 12 External views of seed coating/pelleting machine. (Source: http://www. seedpelletingequipment.com/coatingpan.htm)

10   Seed Packaging

The main functions of packaging are to contain the product and protect it against a range of hazards which might adversely affect its quality during handling, distribution and storage. Seed packaging is the last unit operation, but also of utmost importance after physical upgradation of seed lots to keep it safe until its use at planting. Selection of packaging material has to be done considering its tensile strength, bursting strength and tearing resistance to withstand the normal handling procedures, type of the seed, seed moisture content, expected duration of storage, threat of pathogens and storage environment. Most commonly used packaging materials are cotton cloth, paper film, hessian cloth, laminated polythene, metal, glass etc. Jute/cotton cloth bags are used when the seeds are to be stored under controlled conditions. Pulse seeds are packed in smooth but tough surface packaging material, as the bruchids (major storage insect pest) if infested, cannot lay eggs on it (Sinha 2000). Metallic cans, glass bottles or laminated poly-lined bags are used for smaller amounts of seed where high moisture proofing is required or low moisture seed is to be stored for long periods, such as the germplasm. The ultra-dry seed is only stored in hermetically sealed container. Contrarily, high moisture seeds should not be packed in moisture-proof containers (see chapter on Seed Storage and Packaging for more details). Special attention is required regarding mechanical mixing at the time of packaging operations. The mechanical mixing must be avoided by packaging only one lot at a time and cleaning the total processing area after completing each lot. The packed materials should also be labelled properly, with full lot identification information, e.g. crop, variety, lot no, production year, moisture content, seed quality parameters, weight etc.

References

Agrawal RL (1996) Seed technology. Oxford and IBH Publishing Co., New Delhi, India, p 829

Jorgensen KR and Stevens R(2004) Seed collection, cleaning, and storage, Chapter. 24. In S. B. Monsen, R. Stevens, and N. Shaw, eds. Restoring western ranges and wildlands. Ft. Collins, CO:USDA Forest Service Gen. Tech. Rep. RMRS-GTR-136

McDonald MB, Copeland LO (1997) Seed production: principles and practices. Chapman and Hall, New York

Sinha JP, Modi BS, Nagar RP, Sinha SN, Vishwakarma MK (2001) Wheat seed processing and quality improvement. Seed Res (29):17–178

Sinha JP, Vishawkarma MK, Atwal SS, Sinha SN (2002) Quality improvement in rice (Oryza sativa) seed through mechanical seed processing. Indian J Agril Sci 72(11):643–647

Sinha JP, Vishawkarma MK (2001) Effect of electro-mechanical colour sorting on seed quality in pigeonpea (Cajanus cajan) cultivar 'Pusa 33'. Indian J Agril Sci 71(4):287–289

Sinha JP, Srivastava AP (2003) Seed Processing. In: Singhal NC (ed) Hybrid seed production in field crops. Kalyani Publishers, India, pp 269–296

Sinha SN (2000) Safe storage of seeds: insect pest management. Indian Agricultural Research Institute Regional Station, Karnal, p 29

Sinha JP, Sunil Jha SS, Sinha ASN (2010) Post harvest management of paddy seed. Published by IARI Regional Station, Karnal. TB-ICN 77(2010):1–48

Sinha JP, Sinha SN, Atwal SS, Seth R (2008) Success story (1997–2008): seed Processing & Storage. Published by IARI Regional Station, Karnal, pp 1–62

Sinha JP, Modi BS (2000) Colour sorting and seed quality in mungbean. J Inst Eng 81:1–5

Trivedi RK, Gunasekaran M (2013) Indian minimum seed certification standards, The Central Seed Certification Board Department of Agriculture & Co-operation. Ministry of Agriculture, Government of India, New Delhi

Seed Storage and Packaging

Malavika Dadlani, Anuja Gupta, S. N. Sinha, and Raghavendra Kavali

Abstract

Storage is an essential component of seed programmes, which primarily aims at maintaining the high-quality standards of the seed from harvest till the time of sowing the crop in the next or successive seasons. In addition to this, seeds are also stored for longer durations to maintain stocks for seed trade at national and international levels as per market demands and as a buffer against crop failures in times of natural calamities or other exigencies, to maintain seeds of the parental lines for hybrid seed production in one or more seasons, to conserve active genetic stocks for breeding purposes, and to maintain germplasm for long term use. Seeds of most of the agriculturally important species are categorised as orthodox or desiccation-tolerant. Their longevity increases with decrease in storage temperature and the relative humidity of the storage environment (or seed moisture content). However, notwithstanding the constitutional differences among plant species concerning seed longevity, being a living entity, every seed undergoes deteriorative changes during storage, even in dry stores, primarily in terms of germination and vigour due to physiological deterioration, and changes brought by the presence of the pests and pathogens. A good seed programme aims at maintaining the high planting value of the seed in terms of purity, germination, vigour, and seed health during storage by taking care in seed handling, controlling the temperature and relative humidity of the store (or seed moisture in case of hermetically sealed containers), and following good sanitation

M. Dadlani (✉)
Formerly at ICAR-Indian Agricultural Research Institute, New Delhi, India

A. Gupta · S. N. Sinha
Formerly at ICAR-Indian Agricultural Research Institute, Regional Station, Karnal, India

R. Kavali
Indo—German Cooperation on Seed Sector Development, ADT Consulting, Hyderabad, India

© The Author(s) 2023
M. Dadlani, D. K. Yadava (eds.), *Seed Science and Technology*,
https://doi.org/10.1007/978-981-19-5888-5_11

239

practices. Considering that the facilities for conditioned storage may not be accessible and affordable in many situations, alternative solutions may be considered, especially for on-farm seed storage.

Keywords

Safe storage · Hermetic sealed · Moisture pervious · Safe moisture · Dry chain · Storability factors · Storage diseases and pests

1   Introduction

During seed development, germination and vigour reach their peak at 'physiological maturity'. As the seed moisture is very high at this stage, seed crop is allowed to get sufficiently dried before the seed can be harvested till a stage, commonly referred to as 'field maturity' or 'harvest maturity', when the seed moisture reaches a safe level ($<$15–20% in different species), but not too dry to cause shattering (Verdier et al. 2020). However, in dry season, seed moisture at harvest might go as low as 5–6%, as seen for a number of winter crops (*rabi* season) in the north western states of India. In case of wet season crops, on the other hand, harvested seed often need to be subjected to further drying before processing. Seeds of most of the agricultural and horticultural crops show a typically orthodox behaviour, where seed longevity increases with the reduction in seed moisture, and can be stored at low temperatures for prolonged storability. However, seed being a live entity, the status of viability and vigour achieved at the time of physiological maturity starts declining gradually, though very slowly initially, even on the mother plant, through harvesting and subsequent storage till the seed becomes non-viable (For more details on this please see chapter 'Seed Longevity and Deterioration') losing the ability to germinate and produce a healthy seedling (Ellis 2019). Therefore, the aim of packaging and storage is to maintain desired levels of seed quality, especially in terms of germination, vigour, and seed health, till its intended use. Though in the sequence of a seed programme, packaging is done before the final storage of seed, in order to establish the primary importance of storage, which also determines the type of packaging to be adopted in different situations, the storage requirements are discussed first.

2   Purpose of Seed Storage

Since the beginning of agriculture as an organised activity, farmers laid importance on the safe storage of seeds to plant the next crop. Seeds from the most healthy plants in a crop were harvested separately, cleaned, dried well, and saved in a cool and dry place for sowing the crop next season. With the advancement of agriculture as a specialised science, seed technology has also advanced significantly. Whether farm-saved or stored by the organised seed producers, the primary purpose of storage, accounting for more than 75% of the total produce (personal assessment based on the

Indian seed system), is to make available seed having good germination and vigour for the next planting season in the cycle of cultivation. In addition, some proportion of the total seed stock, especially of the food security crops, is maintained as a buffer stock by the government agencies or by commercial producers to fulfil the seed demand in the event of any calamity, whereas those involved in national and international seed trades also maintain some stocks for later use as per anticipated demands in other regions.

Besides these, seeds of the parental lines of the hybrids are retained mainly by the seed producers in medium-term storage for two seasons or more to save on the cost of production. Small quantities of germplasm seeds and breeding lines are preserved in long-term storage.

Suitable, cost-efficient storage management needs to be planned to meet these purposes.

3   Factors Affecting Seed Storability

Factors affecting the storability of seeds can be broadly divided into two major groups.

3.1   Seed Factors

These refer to all such internal factors which affect the physiological status of seed at the time of harvest and subsequently. These include:

- The genetic makeup of the seed.
- Growing conditions of the seed crop, any biotic or abiotic stress or nutrient deficiencies, especially during seed maturation.
- Initial seed quality, including maturity at harvest, seed size, and physical damage.
- Seed moisture content during processing and storage.
- Pathogen load and internal insect infestation and weed seeds.

3.2   Storage Factors

These are essentially the external factors that significantly influence the storage life of the seed of a species.

- Relative humidity (seed moisture).
- Temperature.
- Gaseous atmosphere.
- Packaging material.
- Structure and sanitation of the seed stores.

Besides the wide genetic variability, seed factors influencing storability need to be monitored during the raising of the mother crops. Care is taken to ensure that the seed crop is grown in an agro-climatically suitable region, with moderate temperatures and dry weather during seed maturation; the soil not be deficient in any vital micro or macronutrients; not affected by any biotic or abiotic stress during crop growth in general and seed maturation in particular; the seed crop is kept free from seed-transmitted diseases and pests; and harvested at the right stage of maturity (For more details on these please see chapters 'Seed Development and Maturation', 'Seed Storage and Packaging', and 'Emerging Trends and Promising Technologies'). Abiotic stress caused by high temperature and soil moisture and biotic stress caused by Phomopsis during seed-fill and pre-harvest stages lead to hard seeds, low seed viability, and vigour in soybean (TeKrony et al. 1980; TeKrony et al. 1996). Physiological deterioration of seed during storage is an inevitable and irreversible process, bringing down individual seeds' ability to germinate, thus lowering the germination per cent of a seed lot (For more details on these please see chapter 'Emerging Trends and Promising Technologies'). In addition, damage caused by pests and pathogens also brings down the planting value of the seed. Once the germination of a seed lot falls below the prescribed standards for a given crop species or is heavily insect-infested, such seed lots are not acceptable. If such seeds were fungicide-treated, those could not even be used for food and feed and are discarded as waste. Thus, loss of germination and damage due to pests and pathogen causes a significant loss of resources. McDonald and Nelson (1986) estimated about >25% loss of seed inventory annually due to a fall in seed quality due to poor storage which could be even more in the tropical and sub-tropical regions (Champ 1985). Since the process of deterioration can neither be halted nor bypassed entirely, efforts are directed to slow down the speed of deterioration by adopting the best storage and packaging practices as required, fulfilling the purpose of storage.

4   Management of Storage Factors

Of all the factors mentioned above, relative humidity (RH), temperature, and gaseous composition of the storage environment are the three most crucial ones that influence seed storability through direct and indirect effects and interactions between the biotic and abiotic factors. Under ultra-dry (0.3–3.0%) and low temperature (−5.0–5.0 °C) conditions, orthodox seeds retained their germinability for prolonged periods (Pérez-García et al. 2007), and such conditions are generally recommended for long-term storage of small quantities of seed, as in the gene banks. The gaseous composition of the storage space also influences seed storage, especially under hermetic conditions (Groot et al. 2015). Dry seeds usually have water potentials between −350 and −50 MPa. Being highly hygroscopic, seeds absorb moisture from the humid atmosphere till a state of equilibrium is attained. However, the chemical constitution of seeds influences the moisture absorption pattern in a given atmosphere. Thomson (1979) found that the moisture content (MC) of most cereal seeds and pulses reached safe moisture contents of 11–13% at 50–60% RH. In contrast, the oilseeds attained <10% moisture even at 70%. Malek et al. (2020) observed that the moisture content at equilibrium RH (ERH) varied in canola seeds at different temperatures and relative humidity. At all temperatures, there was a difference between water desorption and absorption curves, desorption curves being higher than the absorption curves. Therefore, in unconditioned stores, the RH keeps fluctuating; hence, the seed experiences both absorption and desorption of moisture vapour before reaching an ERH. All these factors need to be considered while planning seed storage.

A series of pioneering work conducted by EH Roberts and his team at the University of Reading, UK, on seeds of different crop species stored at varying ranges of MC/RH and temperature led to the development of viability equations to predict the germination of seeds of a given species at a given set of temperature and moisture/humidity (Roberts 1973; Ellis et al. 1989). The basic equations were further modified to better predict specific species' storage behaviour. However, there are considerable reservations about the validity of these equations, which uses universal temperature constants over the range of -13 °C to 90 °C, and moisture constants which are species-specific (Pritchard and Dickie 2003) but do not take into account critical factors such as the genotypic variations within a species, phase transition of the cytoplasm to the glassy state at MC of 5% and below, and gaseous compositions in sealed storage. Nonetheless, it does provide broad indicators for the storability of seeds of different species and emphasises the influence of temperature and humidity (seed moisture) in seed storage.

Though both the factors are essential and interact in influencing seed longevity, moisture plays a more significant role, and its management can be more energy-efficient than cooling, particularly in warm environments. High humidity is considered a bigger problem than temperature for storing any kind of seed or grains (Omobowale et al. 2016). Hence, it is important that even small reductions in ERH can equal the effect of reduced temperature without the infrastructural investment or energy input required for refrigerated storage (Dadlani et al. 2016; Pérez-García et al. 2007).

Bradford et al. (2018) suggested a relationship between the survival and growth of various microflora and insect pests on seeds with decreasing moisture content and summarised that though active respiration of seeds stops below about 95% ERH (equilibrium relative humidity refers to the MC attained by a seed lot equilibrated at a given RH), the bacteria can grow up to about 90% ERH. In contrast, at ERH below 65%, fungi cannot remain active, though storage insects are active up to below 65% ERH due to their ability to limit water loss and generate water from their food through metabolic pathways. However, they are unable to survive at ERH values less than 35% (Roberts, 1972). However, achieving ERH values <35% is often difficult, especially for seeds of grain crops with large volumes in humid environments. Seeds of some species, particularly legumes, are often prone to damage due to excessive drying, hence need care at drying. Therefore, appropriate and cost-effective drying methods must be adopted before storing seeds. Upon drying, seeds need to be stored in a sealed or dry environment to maintain germination and vigour and protect against insects. Very low ERH levels or low seed MC usually are adopted only for storage of seeds such as onion (*Allium cepa L.*), which are known to be poor storers. Rao et al. (2006) found that storage of onion seeds for 18 months or more could be achieved upon desiccation of seeds to 6 ± 1% seed moisture, sealed in moisture impervious containers, and stored at ambient temperatures or 25 °C. Low moisture storage is suitable for medium- to long-term storage of important plant species with short storage life (Ellis and Hong 2007; Hong et al. 2005).

Harrington (1972) emphasised the effect of seed moisture (and RH) and temperature and suggested two thumb rules that with every 1% reduction of seed moisture and with every 5 °C decrease in temperature, the storage life of seed doubles and the effect of these two factors are independent of each other. He further stated (Harrington 1973) that for safe seed storage of most crop species for 1–3 years, the sum of the RH and temperature at °F should not exceed 100. However, seeds can safely be stored at conditions even when the sum of the RH and temperature (in °F) is 120, provided the temperature does not contribute more than half (Justice and Bass 1978). To varying levels, management of storage factors can be achieved by natural or artificial means.

4.1 Management of Moisture

This can be achieved by (a) drying—bringing down the seed moisture to a desirable low level first, and then storing in a hermetically sealed and moisture impervious container or (b) storing seed in equilibrium with a dry environment maintained at a low equilibrium relative humidity (ERH), where it maintains a desirable moisture content. Scientific studies have established that with the reduction of equilibrium relative humidity (ERH), the storability of food products improves as the metabolic activity of spoilage bacteria, fungi, and insects slow down upon desiccation (Crowe et al. 1992). However, seeds (or grains) with higher oil content will have lower MC at a given ERH than those with lower oil content. This is because water is excluded from the hydrophobic oil bodies in the cells, reducing the water content relative to the total product weight. The same applies to seeds. The relationship between MC and ERH at a given temperature (termed an 'isotherm') is consistent for the seeds of a given species. Though the seed moisture content in a given environment is directly proportional to its ERH and reflective of the water activity of the product (Chen 2001), in case of the storage biology of seeds, use of ERH rather than MC is preferred because the effect of ERH on spoilage organisms is consistent across products regardless of their composition (Bradford et al. 2018). In the case of farm-saved seed, natural sun and wind are used (for sun or shade drying) to reduce seed moisture, and the seed thus dried is stored in suitable containers in the coolest and driest available spot. With some plant-protective measures, it was found that it may be safe to keep seeds of most grain crops for at least one crop season under uncontrolled conditions if the maximum ambient RH and temperature do not exceed 70% and 30 °C, respectively, for more than three months in a year (Agrawal 1982). For maintaining germination in commercial seed lots, care is needed for safe drying and equilibrating at a safe RH.

4.1.1 Seed Drying

Drying is the process of removal of moisture from the seed. The extent of drying operation is determined by the initial seed moisture content and the level to be achieved, which in turn will be determined by the following factors:

- Type of seed: This will consider seed structure, composition, and longevity behaviour (whether the species is a poor, moderate, or good storer).
- The abundance of natural sunshine and wind at the location and availability of infrastructure and machinery to use artificial drying economically.
- Conditions of storage to be followed: It is crucial to know if the seed will be (a) packed in vapour permeable containers and stored under ambient conditions; (b) packed in vapour permeable containers and stored in RH and temperature conditioned stores; (c) packed in hermetically sealed containers and stored under ambient conditions; or (d) packed in hermetically sealed containers and stored in conditioned stores.
- Value and volume of operations: The quantity of seed to be dried and the value of the seed per unit weight are the other important considerations in commercial seed operations.

4.1.2 Natural Drying

Several studies have shown that seeds of field crops and vegetables equilibrated at 60–70% RH attain equilibrium moisture content (EMC) of ~ or <12%, which is considered safe for post-harvest handling (Harrington 1960; Leopold and McDonald 2001). Hence, seed crops harvested in favourably dry climates seldom need much drying before threshing and conditioning either for on-farm or commercial storage. When the MC of the harvested seed is high, it may need to be pre-dried before threshing and further dried before storage.

Open-air sun-drying is an effective and practical method of bringing seed moisture to a safe level. Care is needed to ensure that the seed is laid in middle layers and turned periodically for uniform drying. As high temperatures can be detrimental to seeds having high MC (Brooker et al. 1992), the temperatures at drying should preferably be kept < or not above 40 °C. Starchy seeds generally withstand higher drying temperatures than oil-rich seeds (Thomson 1979).

In the subtropical and tropical regions with ambient temperatures of 40^0 C and above, shade-drying with assisted aeration (using domestic-purpose pedestal fans) is often preferred. It also protects seeds from the possible damage caused by the UV radiation of sunlight. Hanging porous seed bags or earheads upside down from the ceiling of the storage structure is also an effective method of on-farm natural seed drying.

4.1.3 Commercial Drying

Depending upon the initial seed moisture at harvest, seeds are dried using single- or two-stage drying methods before storage. Commercial seed drying is of three types (For more details on this see 'Seed Processing for Quality Upgradation'):

1. Layer drying.
2. Batch drying.
3. Continuous flow drying.

Following the main principles of each type of drying, additional features may be added as per needs.

Layer drying (also referred to as bin layer drying) is mainly used for in-storage drying in which seed is dried in layers, and the airflow ducts are built at the bottom of the bin. Once the first layer is dried up to a certain level, a second layer is added; thus, each seed layer is partially dried before the next is added. Several layers will be added, and these will reach moisture equilibrium over a period of time. A stirring and mixing device improves moisture spread across the layers. This type of dryer is more suitable for on-farm storage of seeds.

Batch drying, suitable for both stationary and portable units, dries seeds in batches. Slightly heated air is passed through the seed layers from bottom to top. With the onset of drying, a moisture gradient is created from bottom to top, and moisture is removed from the seed till an equilibrium is reached. Further drying is possible only by raising the drying temperature or the rate of the airflow, or both. Like layer dryers, a stirring device is also used to improve drying in batch dryers. In a batch-type dryer, a fixed amount of seed can be dried at a time.

Besides forced air, both layer and batch driers may have supplemental heating, but since the seed is not moving, it requires a cooling period after drying once the heating is turned off.

In *continuous flow dryers*, unlike layer and batch dryers, the seed moves continuously from the inlet to the outlet through the drying and cooling chambers. The heated air is forced through the seed column from a heated air plenum. This method is suitable for drying a large quantity of seeds. The drying efficiency is highest in continuous flow type dryers, followed by the batch type (better in two-pass type than the single pass) and layer type dryers, respectively.

For better drying efficiency, seed dryers have the option of fitting a dehumidifier to dry the air before it is heated and passed through the seeds. The air dryer lowers the relative humidity of the air by decreasing the total moisture in the air. As a result of dehydration, the temperature is raised; thus, the need for heat to be provided by the heaters is lowered, reducing the fuel cost (Justice and Bass 1978).

In commercial seed production units in warmer and humid regions, dealing with large quantities of field crops during peak seasons, two-stage drying may also be adopted. In this, harvested seed with high MC (~20%) is first dried to a moderate level (~15 to <18%) and stored. Once all seed lots or the whole quantity of seed is dried to such a moisture level, which is safe for short storage, the seed lot is dried to the moisture level recommended for safe storage (Khare and Bhale 2014).

In addition to the above-mentioned basic dryer designs, there are dryers for special purposes, e.g. rotary dryers, used for dying small batches of vegetable seeds; ear-corn dryers for drying seed-in-the-cob of maize; box dryers for drying limited quantities of pre-basic/breeder or basic/foundation seed. Fluidised bed seed dryers can also be used for small-seeded species.

4.1.4 Management of Temperature

The independent yet interactive influence of temperature and humidity on seed longevity is well documented, recognised, and reflected in early prepositions of Harrington's thumb rules, as well as Robert's viability equations. However, in later studies, it has become evident that the humidity of the storage environment plays a bigger role in seed longevity than the temperature. When seeds are dried to a reasonably low MC, their storability gets significantly enhanced, compared to the longevity behaviour of seeds held at higher moisture even at temperatures above 25 ° C. Contrarily, high moisture seeds can only be stored at temperatures below 10 °C, but not at sub-freezing temperatures. At temperatures above 25 °C, seeds with moisture above 12–14% experience higher physiological and pathological deterioration, whereas at subfreezing temperatures, seeds with high MC may experience freezing injury. Comparing the storability of the primed seed of rice (dried to 10.5% moisture, under different conditions of temperature, RH, and availability of air), Wang et al. (2018) observed that the viability of primed rice seeds did not reduce when stored under low temperature (LT <5 °C)—vacuum (V); room temperature (RT ~30 °C) and vacuum (RT-V), or room temperature, aerated and low humidity (LH 20–26%) conditions (RT-A-LH), but was significantly reduced at room temperature-aerated-high humidity (60% and more) conditions (RT-A-HH). Hence, the detrimental impact of storage temperature was evident when the seeds were stored under high RH and aerated conditions. In a vacuum, the increase in temperature (30 °C) did not reduce the longevity of primed seeds in a short storage study, indicating the vital role of air (or its absence) in determining seeds' storability.

4.1.5 Gaseous Composition of the Storage Environment

It is well known that seeds dried to low MC (<5%) and stored in hermetically sealed containers (laminated and poly-lined/polythene, metal cans, glass, etc.) can store well for more extended periods at ambient, lower or higher temperatures as compared to those packed in non-sealed containers (Grabe and Isley 1969). The impact of gaseous components in the storage environment on seed longevity has been studied extensively in seeds maintained in hermetically sealed containers (Bockholt et al. 1969; Lougheed et al. 1976; Justice and Bass 1978; Bennici et al. 1984). Storing seeds in a vacuum, CO_2, or other inert gases has been tested for extending longevity, and variable results were reported. Groot et al. (2015) observed that the ageing of dry seeds was accelerated by the presence of oxygen in the environment and suggested storing seeds in anoxia for prolonged germplasm storage.

4.2 Seed Packaging

After the seed is dried to an optimum moisture level, it needs to be packed in such containers that the moisture content does not exceed the safe level for the kind of storage. Seed moisture below 12% is considered safe for storing most seeds of field crops in moisture vapour permeable bags for at least one season under ambient conditions. For vapour, impermeable containers seed moisture in the range of 6–8%

is considered safe (Anonymous 2013). Seeds in bulk are commonly packed in gunny sacks/burlap/cotton/multiwall paper/high-density polyethylene (HDPE) bags or moisture-resistant bags and cocoons, whereas seeds in smaller quantities can be stored either in flexible or rigid materials as per needs.

Packaging materials are characterised by the degree to which they can resist the passage of gases and vapour. In different types of flexible packaging materials, the gases and moisture vapour can permeate (a) through the macroscopic pores and canals (e.g. kraft paper and parchment paper); (b) by the process of diffusion of the gas through the main bulk, and evaporation from the outer surface (e.g. uncoated cellulose, polyethylene, and cellulose acetate); and (c) through the pinholes on the surface of the packaging material, as in case of aluminium foils (Ranganna 1986).

Introduction of ultra-hermetic flexible UV-resistant cocoon made of polyvinyl chloride (GrainPro®), which have low permeability to air and moisture prevents the exchange of air and moisture into the bag and reduces insect activity considerably. Though more suitable for grain storage, these have shown promising potential for chemical-free, safe seed storage without cooling if the initial seed moisture is brought down to a safe level (<10%). Developed by the International Rice Research Institute (IRRI) and manufactured by GrainPro Inc., Super Grain bags maintained germination for 9 to 12 months under natural conditions (Asian Scientist, 30 July 2012; https://www.asianscientist.com/2012/07/tech/irri-grainpro-supergrainbag-2012). These bags come in 25 or 50 kg capacity and also have ports for fumigation of seeds or drawing seed material.

Moisture-proof packs (having low moisture vapour transmission rates) made of films of polyvinyl, polyethylene, cellophane, polyester, aluminium foil, etc. are generally used for smaller and sealed packaging. Some of these, e.g. polyester, also have low permeability rates to CO_2 and O_2. These are also suitable for vacuum sealing or filling with neutral gases to slow down the process of ageing. A combination of two or more different layers can be fused for better tensile strength and resistance against moisture vapour and other gases, e.g. CO_2 and O_2. However, while packing seeds in moisture-proof containers, caution is needed to check the seed moisture content. The Indian Minimum Seed Certification Standards (Anonymous 2013) provide specific requirements for keeping seed MC in moisture vapour-proof sealed containers. Storing chilli seeds at ~7% MC in partial vacuum (Doijode 1993) sealed packaging in laminated paper-Al foil-polythene bags under ambient conditions was found to maintain satisfactory levels of germination for three years.

4.3   Storage Structures

The important factors determining seeds' longevity during storage are the seed moisture, type of storage container, and storage environment. These factors generally interact, leading to many physiological and biochemical changes in the stored seeds, which result in deterioration of seeds both in quality and quantity, especially in tropical and sub-tropical countries.

Seed is stored in bulk for varying periods from harvest to its use for the next sowing. When the seed crop is harvested, it is often stored in drying sheds or stacked in large heaps covered with waterproof material, e.g. tarpaulin, for short periods before threshing, conditioning, and storage. Such structures only protect against rains, birds, and mechanical damage. However, after drying and processing, the seed needs to be stored in bulk in structures preferably built with waterproof walls and roofs and sealable openings for controlled ventilation and periodic fumigation. It must not allow water entry by seepage from the ground or walls. Columns must be widely spaced to permit the easy use of forklifts and movement of bags in stores, and ceilings should be high to stack seed bags up to 5 m high. Seed stores in sub-tropical and tropical regions are built at least 1 m above the ground, and rat traps are installed at possible entry points to check rodents. Seed warehouses with metal roofs must be suitably insulated and usually painted white for maximum reflectance.

Seeds stored in bulk cause heat build-up and need to be shifted or upturned periodically to break any hot spots. Aeration inside the stores with pedestal or moving fans help in dissipating heat. In modern seed stores, multiple sensors are fixed at different points, monitoring the temperature and humidity inside the stores, which can turn on aeration and cool the seeds when required (Desai et al. 1997). It is also important to ensure that the seed material for storage is almost free from any seed-borne pathogens and pests. The seed material in bags and sacks should be stacked on wooden pallets maintaining proper distance from the walls and the ceilings.

Depending on the type of seed material and purpose, different storage structures are considered for maintaining seed quality in commercial stores.

1. Storage under ambient conditions with or without ventilation: Commercial-scale unconditioned seed storage structures rely much upon natural cooling and ventilation to minimise the adverse effects of heat build-up. In temperate climates, bulk seeds can safely be stored by controlled natural ventilation, whereas in warm, humid environments, ventilation with outside air could be counter-productive (Copeland and McDonald 2001). Seed sacks/bags/piles/single layers/open containers can be kept for a short period, such as pre-processing or before final packing and storage, or for 6–8 months from harvest till the next cropping season in moderate environments. Measures are taken specially to protect seeds from rodents and birds.

2. Storage with only the control of moisture: This kind of storage is more suitable under temperate conditions where the temperature remains relatively cool throughout the year. Seed moisture is controlled either by packing pre-dried low moisture (6–8% moisture) seeds in vapour-resistant bags/containers or by packing seeds in vapour-permeable bags/sacks and using dehumidifiers. Since control of storage temperature in warmer regions, requiring uninterrupted power supply for refrigeration often poses a big challenge, the use of in-the-bag desiccants and frequent use of dehumidifiers, especially during the wet season, to maintain the RH of storage environment <70% maintains seed quality during

storage at least for one season, and also to carry over the unsold stocks for the next season.

3. Storage in low-temperature conditions: Low-temperature storage modules are often maintained by seed companies in warmer regions, dealing with seeds of a variety of crops. High value, low volume seeds of vegetables and flowers, having poor storability are dried to a reasonably low MC, packed in hermetically sealed moisture vapour impervious containers, and stored in low-temperature storage units (modules). However, care should be taken for seeds in pervious moisture containers stored under low-temperature conditions. RH of the ambient air increases with lowering the temperatures raising the seed MC. Thus, when brought to higher temperatures, seeds exiting such storage will be more susceptible to rapid deterioration.

4. Storage under controlled conditions of RH and temperature: This type of storage is used for storing seeds in bulk for at least two or three years and is cost-effective only in the case of high-value seeds, e.g. hybrids; inbred parental lines of the hybrids, especially in case of double or three-way crosses (e.g., maize); or very poor storers, e.g. onion. This is also commonly used for most vegetables (especially the hybrids or small-seeded ones) and flower seeds, owing to their high value per unit weight and smaller bulk. The temperature and RH are typically maintained at 20 °C and 50% (Copeland and McDonald 2001). The operation of this type of store requires an uninterrupted electricity supply.

5. Moist cold storage with control of temperature for limited period storage of desiccation-sensitive seeds from temperate regions: Seeds of recalcitrant seeds with high MC may be stored at low temperatures (just above freezing) for research purposes. The stores are maintained near saturation RH, and the seeds are moistened periodically. However, seeds are seldom stored in this way for commercial use.

4.4   On-Farm Safe Seed Storage

For on-farm seed storage, which accounts for >50% of seed used in subtropical/tropical regions, seeds are sun-dried, air-cooled, and then stored in cool and dry conditions in clean storage containers or structures. The containers of various kinds can be used for the purpose, ranging from jute/hessian bags, HDPE (interwoven) bags, cloth bags, clay pots, metal bins, mud-plastered bamboo baskets, etc. Mud-plastered structures are also used for larger quantities of grains and seeds. Many plant parts having insect repelling properties can be used in these containers/structures (Francis et al. 2015). However, for safe storage in a humid climate, it is not always enough to only dry the seed, unless it is also packed in an atmosphere that will not allow entry of moisture vapour (as in moisture vapour impervious containers). At times, it is difficult to dry the seed below 12% MC, especially for the rainy season crop, which is harvested when the ambient RH is still fairly high (~70%). For storing seeds of high-volume cereal crops, held at 13–14% MC, moisture vapour impermeable bags or bags made of sheets impermeable to gaseous exchange are useful. Storage systems such as Grain Pro™ Superbags or hermetic cocoons (De Bruin et al. 2012; Murdock et al. 2012; Rickman and Aquino 2007) are both moisture-proof and impervious to oxygen, which do not allow the insects to survive (www.grainpro.com) and can be used for one to two season storage. These bags and larger-scale cocoons can significantly improve commodity storage when properly utilised (Afzal et al. 2017).

Another cost-effective way, very promising for on-farm conservation of seed, is the use of zeolite beads, also known as 'drying beads' inside seed bags or drums to dry seeds safely at ambient temperatures maintaining seed viability and also protecting against storage insect pests (Bradford and Asbrouck 2011; Kunusoth et al. 2012; Sultana et al. 2021). As the zeolite beads can be regenerated thousands of times by heating, they can be used repeatedly. Bradford et al. (2018) proposed the dry chain concept to store food products, grains, and seeds safely. Analogous to the cold chain, in which products are maintained at low temperatures throughout storage and distribution, as in the case of fresh foods, in the dry chain, products are dried to low MC and stored in water-impermeable packaging. The dry chain using drying beads and containers resistant to gaseous exchange offers safe storage of seeds for the short to medium term, at relatively low energy consumption and cost, as refrigeration is not required. This approach can be practised effectively for on-farm seed storage by the farmers or for maintaining moderate amounts of seeds in the community seed banks, especially in humid locations (Dadlani et al. 2016). On the other hand, in-silos drying and storing are common with farmers with large holdings.

5   Management of Diseases and Insect Pests During Storage

While the problems of rodents and birds are mainly managed with modification in storage structures and mechanical devices, fungi and insect pests can be controlled by managing seed moisture, relative humidity of the storage environment, and treatment with needed pesticides (fungicides and insecticides). This is also known as integrated pest management.

5.1   Seed Health Management During Storage

Seed health is an important criterion of seed quality, which mainly refers to the presence or absence of disease-causing organisms, such as fungi, bacteria, viruses, animal pests such as nematodes and insects, or physiological disorders due to deficiency of trace elements. The problems of diseases and pests during storage are more aggravated in the warm and humid regions, making it difficult to maintain the prescribed levels of seed vigour and viability from seed harvest till the next sowing season. As the number of small farmers is high in the sub-tropical and tropical countries where farmers use about 60 to 70% of farm-saved seed for sowing, these are highly prone to damages caused by pests and pathogens. According to some rough estimates, nearly 10% of the food grain is lost in storage due to microbial spoilage and insect attack, which also include grains saved for seed purpose.

5.2   Microbial Damage

Seeds become vulnerable to various types of pathogens and saprophytes during storage, especially under humid and warm conditions. Fungi form the major group of microbes causing seed damage, which are classified as storage fungi. Nearly 150 species of fungi have been found associated with grains and seeds in storage (Dharam Vir 1974). Mechanical damage in the seeds, cracks, breaks, or scratches in pericarp or seed coat developed during threshing and processing substantially facilitate invasion of fungi, which find their way to the storage go downs.

The storage fungi can grow without free water, at ERH of 70–90%. Some common storage fungi infecting seed include *Aspergillus, Penicillium, Rhizopus, Fusarium, Cladosporium, Alternaria, Mucor, Chaetomium, Epicoccum*, etc. As most fungi spp. primarily invade the embryo, the infected seed may appear normal during the early stages of infection. However, in later stages, discolouration and distortions of seeds including reduction in seed size, shrivelling, and seed rots are some of the common manifestations of seed-borne pathogens, besides causing seedling decays, pre- and post-emergence mortality, and seedling abnormalities. As the pathogens are well-established, embryos become ungerminable, and the seeds appear discoloured, reducing the seed quality, as well as making the grains unfit for human and animal consumption. Thus seed-borne fungi decrease the market value, germinability, and nutrition of the produce. Excessive fungal growth may also result in heating, caking, and decay. Besides losing viability and vigour, this brings about biochemical changes leading to the production of toxins and reduction in seed weight.

Studies have shown that the invasion of fungi leads to both physical and chemical changes in the seeds. Misra and Dharam Vir (1992) observed an increase in milling losses ranging from 34.0 to 58.6% in discoloured rice grains. Prasad et al. (1990) observed changes in fatty acids, glycerols, sugars, and amino acids in radish seeds infected with *A. flavus*. Similarly, Dube et al. (1988) observed changes in starch, fatty acids, and sugars in wheat grains infected with *A. flavus* and *A. niger*. Joshi et al. (1988) reported a 73% reduction in starch content and an increase in the amount of reducing sugars and phenolic contents in pearl millet seeds infected with storage fungi. Aflatoxins were reported by Bilgrami and Sinha (1983) in infected seeds of maize, groundnut, and other agriculturally important crops used as foods and feeds. Vaidehi (1997) showed that the biochemical changes brought by storage fungi lowered the quality of maize grains. These fungi may be present as dormant spores or mycelium on the seed surface or below the pericarp, which activate and multiply at a fast rate under favourable growing conditions of storage.

Though initially the incidence of field fungi is higher in seeds and that of the storage fungal flora associated with different seeds are initially low, it increases with an increase in the storage duration, whereas the field fungi reduce with the advancing

Table 1 Influence of seed mycoflora on soybean seed germination during ambient storage (Gupta and Aneja 2004)

Storage period (months after seed treatment)	Seed germination[a] (%)	Seed mycoflora[a] (%)	Seed moisture[a] (%)
0	92.0	1.05	7.8
2	86.5	2.4	5.9
4	92.8	3.4	6.6
6	85.3	2.7	9.9
8	50.8	5.9	8.6
10	44.3	4.6	8.9
12	38.3	5.0	8.8
Correlation coefficient (r)	−0.872		0.35

[a] Average of 20 treatments

storage. Proliferation of storage fungi leads to several pathogenic and biochemical changes, resulting in the decrease of seed viability. A significant negative correlation ($r = -0.793$) has been observed between seed viability and the load of mycoflora with the advancement of storage (Table 1).

5.3 Storage Insect Pests

More than half a dozen insects are commonly found associated with stored seed/grain in tropical and sub-tropical environments. These can be grouped into two major categories as internal and external feeders.

Primary or internal feeders: These insects mostly lay eggs inside or on the seed and complete almost entire larval and pupal life inside the seed, only to emerge as adults. They cause significant damage to the seed, which is not visible from the outside. It leads to loss of germination and vigour. The most common of these are:

Rice weevil, *Sitophilus oryzae* **(Coleoptera: Curculionidae)**.
Lesser grain beetle, *Rhyzopertha dominica* **(Coleoptera: Bostrichidae)**.
Pulse beetle, *Callosobruchus maculatus***(Coleoptera: Bruchidae)**.
Angoumois grain moth, *Sitotroga cerealella* **(Lepidoptera: Gelechiidae)**.

Secondary or External feeder: This group of insects feeds on embryo/germ and endosperm from outside. They may attack the whole seed and damage the embryo if the seed moisture content is higher than the recommended, or feed on the damaged/infested or mechanically broken seed. These insects in their different stages of development and growth are generally visible on the seeds. The most common of these are:

Red rust flour beetle, *Tribolium castaneum* **(Coleoptera: Tenebrionidae)**.
Saw-toothed beetle, *Oryzaephilus surinamensis***(Silvanidae: Coleoptera)**.

Rice moth, *Corcyra cephalonica* (**Lepidoptera: Pyralidae**).
Almond moth, *Cadra cautella* (**Lepidoptera: Phycitidae**).

It also includes insects and mites that develop after the infestation of other pests as they feed on cut and broken seeds, moulds and debris, dead insects, and animal wastes such as common grain mite and cheese mite psocids.

Knowledge of the various sources of infestation is crucial for better managing insect pests in the store. These are:

- **Field**: Some insects like bruchids, *Sitophilus oryzae*, and *Sitotroga cerealella* infest seed crops at the reproductive stage in the field. They come along with the harvested produce and multiply during the pre-storage or storage. The infestation is usually detected at the time of the emergence of adults.
- **Stores/godowns**: Insects or their stage(s) hiding in the cracks and crevices, electrical fittings, spillage, filth, etc. are the primary source of the infestation in the seed stores and godowns.
- **Reused old bags/containers and seed transport system:** Adult insects as well as those in developmental stages hide in the weavings, nooks, and corners; infest the seed when stored in such contaminated bags/containers, especially when reusing old bags or in the vehicle while being transported.

Essential components of Integrated Pest Management (IPM) against storage insects are:

- **Prevention:** includes proper seed drying, scientific method of storage, management of storage insect pests, and prophylactic treatments.
- **Inspection and monitoring:** regular inspection and monitoring of insect infestation in the godown and seed lots to make timely and correct decisions on the need and type of control measures.
- **Intervention:** management of insect infestation through appropriate control measures.

5.3.1 Monitoring and Detection of Insect Infestation

Considering the rapid multiplication of insects, regular monitoring and detection of insects in seed stores are necessary for early warning and in taking appropriate and timely control measures. Detection delays may result in pest outbreaks, causing severe contamination of seed materials and quantitative loss. It also helps in achieving better effectiveness of fumigation and other pesticide treatments.

5.3.2 Monitoring of Insect Infestation in Seed Stores and Bulk Godowns

- **Visual Inspection**: It includes inspection of the stores and godowns (both before and after processing) for live, flying, or crawling insects in every season, particularly the warm and humid pre-monsoon, monsoon, and rainy months. Detection of live insects or their castings in sweeps and the presence of flour deposits on bags caused by lesser grain borers indicate insect infestation. The presence of the web in undisturbed places is also a sign of lepidopteran infestation.

- **Light traps**: Most insects are nocturnal and phototropic. Light traps detect the presence of insects and their build-up. Light traps with an electrocution net kill insects attracted to them and help control insects. Mohan et al. (1994) used a 4 W ultraviolet light (peak emission at 250 nm) set at 1.5 m above ground level in the alleyways and corners of godowns. This detected the presence of *R. dominica* accurately.
- **Sticky traps**: These help in the early detection of insects, especially when placed at the top of bins.
- **Traps for crawling insects**: It provides a hiding place and is available in various designs. It can be used with pheromone lures for specific insects or food baits to enhance the capture of multiple species.
- **Pheromone traps**: Unlike light traps, pheromone traps are baited with synthetic chemicals that influence insect behaviour. These chemicals are species-specific and help in better monitoring of particular insect pests. Traps have also effectively detected insects at low population levels early. Pheromone traps are now available with adhesive glue, to which insects get stuck, which helps in removing a proportion of the population (mass trapping).

Detection of External Infestation in Seed Lots
(a) **Direct examination:**
  - Two samples of 200 seeds each are visually examined using a magnifying glass (10X) or stereoscopic microscope with light.
  - Live and dead adult insects, larvae, grubs, etc. are separated, counted, and recorded as numbers per weight of the sample.
  - Insect-damaged seeds are separated and counted, including those whose germ (embryo) has been damaged or have an escape-hole (s) or eggs adhered to them.
  - Other seeds with no visible symptoms of insect injury are subjected to further tests to detect an internal infestation.
  - The number of internally infested seeds is added to the number of seeds found externally damaged by an insect for the final calculation of insect-damaged (ID) seeds.

- There are different methods for detection of internal infestation:
  (a) Dissection method
  (b) Translucent method
  (c) Flotation or specific gravity method
  (d) Staining method
  (e) X-ray or radiography technique
  (f) Acoustic (sound) detection system

(g) CO_2 detection method

(h) Breeding out method

---

6   Good Storage Practices as a Preventive Measure

It requires the maintenance of storage facilities to an adequate standard and efficient control and handling of stocks. Regular and critical inspection of stores and stocks should be done to maintain good storage hygiene. Moreover finally, the chemicals should be the last option on a cost and need basis. The application of pesticides to stored products should be kept minimum. It has two components of pre-storage and in-storage measures.

6.1   Pre-Storage Preventive Measures

6.1.1   Preparation of Seed Stocks

- The seed stock should be clean and free from broken or damaged seeds.
- Ensure drying of seed to moisture content (MC) <10% and for paddy (≤13%) before storage.
- Pulse seeds may carry insect infestation from the field, which is detected when adult insects emerge from the seed in the pre-processing hall before processing. Therefore, pulse seed should be dried in the sun to kill all internal infestation or fumigated immediately after arrival in the godown to avoid insect multiplication.
- Ensure new harvests do not carry field infestation in other crop seeds. Fumigate if the live insect is detected.
- Mix premium-grade Malathion 5%D @200 g/t of seed or Deltamethrin (K-Othrin 2.5 SC) @ 40 mg/kg seed. Treated seed should never be used as food or feed.

6.1.2   Preparation of Seed Store/Shed

- Clean all the structures from debris, webs, and spillage. Disinfect concrete floor and walls with malathion 50 EC or fenitrothion 50 EC @ 50 ml in 5 L water/100 m^2 floor using a knapsack sprayer. Wet all surfaces and fill the crack and crevices to kill hiding insects.
- For non-commercial purposes, treat old seed bags with hot water (>50 °C) for 15 min and dry them before use or treat with malathion @ 10 mL/L water or Deltamethrin @ 02 mL/L water per 20 m^2 bag surface area or fumigate before re-use.
- Seed treatment with Spinosad (Tracer 45 SC) and Indoxacarb (Avant 14.5 SC) @ 2 ppm provide effective (ID: 0.10% and 0.13%, respectively) control of storage insects infesting rice seed up to 12 months of storage under ambient conditions, whereas in untreated lots, seed damage was 7.4% control (Padmasri et al. 2017: https://www.researchgate.net/publication/324476766).

- Treatment of wall or empty surface before storage of seed with following insecticides, if required:

Malathion (50 EC) at 10 mL/L of water @ 5 l/100 m².

Deltamethrin (K-Othrin 2.5 SC) at 40 mg/L water @ 5 L/100 m².

Use of insecticides will be as per the prevailing regulations in a country.

6.1.3 Thermal Treatment

Heat treatment has been used to control the development of pathogens and insects. Both high and low temperatures are reported to be effective against most storage insects (Table 2).

6.1.4 Solar Heat (Solarization)

A solar heater is made of dark cloth or black plastic sheet, which can trap natural ambient temperature to destroy pests. It is particularly effective in sub-tropical and tropical regions where ambient temperature can exceed 40 °C during summer. A solarisation cover can elevate the temperature to >60 °C in about two hours, lethal to many storage insect pests like pulse beetles. Chauhan and Ghaffar (2002) found that pulse beetles at all stages, present in pigeon pea seeds, were killed upon solarisation in polyethylene bags at ICRISAT, Hyderabad, upon reaching the temperature up to 65 °C. Moreover, seeds, thus solarised, remained protected from insect damage even after 41 weeks of storage under ambient conditions of seed store. There was no adverse effect of solarisation on the germination of seeds. This technique is particularly useful for on-farm seed storage. On-farm testing of Sunning and Sieving (S & S), that is, removal of killed insect debris by sieving after solarisation, showed it was as effective in combination with seed treatment with insecticide 1.6% Pirimiphos methyl or 0.3% permethrin after four months of storage. Solarisation of seeds in clear polythene (700 gauge) packets for six days (3 h on each day) was found to be

Table 2 The response of stored-product insects to temperature (Fields and Muir 1996)

Stage	Temperature zone (°C)	Effects
Lethal	>62	Death in less than 1 min
	50 to 62	Death in less than 1 h
	45 to 50	Death in less than 1 day
	35 to 42	Populations die out
Sub-optimum	35	Development stops
	33 to 35	Slow development
Optimum	25 to 33	The maximum rate of development
Sub-optimum	20 to 25	Slow development
	13 to 20	Development slow or stops
Lethal	03 to 13	Death in days (un-acclimatised), and movement stops
	−10 to −5	Death in weeks to months if acclimatised
	−25 to −15	Death in minutes, insects freeze

an effective treatment for reducing insect damage at most of the National Seed Programme (NSP) centres in India (S.N. Sinha, personal communication).

6.1.5 Cold Storage

Low temperature (<20 °C) storage not only slows down physiological seed deterioration but also restricts the growth and development of storage insect pests in the godowns, thereby improving their shelf life. Most storage insects' life cycle gets prolonged, but they are not killed. The relative humidity (RH) in the enclosed space significantly affects the survival of insects under low temperatures. RH <30% is ideal for safe storage of seeds for medium-term period storage (4–5 years). Various factors such as insect species and its stage, density, and distribution in the seed lots, air temperature and relative humidity, length of the exposure period, type of seed and its moisture content, and the initial health status of seed (field infestation, if any) determine the efficacy of insect control in cold storage.

6.1.6 Controlled Conditioned Seed Storage

The hot and humid environment is ideal for insect activity. It also affects seed quality and its shelf life. Hence, low temperature (<20 °C) and low RH (<50%) environment in the seed godown improves the shelf-life and seed quality for a more extended period. Cold storage technology is expensive; therefore, it is suitable for storing low-volume, high-value (LVHV) seeds. In high volume storage of seeds of wheat, paddy, etc., such facility would be uneconomical. The cold storage facility is commonly used with the dehumidifier to store the carry-over seed stock at <20 °C and <50% RH for 1–2 years, or at 15 °C and 30% RH for storage of vegetable and nucleus seeds for 3–5 years period. The dry chain approach can also be used in place of a dehumidified atmosphere storage to control insect pests.

6.1.7 Fumigants

A fumigant is a chemical in vapour or gaseous state that when released penetrates the objects or enclosed areas in such concentrations that are lethal to targeted pest organisms. This excludes aerosols, particles suspended in the air, often called smokes, fogs, or mists. The most important and useful properties of the fumigants are that these penetrate the fumigated materials, neutralising the target organisms and diffuse afterwards.

There are many chemical fumigants recommended for the control of storage insects. A list of such fumigants and their properties are described below, though many of these are now not in use in most countries due to environmental and human health hazards. Many insects have also been reported to have become resistant to some of these. Only phosphine (PH_3) was found safe for all kinds of seeds with up to four fumigations in a multilocation NSP trial (S.N. Sinha, personal communications, ICAR-IISS, Mau) (Table 3).

Table 3 Important fumigants, their dosages, and exposure period

Fumigant	Dose mL or g/m³ space	Dose mL or g/t seed	Exposure period (h)	Repetition (number)	Ovicidal toxicity
ED: CT mix.	480	740	24	2–3	Low
EDBr	32	56	24	02	Normal
CS₂	480	740	24	01	Low
MBr	32	56	12	02	High
PH₃	**02**	**03**	**5–7**	**3–4**	**Moderate**

6.1.8 Hydrogen Phosphide or Phosphine (PH3)

- Aluminium phosphide (AlP) tablets are available for specific use in the names of 'Celphos, Quickphos', etc. It weighs 3 and 1 g of Pellets (used against rodents) and liberates 1/3 phosphine gas of its weight.
- Ammonium carbonate, ammonium bicarbonate, urea, and paraffin are also added. The chemical reaction is.
- $AlP + 2NH_4 OC (O) NH_2 + 3H_2O = \uparrow PH_3 + Al (OH)_3 + \uparrow 4NH_3 + \uparrow 2CO_2$
- CO_2 suppresses the flammability of PH_3 while diffusing from the tablet in the presence of moisture. Ammonia is a warning gas, and it reduces fire hazards.
- Aluminium phosphide produces a carbide or garlic-type odour. It is heavier than air and has low water solubility. It is highly inflammable per se, a safe and convenient method to evolve gas.
- The larvae and adults succumb more easily. In comparison, the eggs and pupae are usually the hardest to kill. The tolerance of eggs and pupae can be overcome by relatively long (10-day) exposure periods.
- Phosphine does not affect the germination of cereal seeds, and legumes with one or two fumigations at comparatively high concentrations. Up to four repeated applications showed no adverse effect on the viability and vigour of different crop seeds in a multilocation trial under the National Seed Project in India (S.N. Sinha, personal communication).

Thus, Phosphine is an effective fumigant for controlling storage insect pests which may be used if permitted by the concerned regulation.

7 Pre- and Post-Harvest Strategies for Disease Management

Storability of seed and incidence of diseases during storage are much influenced by the health status and quality of the seed at the time of harvest. Seed production, therefore, should be planned in safe areas and in appropriate seasons where the occurrence of major seed-borne diseases is known to be minimal or nil. Pre-harvest sprayings with suitable pesticides or bio-control agents and harvesting the crop at the proper maturity stage also help maintain the seed quality during storage. Exposure of the seed crop to biotic and abiotic stresses, especially during seed maturation, influences seed vigour and longevity (Siti et al. 2019).

Discolouration of seed due to pre-harvest rains and the occurrence of diseases reduces the planting and market value of grain seeds, which are mostly sold without any coating or colouring. Therefore, in regions where diseases or discolouration are expected, a pre-harvest prophylactic spray may be applied. Govindrajan and Kannaiyan (1982) observed a reduction in grain discolouration of rice through pre-harvest spraying with copper oxychloride. Seed discolouration in paddy was also reported to increase with higher doses of nitrogen and phosphorus, whereas wider spacing in the field resulted in lesser discoloration (Mishra and Dharam Vir 1991). According to Deka et al. (1996), application of Maneb at the boot leaf stage, followed by a spray with common salt, was effective in reducing the discolouration in paddy grains. However, in some crops, mainly soybean, peanut, and other legumes, the darkening of the seed coat is reported to be indicative of oxidative reactions and not associated with disease (Marzke et al. 1976; Siao et al. 1980).

The association of fungi is reported to be more in the seed crops grown and harvested in the wet season. A higher percentage of storage fungi is generally observed in samples obtained from areas with high humidity (Indira and Rao 1968). There are significant variation in the propensity of storage fungi on seeds of different species and varieties, due to their chemical constituents, presence of alkaloids and antifungal substances in the seed coat, etc. (Misra and Kanaujia 1973). Nair (1982) reported less number of fungi on seeds of *Luffa acutangula* during storage, as these absorb less moisture because of their thick and hard seed coat. Varietal differences concerning the susceptibility to fungal attack during storage have also been observed by Sheeba and Ahmed (1994), who recorded higher fungal incidence on seeds of fertiliser-responsive, high-yielding paddy cultivars than the traditional varieties.

Initial seed moisture, storage temperature, and RH play important roles in maintaining seed health and germination. Hence, harvesting the seed at the right stage of maturity is the first and most crucial step in maintaining its successive storability. As mechanical injuries promote invasion of pests and pathogens, care needs to be taken during harvesting/ threshing and all other post-harvest handlings to minimise any physical damage to the seed. Pre-storage seed treatments also help improve the storability of seed by protecting the seed from microbial attacks during storage and in the field upon sowing, thereby resulting in better seed germination and field stand. Lal et al. (1976) reported that the application of propionic acid and potassium metabisulphite was effective against *A. niger, A. flavus, Penicillium oxalicum,* and *Alternaria alternata* on wheat and maize grains, whereas acetic acid and propionic acid were effective against *A. flavus* and *Curvularia lunata* on groundnut kernels. According to Vaidya and Dharam Vir (1986; 1987), sodium metabisulphite and propionic acid checked the growth of *Aspergillus* and *Penicillium* spp. on groundnut kernels. Besides ensuring the seed quality at the time of storage, it is also important to maintain sanitation of the seed godowns by keeping it clean, sanitised, dry, cool, free from cracks and crevices, and adequately ventilated. The seed material should be packed in clean containers. Gunny (hessian) and cloth bags are sometimes reused for packing grain seeds in bulk. These must be disinfected, or appropriately fumigated to avoid contamination by the carry-over pathogens. The seed stores should be regularly checked for the development of any pests, and efficient remedial measures, such as fumigating, must be employed immediately to keep these under control. Disease and pest management of stored seeds, thus, require optimum storage conditions and deployment of such treatments (of seeds, godowns, and bags) which do not pose any health hazards to the seed handlers and users. (See chapter 'Seed Health: Testing and Management').

Planting material of horticultural crops, such as stems, roots, leaves, tubers, corms, rhizomes, suckers, grafts, etc., may carry many pathogens due to their high moisture content. These may cause diseases reducing their planting value. The pathogens present in the soil may also hamper the field establishment of such propagules. Adoption of an appropriate seed treatment technology can help reduce most of these problems with greater efficacy in disease control, less environmental pollution, low health hazards for the operators, and lesser use of workforce compared to later stage spraying, besides reducing the wastage of the chemicals.

8   Long-Term Seed Preservation

Long-term seed storage is most commonly used for conserving plant germplasm for future use and to maintain the wealth of biodiversity. For this, seeds are preserved at very low MC and low to very low temperatures in the Genebanks. Genebanks play a vital role in the conservation, availability, and use of plant genetic diversity for crop improvement. Maintaining viability, genetic integrity, and quality of stored seed samples and making them available for use even after decades of storage is the primary objective of genebanks (FAO 2014). Hence, seeds are dried to attain a glassy state (See chapter 'Seed Longevity and Deterioration' for more details on glassy state) and stored at low to sub-freezing temperatures to preserve germination for prolonged periods. Stocks are regenerated when germination falls below the stipulated standards. Hay and Whitehouse (2017) suggest that instead of planning regeneration of seed stocks based on initial germination, these may be based on seed storage experiments to identify which seed lots to test first and use sequential testing schemes to reduce the number of seeds used for viability testing, besides using tolerance tables. Different methods can be used for seed drying, such as equilibrating in a dehumidified atmosphere by storing seed with desiccants or using a dehumidifier chamber. The methods chosen will depend on the available equipment, number and size of the samples to be dried, local climatic conditions, and cost considerations. However, there is a limit to which drying can increase longevity. For long-term storage (Cromarty et al. 1982), therefore, seed samples should be dried to an equilibrium of 10–25% RH in a controlled environment of 5–20 °C, so the MC of seeds reaches <5% and are sealed in a suitable airtight container. These are stored at -18 ± 3 °C and relative humidity of 15 ± 3% for long-term storage of Base Collection (usually stored for up to 50 years). For longest-term storage, seeds in hermetically sealed containers are stored in liquid nitrogen canisters. For medium-term storage (10–15 years), samples are stored at 0–10 °C. Working collections can also be stored for 3–5 years at 5–10 °C. These collections are used for evaluation, multiplication, and distribution of the accessions for use. Active collections are usually maintained by multiplying the seeds of their accessions or periodic regeneration of the base collection. However, significant variations can be seen in longevity of seeds of different species maintained under similar conditions of long term storage (Walters et al. 2005).

Ultra-drying reduces seed moisture to 1–3% using desiccated forced air drying, heated drying, or freeze-drying. Kong and Zhang (1998) demonstrated that there was practically no difference in longevity when seeds were dried over silica gel by freeze-drying or heating to 50 °C, as long as the seeds were not over-dried below 1.5%. However, freeze-drying and heating treatments were more advantageous than drying over silica gel due to the faster speed at which seeds could be dried (10 times faster).

Acknowledgement The authors thank Dr. Kent J Bradford, Distinguished Professor and Director of the Seed Biotechnology Center, UC Davis, USA, for reviewing the chapter and providing valuable input.

References

Afzal I, Bakhtavar MA, Ishfaq M, Sagheer M, Baributsa D (2017) Maintaining dryness during storage contributes to higher maize seed quality. J Stored Prod Res 72:49–53. https://doi.org/10.1016/j.jspr.2017.04.001

Agrawal PK (1982) Viability of stored seeds and magnitude of seed storage in India. Seed Tech News 12(1):47

Anonymous (2013) Indian minimum seed certification standards. The Central Seed Certification Board, Department of Agriculture and Cooperation, Min. of Agriculture, GOI, p 569

Beninci A, Binoti MB, Floris C, Gennai D, Innocenti AM (1984) Ageing in Triticum durum wheat seeds. Early storage in carbon dioxide prolongs longevity. Environ Exp Bot 24:159–165

Bilgrami KS, Sinha KK (1983) Aflatoxin contamination in agricultural commodities with reference to maize and groundnut. Phytopathol. 73:253–263

Bockholt AJ, Rogers JS, Richmond TR (1969) Effects of various storage conditions on longevity of cotton, corn and sorghum seeds. Crop Sci 9:151–153

Bradford, K. J. & Asbrouck, J. (2011). Desiccant beads for efficient seed drying and storage. In 10th ISSS conference on "seed science in the 20th century", April 10–15, Salvador, Brazil

Bradford KJ, Dahal P, Van Asbrouck J, Kunusoth K, Bello P, Thompson J, Felicia W (2018) The dry chain: reducing post harvest losses and improving food safety in humid climates. Trends Food Sci Technol 71:84–93

Brooker DB, Bakker-Arkema FW, Hall CW (1992) Drying and storage of seeds and oilseeds. Van Nostrand Reinhold, NY

Champ BR (1985) Occurrence of resistance to pesticide in grain storage pests. In: Champ BR, Highly E (eds) Pesticides and humid tropical grain storage pests, vol 14. Australian Centre for Intl. Agric. Res., Proceedings, pp 229–255, cited in Salunke, B.K., Prakash, K., Vishwakarma, K.S. et al. (1985). Plant metabolites: an alternative and sustainable approach towards post harvest pest management in pulses. Physiol Mol Biol Plants 15, 185 (2009). https://doi.org/10.1007/s12298-009-0023-9

Chauhan Y, Ghaffar MA (2002) Solar heating of seeds - A low cost method to control bruchid (Callosobruchus spp.) attack during storage of pigeonpea. J Stored Prod Res 38:87–91

Chen C (2001) Factors which affect equilibrium relative humidity of agricultural products. Trans Am Soc Agric Eng 43(3):673–683

Copeland LO, McDonald MB (2001) Principles of seed science and technology, 4th edn. Kluwer Academic, Hingham, MA

Cromarty AS, Ellis RH, Roberts EH (1982) The design of seed storage facilities for genetic conservation. International Board for Plant Genetic Resources, Rome

Crowe JH et al (1992) Anhydrobiosis. Annu Rev Plant Physiol 54(1):579–599

Dadlani M, Mathur P, Gupta A (2016) Community seed banks. Agric World 2:6–9. https://issuu.com/krishijagran/docs/krishi_jagran_agriculture_world-jan

De Bruin T, Navarro S, Villers P, Wagh A (2012) Worldwide use of hermetic storage for the preservation of agricultural products. In: Navarro S, Banks HJ, Jayas DS, Bell CH, Noyes RT, Ferizli AG, Emekci M, Isikber AA, Alagusundaram K (eds) Proc 9th. Int. Conf. on Controlled Atmosphere and Fumigation in Stored Products, Antalya,Turkey. 15 – 19 October 2012. ARBER Professional Congress Services, Turkey, pp 450–458

Deka B, Ali MS, Chandra KC (1996) Management of grain discoloration of rice. Indian J Mycol Plant Pathol 26:105–106

Desai BB, Kotecha PM, Salunkhe DK (1997) Seeds Handbook. Marcel Dekker, Inc., New York, pp 531–545

Dharam Vir (1974). Study of some problems associated with post-harvest fungal spoilage of seeds and grains, In: S.P. Raychaudhury and J.P. Verma (eds), Current Trends in Plant Pathology, pp. 296-304, Botany Department, Lucknow University, Lucknow.

Doijode SD (1993) Influence of partial vacuum on the storability of chilli (Capsicum annum) seeds under ambient conditions. Seed Res Special 1:288–293

Dube S, Shukla HS, Tripathi SC (1988) Changes in sugar and protein content of wheat due to Aspergili. Indian Phytopath. 41:633–635

Ellis RH, Hong TD (2007) Seed longevity—moisture content relationships in hermetic and open storage. Seed Sci Technol 35(2):423–431

Ellis RH (2019) Temporal patterns of seed quality development, decline, and timing of maximum quality during seed development and maturation. Seed Sci Res 29(2):135–142

Ellis RH, Hong TD, Roberts EH (1989) A comparison of the low-moisture-content limit to the logarithmic relation between seed moisture and longevity in twelve species. Ann Bot 63(6):601–611

FAO (2014). Genebank standards for plant genetic resources for food and agriculture. Rev. ed. Rome

Fields PG, Muir WE (1996) Physical control. In: Subramanyam B, Hagstrum DW (eds) Integrated management of insects in stored products. Marcel and Decker, New York, pp 195–221

Francis O, Ogu E, Ikehi M (2015) Use of neem and garlic dried plant powders for controlling some stored grains pests. Egypt J Biol Pest Control 25:507–512

Govindrajan K, Kannaiyan S (1982) Fungicidal control of grain infection. Int Rice Res News 7:1

Grabe DF, Isley D (1969) Seed storage in moisture resistant packages. Seed World 104(2):4

Groot SPC, de Groot L, Kodde J, van Treuren R (2015) Prolonging the longevity of ex-situ conserved seeds by storage under anoxia. Plant Genet Resour 13:18–26

Gupta A, Aneja KR (2004) Seed deterioration in soybean [Glycine max (L.) Merrill] cultivars during storage-physiological attributes. Seed Res 32(1):26–32

Harrington JF (1960) Drying, storing and packaging seed to maintain germination and vigor. Seedsmen Digest 11(1):16

Harrington JF (1972) Seed storage and longevity. In: Kozlowski TT (ed) Seed biology, vol 3. Academic Press, NY, pp 145–240

Harrington JF (1973) Biochemical basis of seed longevity. Seed Sci Technol 1:453–461

Hay FR, Whitehouse KJ (2017) Rethinking the approach to viability monitoring in seed genebanks. Conserv Physiol 5:cox009. https://doi.org/10.1093/conphys/cox1009

Hong TD, Ellis RH, Astley D, Pinnegar AE, Groot SPC, Kraak HL (2005) Survival and vigour of ultra-dry seeds after ten years of hermetic storage. Seed Sci Technol 33(2):449–460

Indira K, Rao JG (1968) Storage fungi in rice in India. Kavaka 14:67–76

Joshi S, Williamson D, Sharma M, Iyer SR (1988) Combined effect of fungi on hexoses, pentoses and phenolic contents of *Pennisetum typhoides* (Bajra) grain during storage: A biochemical estimation. Proc Nat Acad Sci India 58B:149–153

Justice OL, Bass LN (1978). Principles and practices of seed storage. USDA Agricultural Handbook, 506

Khare D, Bhale MS (2014) Seed technology. Scientific Publishers, Jodhpur, India, pp 180–192

Kong X-H, Zhang H-Y (1998) The effect of ultra-dry methods and storage on vegetable seeds. Seed Sci Res 8(1):41–45

Kunusoth K, Dahal P, Van Asbrouck JV, Bradford KJ (2012) New technology for post-harvest drying and storage of seeds. Seed Times (New Delhi) 5:33–38

Lal SP (1975) Studies on storage fungi of wheat and maize. Ph.D thesis, IARI- New Delhi

Leopold, L.O. M.B. McDonald (2001), pp. 193-230, Springer (India) Pvt. Ltd., New Delhi

Lougheed EC, Murr DP, Harney PM, Skyes JT (1976) Low pressure storage of seeds. Experientia 32:1159–1161

Malek M, Ghaderi-Far F, Torabi B, Sadeghipour H (2020) Quantification of changes in relative humidity and seed moisture contents of canola cultivars under different temperatures using hygroscopic equilibrium curve. Iranian J Seed Res 7:39–52. https://doi.org/10.29252/yujs.7. 1.39. (English Summary only)

Marzke FO, Cecil SR, Press AF, Harein PK (1976) Effects of controlled storage atmospheres on the quality, processing, and germination of peanuts. USDA ARS 114:1–12

McDonald MB, Nelson CJ (eds) (1986) Physiology of seed deterioration. Crop Science Society of America, Madison, WI

Mishra AK, Vir D (1991) Assessment of losses due to discoloration of paddy grains. I. Loss during milling. Indian J Mycol Pl Pathol 21:277–278

Misra AK, Vir D (1992) Effect of different agronomic practices in incidence of seed discolouration of paddy. Indian J Plant Pathol 22:44–48

Misra RR, Kanaujia RS (1973) Studies on certain aspects of seed borne fungi II. Seed borne fungi of certain oilseeds. Indian Phytopath 26:284–294

Mohan S, Gopalan M, Sundarababu PC, Sreenarayanan VV (1994) International J Pest Management 40:148–154

Murdock L, Margam V, Baoua I, Balfe S, Shade R (2012) Death by desiccation: effects of hermetic storage on cowpea bruchids. J Stored Prod Res 49:166–170

Nair LN (1982) Studies on mycoflora of seeds of some cucurbitaceous vegetables. J Indian Bot Soc 61:343–345

Omobowale MO, Armstrong PR, Mijinyawa Y, Igbeka JC, Maghirang EB (2016) Maize storage in termite mound clay, concrete and steel silos in the humid tropics: Comparison and effect on bacterial and fungal counts. Trans Am Soc Agric Biol Eng 59(3):1039–1048. https://doi.org/10. 13031/trans.59.11437

Padmasri A, Srinivas C, Lakshmi K, Pradeep T, Rameash K, Anuradha C, Anil B (2017) Management of rice weevil (Sitophilus oryzae L.) in Maize by botanical seed treatments. Int J Curr Microbiol App Sci 6:3543–3555. https://doi.org/10.20546/ijcmas.2017.612.412

Prasad BK, Rao RN, Narayan N, Singh AN, Rahman A, Singh SP, Daya SP, Shankar U (1990) R.N Changes in sugar and amino acid contents in stored radish seeds due to *Aspergillus flavus*. Indian Phytopath 43:457–460

Pérez-García F, González-Benito ME, Gómez-Campo C (2007) High viability recorded in ultra-dry seeds of 37 species of Brassicaceae after almost 40 years of storage. Seed Sci Technol 35:143–153

Pritchard, Hugh & Dickie, John. (2003). Predicting Seed Longevity: the use and abuse of seed viability equations; https://www.researchgate.net/publication/249968798_Predicting_Seed_Longevity_the_use_and_abuse_of_seed_viability_equations

Ranganna S (1986) Manual of analysis of fruits and vegetable products. Tata McGraw Hill Publishing Co. Ltd., New Delhi, pp 441–495

Rao RGS, Singh PM, Rai M (2006) Storability of onion seeds and effects of packaging and storage conditions on viability and vigour. Sci Hortic 110(1):1–6

Rickman JF, Aquino E (2007) Appropriate technology for maintaining grain quality in small- scale storage. In: Donahaye EJ, Navarro S, Bell C, Jayas D, Noyes R, Phillips TW (eds) Proc. Int. Conf. Controlled Atmosphere and Fumigation in Stored Products, Gold-Coast Australia 8-13th August 2004. FTIC Ltd. Publishing, Israel, pp 149–157

Roberts EH (1972) Storage environment and the control of viability. In: Roberts EH (ed) Viability of seeds. Chapman and Hall Ltd, Syracuse, NY, pp 14–58

Roberts EH (1973) Predicting the storage life of seeds. Seed Sci Technol 1:499–514

Sheeba T, Ahmed R (1994) Variation in seed borne mycoflora of paddy in Mangalore taluk of Karnataka. In: Singh T, Trivedi PC (eds) Vistas in seed biology. Printwell, Jaipur, pp 220–228

Siao K, Nikkuni I, Ando Y, Otsuru M, Terayuchi Y, Kito M (1980) Soybean quality changes during model storage studies. Cereal Chem 57:77–82

Siti M, Rahman A, Ellis RH (2019) Seed quality in rice is most sensitive to drought and high temperature in early seed development. Seed Sci Res 29(4):238–249. https://doi.org/10.1017/S0960258519000217

Sultana R, Kunusoth K, Amineni L, Dahal P, Bradford KJ (2021) Desiccant drying prior to hermetic storage extends viability and reduces bruchid (Callosobruchus chinensis L.) infestation of mung bean (Vigna radiata (L.) R. Wilczek) seeds. J Stored Prod Res 94:101888

TeKrony DM, Egli DB, Balles J, Tomes L, Stuckey RE (1980) Effect of harvest maturity on soybean seed quality and Phomopsis sp. seed infection. Crop Sci 24:189–193. https://doi.org/10.2135/cropsci1984.0011183X002400010045x

TeKrony DM, Grabau LJ, DeLacy M, Kane M (1996) Early planting of early-maturing soybean: effects on seed germination and Phomopsis infection. Agron J 88:428–433. https://doi.org/10.2134/agronj1996.00021962008800030011x

Thomson JR (1979) An introduction to seed technology. Leonard Hill, London

Vaidehi BK (1997) Seed mycoflora in maize- an appraisal. In: Tewari JP, Saxena G, Mittal N, Tewari I, Chamola BP (eds) New approaches in microbial ecology. Aditya Books Pvt. Ltd., New Delhi, India, p 337

Vaidya A, Vir D (1987) Efficacy of fungicides XLVII. Evaluation of antifungal chemicals for control of post-harvest spoilage of groundnut caused by Penicillium spp. Indian J Mycol Pl Pathol 17:241–244

Vaidya A, Vir D (1989) Changes in the oil in stored groundnut due to Aspergillus nigerand A. flavus. Indian Phytopathol 42:525–529

Verdier J, Leprince O, Buitink J (2020) A physiological perspective of late maturation processes and establishment of seed quality in Medicago truncatula seeds. In: de Bruijn FJ (ed) The Model Legume Medicago truncatula. John Wiley & Sons, New York, pp 44–54

Walters C, Wheeler LM, Grotenhuis JM (2005) Longevity of seeds stored in a gene bank: species characteristics. Seed Sci Res 15:1–20

Wang W, He A, Peng S, Huang J, Cui K, Nie L (2018) The effect of storage condition and duration on the deterioration of primed Rice seeds. Front Plant Sci 9:172

Role of Seed Certification in Quality Assurance

Manjit K. Misra, Adelaide Harries, and Malavika Dadlani

Abstract

High-quality seed is a *"sine qua non"* condition to achieve maximum crop productivity and good returns. The national seed sector, composed of the public and private players, plays a key role in making available high-quality seed in sufficient quantity by following its regulatory framework and adhering to the quality standards stipulated in it. The seed laws of different countries operate on the basic philosophy of encouraging breeding and variety development to meet the demand for quality seeds of many superior varieties of different crops suitable for various agro-ecologies and discourage unscrupulous practices in the seed supply system. However, the mechanism of applying the laws, operating procedures, classification, and quality standards is formulated keeping in view the seed scenario, availability of infrastructure, and trained human resources to handle the seed system. At the same time, given the fast-expanding global seed trade, it is desired for all participating countries to harmonize their national regulations with the international conventions and treaties that provide a regulatory framework to guide and oversee the interests of breeders, seed producers, and consumers globally. It is equally important that the said national regulatory frameworks facilitate farmers' access to seeds of the best quality of superior/

This chapter has been adapted from the "Reference Handbook on Seed Laws, Regulations, Policies and Systems" by the Seed Science Center at Iowa State University.

M. K. Misra (✉)
Seed Science Center, BIGMAP, GFSC, IOWA State University, Ames, IA, USA
e-mail: mkmisra@iastate.edu

A. Harries
Seed Net Inter-American Institute for Cooperation on Agriculture (IICA), Miami, FL, USA

M. Dadlani
Formerly at ICAR-Indian Agricultural Research Institute, New Delhi, India

improved varieties while promoting competitive seed markets with only the essential checks and lesser barriers to seed trade.

Keywords

Seed certification · Seed quality · Quality assurance system · International seed movement · AOSA · ISTA · OECD and Indian seed certification

1   Seed Quality Assurance

Seed quality assurance is one of the basic requirements for the development of an effective seed industry, based on the confidence that farmers place in the seed they acquire. The establishment of a quality assurance programme from field-level production to marketing/distribution provides an adequate environment for securing high-quality seeds in the marketplace.

Seed certification is the seed quality control system conducted by government-designated agencies, where seed standards are established in the national seed regulations, and checks and controls are put into place to ensure that the quality of seeds in the field and market comply with the standards. It is a process designed to secure, maintain, and make available seeds (and vegetatively propagating materials) of superior varieties so handled as to ensure desirable levels of seed quality.

Seed certification system helps in accomplishing the following objectives:

- Release, Registration, and Notification of superior varieties.
- A rapid increase in the production and availability of quality seed of newly released/registered varieties repeatedly, for a long time, on a large scale- in their original constitution.
- Maintenance of identity and purity of varieties.

The principal components of seed quality, which are central to the certification process, are:

- Variety identity/purity.
- Genetic purity.
- Physical/mechanical purity.
- Physiological status (viability, germination, vigour, and longevity/storability).
- Phytosanitary status (seed health).

A variety is defined in Article 1 of the International Convention for the Protection of New Varieties of Plants, 1961 as a plant grouping which can be defined by the expression of the characteristics resulting from a given genotype or a combination of genotypes; can be distinguished from any other plant grouping by the expression of at least one of the said characteristics; and which can be considered as a unit concerning its suitability for being propagated unchanged (UPOV website. https://www.upov.int/overview/en/variety.html).

While the variety identity refers to the official description of its characteristics, resulting from a given genotype or combination of genotypes, genetic purity refers to the proportion of plants or seeds within a population that conforms to the declared (official) description of the variety. Plants or seeds are considered to be varietal impurities (off-types) when they are different from the official description of the variety. Therefore, the genetic purity of a seed lot of a variety is evaluated by the trueness to the variety during various stages of seed production and handling/conditioning.

The physical quality, also known as physical or mechanical purity, is evaluated for the presence/absence of (a) any non-seed inert material such as soil, stones, and dust, (b) non-seed plant parts such as dried stems, leaves, or chaff, (c) seeds of other species (including weed seeds), and (d) ill-filled, undersized, poorly developed, mechanically damaged/injured seed.

The physiological quality refers to the viability, germination, vigour status, and longevity/storability of the seed after harvest and processing (conditioning) at the time of labelling, tagging, and sealing, because the physiological quality may be affected on account of because of poor harvesting, processing/conditioning, and storage conditions.

The phytosanitary quality refers to the absence of insect pests and pathogens in a seed lot that may affect crop performance including yields.

Most countries have developed seed certification regulations that suit their specific needs while in harmony with international norms, and have independent authorities for their implementation. While the system followed in the EU, UK, and other industrialized countries, following the basic OECD certification schemes, are somewhat similar, in many developing economies such as India, other SAARC nations, and the African continent, variety registration, and seed certification are not compulsory, but quality assurance of the seed producer by labelling is mandatory. The seed laws of the USA permit both voluntary variety registration and certification and a quality assurance by truth-in-labelling.

1.1 Seed Certification

Seed certification is the main instrument in the process of quality assurance. Seeds intended for domestic and international markets are controlled and inspected by official sources to guarantee consistent high quality for consumers. The purpose of seed certification is to make available quality seeds of superior/registered varieties, bred, and released for cultivation after a systematic process of evaluation. It ensures the varietal identity and genetic purity of the variety, physical purity, physiological

Fig. 1 Seed value chain

quality, and other quality parameters as per the standards set by the concerned authority under the Seed Laws. The seed certification schemes include minimum quality standards for different classes of seeds of specific crops; inspection processes in both the field (seed crop) and post-harvest stages; representative seed sampling procedures; seed testing, tagging, labelling, and sealing. The inspections carried out should ensure that there is no varietal contamination and that the variety is true to type. This is accomplished by maintaining safe isolation from possible contaminants, by both space [i.e. distance] and time, inspections of the seed crop at critical growth stages, and growing samples in pre- or post-control plots, as required, to verify and confirm that the progeny conforms to the characteristics of the said variety.

Figure 1 shows a broad outline of the activities that are essential in the seed quality assurance system from variety release and registration up to the distribution to farmers.

The actual system implemented by different countries may vary within this general framework. However, the following broad aspects are common:

- The Seed Act should include the general provisions for seed quality assurance, while the specific requirements and standards shall be included in the regulations so that changes may be easily updated. The Seed Act also defines the designation of the government authority responsible for the application of the legislation to the quality and market control.
- The scheme is valid only for registered/notified/released varieties, for which a well-defined system of testing the Distinctness, Uniformity, and Stability (DUS) and the Value for Cultivation and Use (VCU) must be in place.
- The scheme includes the Plant Variety Protection Law which should be based on the UPOV guidelines or an internationally accepted *sui generis* system of Plant Variety Protection (PVP). Its enforcement will encourage the breeding of new

varieties by public and private institutions, and safeguard the rights of the breeders as well as the farmers through the seed production/distribution systems.
- The activity of seed certification includes the filing of the formal application, verification of the seed source (generation class), field/seed crop inspection to verify conformity to the prescribed Field Standards, seed crop approval, harvest clearance, supervision during harvesting, seed lot identity allotment, seed processing/conditioning, seed sampling, seed testing to verify conformity to the prescribed Seed Standards, tagging, labelling, and sealing. When the entire activity of certification is concluded, based on the inspection reports and seed quality results, a seed certification certificate is issued for the entire seed lot with a validity period (concerning germination), and the individual seed containers constituting the seed lot are tagged, labelled, and sealed based on the regulations. The colours of the tag labels for certification are also specified in the regulations to distinguish the generations in the seed chain for the different classes of seed under different systems.

Under the seed law, the government seed authority is responsible for the process of seed certification, carrying out the official task for the purpose. For more efficient utilization of resources, the seed laws of most countries have provided for the "accreditation/authorization/licencing" of agencies or individuals to undertake certification activities, following the national framework for accreditation. In some of these countries, e.g. the USA, part or the entire seed certification scheme may be delegated to third parties, either public or private, to carry out any or all of the activities. Different countries worldwide are using various denominations for this activity, i.e. accreditation, authorization, designation, franchising, or licencing.

The process to conduct/perform seed certification requires skilled personnel with a level of understanding and knowledge of the steps involved. Therefore, there should be a continuous investment in human resource capacity building and exposure to emerging new knowledge.

1.1.1 Seed Certification System Operating in the USA, EU, and the UK

Seed certification is the process to ensure that the *genetic* identity and purity of a plant variety are maintained during multiplication from one generation to the next. The seed certification schemes rely on regulatory standards and procedures implemented at each step of the seed production process for different crops. Seeds put on the market with a label of "Certified seed", means that the seed complies with the quality standards prescribed for certification. Under the seed law, the government seed authority is responsible for the process of seed certification, meaning that they carry out the official task for seed certification.

There are different schemes for seed certification. In the USA, the initial recognition of certification and official certifying agencies was established under the US Federal Seed Act of 1939. In 1969, in the Federal Seed Act Regulations—Part 201 and under the Association of Official Seed Certification Agencies (AOSCA), the standards for land history (concerning the previous crop), field isolation from designated contaminants, and varietal purity in the field and seed stages were incorporated.

The Federal Seed Act also established the seed certification standards and regulated the interstate shipment of agriculture and horticulture seeds. The seed companies are required to use truthful label provisions, with the quality information disclosed on the packages to place seeds in the market. The national certification standards under the AOSCA seed system are used by the seed industry as the base standard, meaning that the quality will fulfil the minimum stipulated standards. However, most seed companies, to establish a market reputation, try to reach for higher standards through their stringent and more elaborate internal quality management programmes.

All the states in the USA have seed laws that are based on truth-in-labelling to provide customers with the identity and quality of the seed put into the market. For interstate seed marketing, the Federal Seed Act must be followed, and the United States Department of Agriculture-Agricultural Marketing Services (USDA-AMS) is responsible for the enforcement of the seed standards. The voluntary certification scheme in the USA requires the application of the DUS criteria, and the seed is produced under a limited generation system that ensures the varietal identity, genetic purity, and the desired levels of quality. In the US seed system, voluntary registration and certification, together with truth-in-labelling, allow the seed companies to have unrestricted market access.

In Europe, variety registration is compulsory, and seeds may only be put on the market if it is certified (except in the case of vegetables). The variety must be listed in an official catalogue, together with completed DUS and VCU testing results. The certified seed put into the market needs to have a certain level of quality, such as germination and disease-free conditions. Europe follows the OECD seed schemes; for non-European countries that want to sell seed in the international market, they must have equivalence with the EU. This signifies that the exporting country must meet the same criteria for seed quality, characteristics, examination, identification, labelling, control, and packaging. Seed sampling, fastening, and labelling of containers can be carried out under the official supervision of the competent authorities (in third countries) based on the OECD rules. All seeds sold must be registered in the National Catalogue or the European Common Catalogue of Agricultural Plant Varieties, to confirm that the variety has passed identity and purity tests. Europe conducts post-control grow-out tests with small field plots to that allow certification inspectors to verify the varietal identity and genetic purity, varietal identity, and genetic purity status of the variety.

Depending on the national seed framework, there are different criteria for accepting varieties for certification. Some countries request that the eligible varieties, to be submitted for certification, should be released and registered in the national list; others accept varieties listed in non-official lists or varieties registered for protection in the List of Protected Varieties. The important condition is the availability of the description of the variety. In general, the certification schemes around the world require that the person who applies for seed certification needs to be registered under the national seed legislation. Usually, there is an application form to be completed with basic information about the variety, such as denomination, botanical description, and characteristics. This is followed by the registration of the seed field which includes information on the location, field ownership, the size of the seed plot, and previous cropping history.

Varieties considered eligible for certification are those that have been approved by the AOSCA National Variety Board, the Plant Protection Office, the Official Seed Certifying Agency, and/or the OECD Seed Schemes. Varieties of foundation and breeder seed may be accepted for inspection if an adequate description is provided with the final certification, subject to later approval, when the breeder, owner, or agent of a variety provides the more information, such as a description of the morphological, physiological characteristics, and any other information that validate the identity of the variety.

Variety Release: Pre-condition for Certification

The variety release and registration is a process that is a mandatory requirement before seed certification. The national, regional, or international list of registered varieties provides the information on Distinctiveness, Uniformity, and Stability (DUS) of the variety, as well as its performance on Value for Cultivation and Use (VCU).

DUS testing is a system of determining whether a new variety is distinct from existing varieties within the same species (Distinctiveness), whether the characteristics are expressed uniformly in a population (Uniformity), and whether the characteristics do not change over the generations (Stability). This morphological description is mandatory for the grant of plant variety protection and is also used for the field inspection in the seed certification scheme to verify the genetic identity.

VCU testing, on the other hand, emphasizes the evaluation of the performance of a variety to be grown in pre-established agro-ecological conditions, dependent on the marketing zones decided by the breeder and/or seed enterprise, together with the use to be made of the harvested crops and the products produced from the variety. Some countries establish that the candidate variety must have superior cultivation value in comparison with the existing varieties published in the national list. The most important aspect of testing for VCU, usually, is the yield performance under a given set(s) of agro-ecological condition(s), but other characteristics must also be checked that may be different and superior from crop to crop, such as insect pest/disease resistance, nutritional value, commercial quality, adaptability, agronomic merits, etc.

The release and registration of a new variety with DUS and VCU tests provide the basis for quality assurance under the seed certification scheme, the registration in the Plant Variety Protection system, and the maintenance of the nucleus and breeder seed.

Seed Certification Phases

As mentioned above, the receipt and checking of the application is the first phase of seed certification, in which the general eligibility requirements are considered. The second phase is the verification of the seed source for compliance with generation limits, by checking the tags, labels, containers, or purchase receipts/records.

The third phase is the conduct of field inspections during critical crop growth stages to ascertain that the expected/prescribed field standards are met and to make sure that factors which can cause irreversible damage to genetic purity or seed health are not present beyond the maximum permissible levels. The certification inspector takes into account various factors:

– Isolation by both space (distance) and time from all the possible source(s) to prevent undesirable/foreign pollen from contaminating the seed crop, by complying with the recommended isolation.
– Off-types: The observation that the field/seed crop has been cleared of contaminants including plants that do not conform to the description of the variety; weeds, and diseased plants). The seed grower must undertake roguing at specific stages of crop growth to remove all undesirable plants at the right stages. The certification personnel would identify and count the number of off-types to decide accepting or reject the seed crop for harvesting and conditioning.
– The number of inspections: Usually 3–4 inspections are recommended—preliminary inspection before sowing/ planting; pre-flowering at the emergence of the flowers/inflorescence; flowering, post-flowering, and pre-harvesting when the seed has reached physiological or harvest maturity.
– Pattern to follow: Certain patterns are used when inspecting seed fields for certification to get maximum coverage while walking a minimum distance, i.e. alternate change of directions and model X with linked ends as shown below.

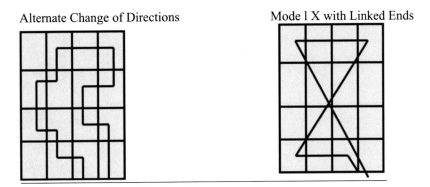

Alternate Change of Directions   Mode l X with Linked Ends

Source AOSCA

To provide some flexibility to the inspector for moving within the seed crop, some other models as shown below are also available. Depending on the field size-dimensions, crop condition, spacing, population, etc., the inspector may opt for any of the models shown, ensuring that the same plant is not counted again.

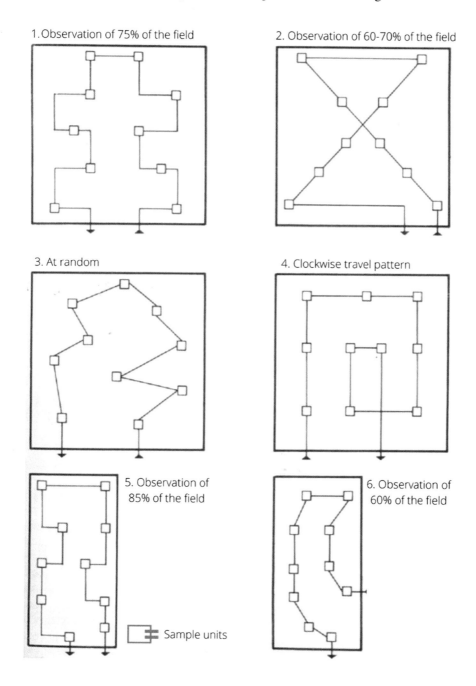

1. Observation of 75% of the field

2. Observation of 60-70% of the field

3. At random

4. Clockwise travel pattern

5. Observation of 85% of the field

6. Observation of 60% of the field

Sample units

Table 1 Equivalence classes of seed

Generation	AOSCA (USA)	OECD (EUROPE)
1	Breeder	Breeder
2	–	Pre-basic
3	Foundation	Basic
4	Registered	Certified 1st
5	Certified	Certified 2nd

OECD seed scheme recommends field inspection following the quadrat method. In this sampling procedure, a detailed examination is made of small areas of the seed crop, called "quadrats". The number and size of these areas are decided on the minimum varietal purity standards of the specific crop. For crops like wheat, barley, and oats, at least 10 quadrats of 10 m^2 (1 m × 10 m) is recommended.

In the seed certification scheme, different classes of seed are designated, such as breeder seed, foundation or basic seed, registered seed, and certified seed.

Breeder seed is outside the ambit of the certification process and is produced under the direct and personal supervision of a qualified Breeder, and it is controlled by the originating breeding institution or person. It is expected to have 100% genetic purity.

Foundation seed is the seed produced from the breeder seed, the registered seed is the class of seed produced from the foundation, and the certified seed is the seed that is the progeny of the foundation, or registered seed.

In all classes, it is necessary to follow the procedures that ensure the maintenance of genetic purity and variety identity.

In addition, the number of generations that the variety may be multiplied will be limited by the breeder or owner of the variety and will not be more than two generations beyond the foundation seed class with some exceptions that, in case of emergency, may be established.

Depending on the seed certification schemes, there are different seed classes as shown in the table below. The class names may differ between the denomination in AOSCA classes of seed and the OECD classes, but the equivalence is the same (Table 1).

Field inspections are conducted by the seed certification agency or the accredited/designated institutions, either public and/or private. Usually, two to four inspections are conducted based on the established requirements for each crop.

Field standards include:

1. Isolation, i.e. minimum distance from other varieties of the same kind/species [other species in some crops] and the same crop not meeting the varietal purity requirements for certification.
2. Off-type, i.e. the plant that deviates in one or more characteristics/features from the one described originally for the variety under seed production/certification. Off-type plants and other varieties, exceeding the standards must be rogued out and removed from the field well before they contaminate the seed crop and prior to inspection.

Table 2 Purity standards for some field crops as per OECD certification schemes

Crops	Genetic purity (% minimum) for different seed Classes		
	Basic	Certified CS1	Certified CS2
Barley, wheat, paddy, oats	99.9	99.7	99.0
Groundnut	99.7	98.0	97.0
Sunflower varieties	99.7	99.0	98.0

Under the Common Rules of OECD, Post-control is obligatory for all samples of Certified Seed, when the lot is to be used for the production of further seed generation. In this case the post-control is also a pre-control of the following generation

3. Pollen shedders (in male-sterile female parents), shedding tassels (in the female parent of maize hybrids), selfed bolls (in the female parent of cotton hybrids), selfed flowers, ears, heads, fruits (in the female parent of rice, pearl millet, sorghum, castor, sunflower, and vegetables).
4. Weed plants and other crop plants, seeds of which are likely to get mixed up with the crop seed and are difficult to remove during the post-harvest conditioning process. Some of them are designated as "Objectionable/Noxious" for certification purposes. They should be rogued out and removed from the seed field before setting seed and the inspection of the seed crop.
5. Plants affected by designated diseases, if any, for the crop under inspection.

Post-control grow-out testing is a robust mechanism for verifying/confirming the varietal purity of the seed produced under certification process most conclusively. It is mandatory for the pre-basic seed of pure line varieties, and also for the basic seed of the parental lines of hybrids, though the post-control of breeder seed is the breeder's responsibility and not that of the certification agency. The permissible number of off-types varies with the seed classes (Table 2).

Sampling: After harvest, the seed should be kept in a single lot to obtain a sample that represents the lot with a unique number assigned to the crop variety and class of seed. The entire seed lot must be made homogenous and processed/conditioned before the sample is drawn, which is then sent to the laboratory to conduct the tests for seed quality. The sample must be drawn properly, following the prescribed procedures for sampling so that it truly represents the seed lot, whether the seed is in small containers, bags, or bulk. The Composite Sample obtained by pooling together all the Primary Samples is divided into two samples, one for sending to the lab as Submitted sample and the other as Guard/Reference sample to be retained by the Sampler.

Usually, before the seed certification agency takes a sample, the seed must be packaged in containers with the seed company's identification. Also, the seed company is authorized for taking the samples and sending them to the official or accredited seed testing laboratory to check the quality. There are also the maximum seed lot sizes and the minimum Submitted and Working sample sizes established by indicated in the ISTA Rules that should be followed. Usually, for species with seed size similar to that of wheat, paddy, and barley, one lot is not more than 30,000 kg; a lot of smaller seed-sized crops such as mustard or Finger Millet (*Eleusine coracana*) will not exceed 10,000 kg, whereas, in maize and other crops having bigger seed size, a lot will be as large as up to 40,000 kg.

Seed testing includes physical purity, germination, moisture content, seed health, and any other quality test that is required by the authority or the producing agency (see chapter "Testing Seed for Quality"). If the results of the laboratory comply with the seed certification standards, the authorized body or the seed certification agency can issue one certificate for the entire seed lot, and certification tags for all the containers in the lot.

All the containers of seed to be sold as certified must be tagged and must be securely sealed in such a manner that they cannot be removed without tampering damage. The certification tag content includes the crop/variety name, class, lot number, name or number of the applicant, net weight, percent pure seed (purity), inert matter, other crop seeds, weed seeds, germination, MC, and test date. The colour of the tag/label is based on the class of seed: white for foundation seed, purple for registered, and blue for certified seed.

1.1.2 Accreditation for Seed Certification System

Under the seed law, the government seed authority is responsible for the process of seed certification. Due to the limitation of resources to comply with these activities, the seed laws of some countries have included the possibility of accreditation/ authorization/licencing of the activities of certification by accredited persons or entities following the national regulations. Accreditation is defined as "the formal recognition of technical competence to carry out official specific tasks". An accredited person or entity is defined as "a public or private body empowered by the Minister to undertake quality control and certification activities". Accreditation allows seed quality control to be performed by individuals, third parties, seed laboratories, or seed entities that shall be allowed to inspect fields, take samples, test seeds, and issue labels.

There are different examples of accreditation in the seed sector, such as the Organization for Economic Cooperation and Development (OECD) Seed Schemes, and the International Seed Testing Association (ISTA) seed testing accreditation programme. The National Seed Health System (NSHS) and the Accredited Seed Laboratories (ASL), in the USA, are also successful examples. The Economic Community of West African States (ECOWAS) Seed Regulation, the South African Seedsmen Association (SANSOR), and the Zambia Seed Certification and Control Institute (ZSCCI) are examples of accreditation at regional and national levels.

1. The OECD seed accreditation is established under the Guidelines for the Authorization of some certification activities under the OECD Seed Schemes (OECD Paris 2012). The OECD accreditation scheme allows third parties to perform certain activities necessary for seed certification on behalf of the Designated Authority. OECD has authorized accreditation for field inspection, sampling (including labelling and sealing), and testing activities.
   - The inspectors, having the required qualifications and expertise, are authorized to inspect certified seeds to carry out the tasks to take care of all the steps involved in certification. The level of check/supervision by the official authority is established at the level of at least 5% of the production, and the designated authority needs to fix penalties for the infringement of the rules.
   - The authorized seed samplers shall have technical qualifications through training courses; use approved sampling methods and equipment; be independent persons; be persons employed by a neutral or legal entity that does not involve in seed activities; or be persons employed by seed companies. The sampler employed with a seed company can take samples only on seed lots produced on behalf of his employer. The Designated Authority will conduct auditing, monitoring, and checking of random sampling in at least a proportion of 5% of the cases.
   - The authorized laboratories shall maintain the conditions required, have staff with necessary qualifications and training; be an independent laboratory; or be a laboratory belonging to a seed company. Official supervision shall be conducted by the Designated Authority through auditing, monitoring, and check analysis in at least a proportion of 5% of the samples (Organization for Economic Cooperation and Development, OECD, website 2020).
2. The ISTA accreditation programme includes the member laboratories that have proven their technical competence in carrying out seed testing following the ISTA Rules and operating an effective quality management system. To be eligible for accreditation an ISTA member laboratory needs to participate in the proficiency tests and establish a quality management system developing a quality manual, after which an audit is conducted by ISTA experts who evaluate and make the decision for the accreditation. The ISTA accreditation is a formal recognition of the technical competence of a seed lab to carry out specific tasks, for which the accredited labs need to clear periodic evaluation of their proficiency. These laboratories are authorized to issue international Seed Analysis Certificates such as Orange Certificates (for details, please see International Seed Testing Association, ISTA website 2021).
3. In seed health management, there is a well-established system for accreditation in the USA. The National Seed Health System (NSHS) is a USDA-APHIS (USDA-Animal Plant Health Inspection Service) programme administered by Iowa State University's Seed Science Center to accredit both private and public entities to perform certain activities needed to issue the federal phyto-sanitary certificates for the international movement of seed.

The activities for which entities can obtain NSHS accreditation include:

- Laboratory seed health testing: A laboratory-based programme to test for plant pathogens in seeds. A comprehensive list of approved NSHS Seed Health Testing labs and methods is published.
- Phyto-sanitary field inspection: Inspection is conducted to detect the diseased plants grown to produce seed in the field, nursery, or greenhouse.
- Seed sampling: Sampling of seeds is done as per the recommended procedure to be submitted to the laboratory for seed health testing.
- Visual inspection: visual inspection of seed shipments at exporter's facility, before issuing phyto-sanitary certificates.

  There are two Reference Manuals to support NSHS. Manual A deals with the administration, procedures, and policies of the NSHS; and Manual B with the Seed Health Testing and Phytosanitary Field Inspection Methods (Iowa State University, Seed Science Center Website 2021).

  Another example of accreditation is the programme for seed labs testing for purity and germination in the USA. This covers species contained in the Association of Official Seed Analysts (AOSA) Rules for seed testing and/or the Federal Seed Act. The eligible laboratories are members of the AOSA and the Society of Commercial Seed Technologists (SCST). These laboratories are required to develop a management system, participate in proficiency testing, and issue seed analysis reports/certificates.

4. One of the Regional Examples is the Technical Agreement of the Economic Community of West African States (ECOWAS) under the Regulation 4/05/2008 for the Harmonization of Rules on Quality Control, Certification and Marketing of Plant Seeds and Seedlings in the ECOWAS Region, Article 13: "*Seed quality control in each Member State shall be carried out by the official quality control and certification authority or any other accredited private body, following the provisions of the regulations*"

Similarly, in South Africa since 1989, SANSOR (South African Seedsmen Association), has been the designated authority (DA) to manage and execute all functions regarding seed certification on behalf of the government. This includes not only the National Seed Certification Scheme but also the international seed schemes such as AOSCA, OECD, and SADC (Southern African Development Community Seed Certification System).

In the case of Zambia, the Plant Variety and Seeds Act has a provision "for licencing seed companies as certifying agencies, and for approval of any person as an official seed inspector, sampler or tester for a certifying agency". Seed Control and Certification Institute (SCCI), Zambia's seed authority may licence any seed company or institution as a certifying agency for inspecting, sampling, or testing seed. According to the African Seed Access Index (TASAI), which monitors indicators that are essential to seed sector development at the national level, Zambia has 118 licenced seed inspectors, including 83 private seed inspectors (Mabaya, Miti, Nwale and Mugoya, Zambia Brief 2017, The African Seed Access Index (TASAI) September 2017). Zambia's Plant Variety and Seeds Act, Chapter 236, regulates seed production, control, sale, import/export, certification, and testing with quality standards.

An accreditation scheme should include different procedures to have an efficient and effective seed certification system in compliance with the national and international seed standards. The Seed Science Center of Iowa State University has developed Accreditation Procedures Manuals for different regions and national authorities (J. Cortes and A. Harries, Procedures Manuals for Seed Certification/ Accreditation, Seed Science Center-Iowa State University 2015).

The necessary accreditation system is for entities, individuals, and laboratories, where different criteria are established and in which a National Seed Authority (NSA) must be satisfied that the Third Party/Seed Entities have been sufficiently trained and are competent to carry out seed certification. The following conditions are suggested for the establishment of a national accreditation system:

- The Accredited Entity (AE) shall establish, document, implement, and maintain a quality management system that ensures that the service conforms to the requirements of National Standards. The AE must continually improve the effectiveness of the quality system.
- The AE should have a documented quality management system that describes its regulations, organization, working procedures, and standards. The AE shall establish and maintain a quality manual that includes an organization chart, the scope of the quality system, documented procedures, activities, references, and a description of the interaction and interlinkages between the procedures.
- The AE should develop and maintain documented procedures for the accredited activities, inspection, sampling, and/or testing activities to verify the specific requirements to be met by the product.
- Records shall be established and maintained to provide evidence of conformity to requirements and of the effective operation of the quality system. The required activities for field inspection, sampling, and testing, and the way to record shall be described in the quality manual.
- The Accredited Individuals (AI) should demonstrate commitment to quality service and meeting requirements of seed regulations and assure the ability to carry out the inspection services, sampling, and/or testing. The system should also include technical training, both theoretical and practical, conducted by the seed authority.
- The AI should keep records of complaints related to field inspection, and sampling, investigate the reasons and take corrective actions. He/she also shall establish and maintain documented procedures for performing, verifying, and reporting that the activities meet the specific requirements.

Similarly, in countries such as the USA, the national seed authority also undertakes accreditation of seed testing laboratories (STLs), based on the competence of the technical staff; necessary infrastructure; quality management system including a quality manual, records of maintenance, and calibration, reference materials, etc. The STL must clear the proficiency tests administrated by the national seed authority before the grant of accreditation, and periodically thereafter.

Having an accreditation scheme at the national level saves time and economic resources, as it is difficult for government inspectors to inspect/check all the sites for field control and timely completion of testing thousands of samples. For seed enterprises to have their seed inspectors and a list of accredited laboratories to conduct seed testing on time, is considered a better practice for enforcing the quality management concept in the production of certified seed.

1.2   Truth-in-Labelling/Truthfully Labelling System

This system may be considered an officially recognized methodology of self-control, through which the seed producers take care of the quality of seed during the entire process of production, processing/conditioning, storage, and marketing, following the official standards and the company's internal quality control standards. This system is used in one or the other form in different countries worldwide, such as India, the USA, Japan, Korea, and Thailand. The international seed trade of vegetable seeds, for example, is based on an effective truth-in-labelling system (ISSD Africa Synthesis Paper-Effective Seed Quality Assurance, Kit Working Papers 2017).

The truth-in-labelling system is based on two major premises:

(a) The seed company has the responsibility and control of the entire quality assurance process. This allows for a speed-up of the process and the reduction of associated costs. Another advantage is that the seed producer needs to maintain the reputation in the market, and therefore, the internal controls and corrective activities are in the best interest of the seed company. There are different opinions on this concept, but in practicality, the scheme is considered an ideal self-regulatory quality assurance system providing a seed production enterprise the opportunity to demonstrate its capacity and professional quality seed system. One disadvantage of this scheme in developing countries is the lack of speedy judicial systems for sanctions against the seed producer when they don't meet the internal quality standards.

(b) Consumers should have the freedom to choose which varieties are best for them so the responsibility for quality control is not under the seed authority. The basic condition regulated in the USA is the truth-in-labelling (Department of Agriculture Agricultural Marketing Service, 7 CFR Part 201 [Docket No. LS–02–12], Enforcement of the Varietal Labeling Provisions of the Federal Seed Act) through the Federal Seed Act that controls the labelling of seed marketed between and among states. The seed is controlled by the State Department of Agriculture with qualified inspectors who may draw samples, and submit the same to a designated STL to test and verify the variety and quality declared on the label. A similar system also prevails in countries that allow the sale and use of the Labelled seed (often referred as Truthfully Labelled seed), as in India (Prasad et al. 2017). Any violation of the labelling provisions, or the seed being sold as labelled and not meeting any of the prescribed quality standards, involves monetary penalties to the seed companies (Federal Seed Act Policy: The Federal Seed Act (FSA) (7 U.S.C. 1551–1611).

1.3   Seed Quality Assurance System in India and Other SAARC Countries

Given a strong association between quality seeds, crop yields, and production, Singh and Jain (2014) noted that developing countries like India and Bangladesh will be in greater need to produce and use high-quality seeds because of their fast-growing population, changing demographic profiles, and also the need for poverty alleviation. Realizing the role of quality seeds in bringing the Green Revolution during the 1960s and 1970s that led to achieving food security, seed legislations were introduced in India, Pakistan, Bangladesh, and Nepal, which adequately provided for seed quality assurance. The Seeds Act, 1966 (India), the Seed Act, 1976 (Pakistan), the Seeds Ordinance, 1977 (Bangladesh) (which was amended in 1997, 2005, and 2007), and the Seeds Act, 1988 (Nepal) were vital instruments introducing a system of seed quality control in these countries, which have many similar features, and recognize the needs of a large proportion of farmers who use farm-saved seed (Koladya and Awal 2018). The Indian system of seed certification, discussed below, presents a general model being followed in these countries with some minor variations.

The organized seed sector, particularly in agricultural crops, in India took its roots with the establishment of the National Seed Corporation, a Government of India undertaking in 1963. With the enactment of the Indian Seeds Act in 1966, and the Seed Rules in 1968, seed certification gained a legal status. Agriculture being a State subject all the 28 States have been given the powers to establish the State level certification agencies either under the Department of Agriculture or as an autonomous body, and govern these under the Seeds Act, 1966. The first official Seed Certification Agency as part of the Department of Agriculture was established by the state of Maharashtra in 1970, whereas Karnataka was the first state to establish an autonomous Seed Certification Agency in 1974. Currently, there are 25 State Seed Certification Agencies in India; and the Central Seed Certification Board (CSCB) at the national level takes care of the standards and procedures for uniform adoption by all the State agencies. Elaborate seed certification standards specifying the crop-wise quality norms, i.e. Field Standards and Seed Standards are prescribed by the CSCB in the Indian Minimum Seed Certification Standards (IMSCS) prescribed by the Government of India. So also the procedures involved in the various phases constituting the certification process are periodically announced by the CSCB. There are 132 notified Seed Testing Laboratories (STL) for testing the seed quality for certification/labelling or other purposes. These are technically supported and guided by a Central Seed Testing Laboratory (CSTL), a Referral Lab, established under the Department of Agriculture and Farmers Welfare, Govt. of India. There are 20 ISTA member STLs in India, of which eight laboratories (6 in private sector and 2 in public sector) are accredited with ISTA which perform seed testing as per ISTA

Rues and are entitled to issue the orange certificate for international seed trade. India is a member of OECD seed schemes. However, like in the USA and many other countries, seed certification is voluntary, while labelling is mandatory for any seed in commerce.

The number of varieties notified and released since the enactment of the Seeds Act in India is ~5300, of which nearly 2000 varieties are in the seed chain. While public research institutions are the major contributors to notified varieties, with a share of 89%, private seed companies also contribute 11% of varieties, mainly hybrids (Yadava and Chowdhury 2021). Varieties can be released by the State Seed Committee or Central Seed Committee (SSC or CSC) based on the Release Proposals presented by the research system. For bringing the released varieties under the ambit of the Act, they are notified at the central level by the CSC. The receipt of indent and monitoring of the Breeder Seed allocation and supply (for the production of Foundation and Certified Seeds) is the responsibility of the Seeds Division, Govt. of India. The steps undertaken for the production of certified seed are similar to those in other countries as discussed above. The Labelled Seed is required to meet the minimum quality parameters prescribed for Physical Purity and Germination for certified seed.

The broad differences/similarities between the Indian and OECD systems of seed certification are presented below (based on Trivedi and Gunasekaran 2015):

Parameters	Indian Seed Certification System	OECD Seed Certification Scheme
Eligible varieties	Only the varieties notified under Section (5) of the Seeds Act, 1966 of the Govt. of India can enter seed certification chain	All varieties in the National List of varieties for OECD seed schemes are eligible
Variety nomenclature	Compulsory	Compulsory
Variety maintenance	Responsibility of the breeders/breeding institutions	Responsibility of the breeders/breeding institutions
Generation system of seed multiplication	Breeder → Foundation → Certified (FS1 to FS2 or CS1 to CS2 are permissible, provided seed multiplication under Certification does not exceed four stages beyond breeder seed)	Pre-Basic → Basic → Certified1 → Certified2 (seed multiplication is permitted for two generations of Certified seed)
Nucleus seed	No specific tag; nucleus seed is maintained by the breeder/breeding institute; used for breeder seed multiplication	No specific label; controlled and maintained by the maintainer/breeder; used for pre-basic seed multiplication
Breeder seed/ pre-basic seed	• Golden yellow label • Monitored by a multi-disciplinary team of experts • Produced by the Breeder Seed Production (BSP) centres as per the indent allocated by the DAC, Govt. of India • Used for producing Foundation Seed (FS1)	• White label with diagonal violet stripes • Controlled by the Designated Authority (DA)/concerned Certification Agency and the Maintainer/breeder/institution • Produced officially by the recognized institutes/organization • Used for producing basic seed • Cannot be commercialized

(continued)

Parameters	Indian Seed Certification System	OECD Seed Certification Scheme
Foundation seed/basic seed	• White label • Monitored by the Seed Certification Agency • Used for producing Certified Seed (CS1) or FS2 seed with the necessary approval • Produced at Institutional Farms/by registered seed growers • Pre-/post-control plot tests are not mandatory • Validity of seed quality: Germination result]valid initially for 9 months from the date of testing, and subsequent validation is for 6 months	• White label • Controlled by the Designated Authority (DA)/concerned Certification Agency • Used for producing Certified Seed (CS1) • Produced officially by the designated institutions • Pre-/post-control plot tests are mandatory if used for further multiplication • No limit to the validity of seed quality
Certified Seed/ Certified 1(CS1) Seed	• Azure blue label • Produced by the registered seed growers • Monitored by the Seed Certification Agency • Can be used for CS2 seed production, provided FS1 seed was used • Validity of seed quality: Germination result is for 9 months from the date of testing for the initial certification. For the subsequent validations, the validity period is 6 months as long as the seed lot is meeting the quality standards	• Blue label for Cs1/red label for CS2 • Produced by registered seed producers • Controlled by the Designated Authority (DA)/concerned Certification Agency • No restriction for the validity of seed quality
Labelled seed/ standard seed	• Labelled seed/truthfully labelled seed: Opal green tag label • Seed is monitored by the producing organization itself • Must meet the minimum standards for physical purity and germination prescribed for the certified seed	• Standard seed: Carries dark yellow label • Seed is declared by the supplier as being true to the variety and of satisfactory varietal purity. It must conform to the appropriate conditions in the scheme
Field inspections, seed sampling, and testing	• Field inspections and seed sampling are performed by the officials of the Seed Certification Agency • Seed testing is conducted by notified STL only, including the STL of the Certification Agency	• For certified seed production, non-official, but licenced inspectors and seed samplers are also allowed • Seed testing may be conducted at non-official STLs as authorized by the DA
Grow-out test and pre- and post-control plots	• Grow-out test is compulsory only for the hybrids that are produced following a system of manual emasculation and pollination; or using chemical hybridizing agents (CHA) • A post-control plot may be compulsory for the Breeder seed of certain species, as identified	• Pre-control test is compulsory for pre-basic and basic seed • A part of every sample of basic seed and 5–10% of the certified seed shall be checked in a post-control test

The quality of certified seeds for use within the country will be regulated by the Seeds Act, 1966 and the Indian Minimum Seed Certification Standards (IMSCS) as amended from time to time, whereas that for OECD certification will be as agreed by the DA. Seed analysis of certified seed is performed by the notified laboratories, following the national Seed Testing Manual approved by the Ministry of Agriculture & Farmers' Welfare, GOI. However, ISTA Rules for Seed Testing serve as authentic reference source on the subject. OECD certification can be performed at the STLs approved by the DA, which may include notified STLs or ISTA-accredited labs.

The norms for field inspection, sampling, and seed testing could be more stringent in specific cases under the national seed regulations than under OECD, and vice versa.

1.4   Quality Declared Seed (QDS) System

The Food and Agriculture Organization of the UN (FAO) established a category of seed referred to as Quality Declared Seed (QDS), to have an alternative quality assurance system for developing countries. One major challenge in designing the QDS system was to promote flexibility in its implementation while retaining the basic principles of quality assurance (Quality Declared Seed, FAO, Rome, Paper 185/2006).

QDS is a quality management system with established standards to provide quality seed to farmers with a label and is considered an intermediate seed system. The implementation of a QDS system has the following requirements:

- A list of varieties eligible for the system should be available.
- Seed producers should be registered under the national seed authority.
- The national seed authority should check at least 10% of the fields registered to produce QDS and 10% of the seed offered on sale.
- The national seed authority should establish seed quality standards that can and must be achieved.
- The registered seed producer has the responsibility to apply for the production of QDS providing all necessary information and details of the seed to be produced and the location of the fields.
- One important premise in QDS is to produce quality seed while keeping the procedure simple, inexpensive, and not bureaucratic so that the smallholder farmers can adopt the same.

Thus, the QDS is seed produced by a registered seed producer that follows the QDS standards and requirements. The varieties for producing QDS include local varieties or landraces, varieties obtained by conventional breeding, and varieties obtained by other systems such as "participatory plant breeding". In all cases, the applicant needs to provide a statement of the origin of the variety, data of the variety, a description of the morphological characteristics, the recommended agro-ecological zone, and a statement of the procedures used to maintain the variety.

For bred varieties, a description of the morphological aspects and other characteristics of the eligible variety is required, together with the agro-ecological zone recommended for cultivation. The procedures for maintaining the variety must also be provided for QDS certification. To maintain the genetic purity for this seed category, the limitation of generations and/or the additional isolation of seed crops may be established.

The production process of QDS includes:

1. Ensuring that the seed production fields have satisfactory previous cropping and the seed used is eligible to produce QDS.
2. Ensuring that the seed crop is well grown and measures are taken, such as roguing of off-types, weeds, and diseased plants to ensure healthy crops.
3. Inspecting fields based on the standards and rejecting those that do not reach the standards.
4. Ensuring that the identity of the seed at harvest is maintained and is delivered for conditioning in identified containers.
5. Ensuring that seed conditioning is performed preserving the identity and varietal purity of the seed.
6. Securing appropriate samples of the lot and submitting them for testing to a laboratory.
7. Keeping records of all activities, inspections, test results, and completing the QDS declaration.

The QDS system helps smallholder farmers who wish to buy improved seeds but have no access to improved certified seeds from any known source. The system also allows farmers with QDS to trade with other farmers. This seed has been used as a "relief seed" in many developing countries during emergencies, natural calamities etc. when shortfalls in seed availability can occur.

Summary of a case study from Tanzania is given below, in which the production of QDS helped farmers considerably.

Quality Declared Seeds Production in Tanzania (Pearl Millet, Sorghum, and Groundnut 2016)

(Source: Eco ACT project, Global Climate Change Alliance)

This was a model where farmers that could not afford to buy certified seeds, produced QDS seeds that was distributed to over 2500 farmers in target communities. Farmers improved their capacity to manage fields for seed production and obtained quality seeds and traded 1 kg of QDS seeds for 2–3 kg of crop produced.

This project also allowed farmers to expand their business.

Summary of another case study carried out in Uganda by ISSD is given below:

QDS Filling the Gap Between Formal and Informal Seed Systems: A Case of Common Beans in Uganda 2013

Farmers involved in producing QDS in a local project produced 4% of quality seed in the region at less cost as compared with certified seed. These farmers became local seed businessmen. Another success of this project was that good quality bean seed increased yield to 670 kg/hectare, resulting in extra income for farmers and good agricultural economic development.

There are also other uses of QDS, such as for relief purposes in climate and hunger emergencies. It also serves as a reference scheme for seed supplies since national seed organizations are often unable to provide comprehensive documentation for rapid international movement. There are other potential suppliers including cooperatives, farmer groups, large private farms, and NGOs for whom QDS could provide a cost-effective entry point to seed quality assurance.

In developing countries, the intermediate seed system may include the QDS scheme that requires seed producers to conduct internal quality assurance and declare the quality of their seed based on limited quality controls established by the regulatory authorities, e.g. an inspection of 10% of the total seed produced instead of undergoing the full inspection and quality testing procedures.

QDS is not proposed as a global scheme that countries would formally recognize or adopt as a basis for trade. However, it may facilitate seed movement at the national and regional levels if no other such scheme is available. Likewise, the standards set out here may provide a basis for regional seed schemes to develop their standards according to their specific trading needs.

2   International Certification Systems for Seed Movement

2.1   Association of Official Seed Certifying Agencies (AOSCA)

The AOSCA is committed to assisting its clients in the production, identification, distribution, and promotion of certified seeds. It was established as an International Crop Improvement Association including all Seed Certifying Agencies from the USA in 45 states. Other countries such as Canada, Argentina, Brazil, Chile, Australia, New Zealand, and South Africa are members of the AOSCA and are governed by AOSCA Rules.

The purpose of the association was to establish minimum standards for genetic purity and identity for the classes of certified seed for the national and international movement of seed. Also, AOSCA cooperates with the regulatory agencies in seed regulations and procedures related to the seed movement intra-state, inter-state, and internationally (Association of Official Seed Certification Agencies, AOSCA, Web-site 2021).

The seed certification agencies from the states are members of AOSCA, applying the standards and procedures of AOSCA and the Federal Seed Act Regulations.

The programme ensures that the seed is produced, harvested, cleaned, and tested under very strict guidelines that include the following:

- The AOSCA classes of seed: Breeder seed, foundation seed, registered seed and certified seed.
- Application: The applicant must complete an application form with a tag or invoice accrediting the class of seed, variety, lot number, and grower number for the seed to be planted.
- Special field requirements: The field selected to produce certified seed must be free of noxious and restricted weeds. The field also should not have been planted in the previous season with another variety of the same crop or class of seed that could produce volunteers that may cross with the variety being planted. The field must be isolated from other varieties of the same kind and same variety not meeting the genetic purity requirements by the prescribed distances so as avoid chances of contamination by cross-pollination/mechanical admixture.
- Eligible crop varieties: Eligible crops are governed by each certifying agency. Typically, a crop is eligible if it has passed the review of one of the following review boards:
  - Plant Variety Protection Office.
  - National Certified Variety Review Board, a division of AOSCA.
  - Member agency of the AOSCA.
  - Organization for Economic Cooperation and Development (OECD) Seed Scheme.

Field inspections: Seed crops are inspected by the state to ensure that they comply with the standards. The agency may reject a field where its conditions do not allow an adequate inspection to verify the genetic identity and purity. The inspectors may also reject the fields if they are excessively weedy, have poor stand development, or the crop has disease, insect damage, or other factors that would affect the quality or genetic purity of the seed.

Harvesting: Certified seed must be harvested with equipment that is not contaminated with other crops or weed seeds.

Conditioning: Certified seed can be cleaned only in the facilities that have been approved during the inspection by the certifying agency. This is to ensure that the conditioning facility and machinery significantly improve the quality of harvested seeds.

Seed sampling and testing: A sample of the conditioned seed is typically drawn under the auspices of a state representative and tested in an officially recognized seed testing laboratory.

If the tested seed meets the minimum requirements for purity (genetic/physical) and germination that is specified by the state, it is eligible to be a certified seed. Each lot of certified seed receives tags that are affixed on the bags. The colour of the tag depends upon the generation of seed produced. The classes and tag colours are as follows:

Breeder Seed (White tag): This is the first-generation seed of the variety produced from Nucleus/ Breeder Seed. This class of seed is directly controlled by the entity releasing the variety. This class is outside the ambit of certification and forms the source for the subsequent generation, i.e. foundation seed class.

Foundation seed (White tag): This is produced from breeder seed. This class of seed is typically a second-generation seed.

Registered seed (Purple tag): This class is produced from foundation seed. It is typically a third-generation seed.

Certified Seed (Blue tag): This is produced from registered or foundation seed. This is typically a fourth-generation seed. It is the class of seed usually sold for commercial crop production purposes and not meant for further certified seed production. Blue-tagged certified seed is not eligible for further seed multiplication.

Substandard certified seed (Blue tag): A seed that has gone through the certification process but has failed the minimum mechanical purity and germination requirements, may be tagged as a substandard certified seed. Doing so is completely up to the state certifying agency and is considered on a lot-by-lot basis.

Source-identified seed (Yellow tag): This is the seed that has been harvested from natural stands or grown in field production but has not been tested for its traits. It is produced under the auspices of the state and if it meets the prescribed quality requirements it is labelled as source-identified seed.

Selected seed (Green tag): This is the class of seed that exhibits characteristics of a variety but has not been definitively proven to have traits that can be inherited by subsequent generations. It is usually material that is undergoing testing and awaiting conclusions. It may be field-produced or harvested from natural stands.

2.2   Organization for Economic Cooperation and Development (OECD) Seed Schemes

This is an international seed certification programme designed for international seed movement/trade. The programme was initiated in 1961 with the aim of facilitating a transparent international seed trade through common quality standards for seed certification. The objective is to encourage the use of certified seeds of high quality, authorize the use of labels and certificates produced for international trade according to agreed standards, and enhance cooperation between importing and exporting countries.

The main instruments of this OECD Seed Scheme are the Rules and Regulations (2022), the OECD list of varieties, and the guidelines for control plot tests and field inspection for seed crops.

The scheme is open to OECD countries as well as non-OECD countries from the United Nations (UN) and the World Trade Organization (WTO). Currently, there are 61 participating countries.

The OECD Seed Scheme is applied for those varieties that are officially recognized as distinct, uniform, and stable and having an acceptable value for cultivation and use in at least one participating country. An OECD list of eligible varieties is published annually. There are over 62,000 crop varieties that are traded internationally (Quality Seeds for World Needs, OECD Seed Schemes, 2021). For a country to issue OECD labels, it is required to register the variety in this OECD List of Varieties.

In addition, satisfactory conditions of production and conditioning of basic and certified seeds must be ensured through field inspections and post-control tests.

The OECD Rules establish quality standards for seven groups of species: (1) grasses and legumes, (2) crucifers and other oil or fibre species, (3) cereals, (4) fodder and sugar beet, (5) subterranean clover, and similar species, (6) maize, sorghum, and (7) vegetables. It also establishes common rules and regulations for seed certification.

The categories or classes of seed in this scheme of certification are the following: Pre-basic seed, basic seed, and certified seed. For the pre-basic seed, the colour of the labels is white with a diagonal violet stripe. In the case of basic seeds, the label colour is white, while for certified seeds, first generation is blue and second generation is red. For not-finally certified seeds, the label colour is grey. While all classes of OECD-certified seed lots must be accompanied by an ISTA OIC, the seed which is categorized as "not finally certified" (grey label), OIC is not required. A category known as "standard seed" refers to the seed that is declared by the supplier as being true to the variety with satisfactory variety purity. It must.

conform to the appropriate conditions in the Schemes. It carries a dark yellow label. This category mainly exists in the vegetable seed scheme (Trivedi and Gunasekaran 2015).

Seed-not-finally certified is the seed that is exported from one country of production after field inspection with labelled containers. The designated authority of the importing country must verify the final certification process with all the information and documents provided by the designated authority from the country of production.

The designated authority should ensure the availability of the description of the variety or of the parental components before the time of the field inspection. The description should be based on the international guidelines developed by the International Union for the Protection of New Varieties (UPOV).

The designated authority is responsible for ensuring that the appropriate labels are affixed to the containers.

The process of OECD seed certification includes the control of the production and carrying out field inspections to verify the varietal identity and purity. The crops standards include minimum requirements of the previous cropping, field isolation, noxious weeds present, number of inspections, species purity, varietal purity, size of the seed lot, and special conditions. One or more field inspections shall be conducted, but at least one after the emergence of the inflorescence/during flowering. The inspectors shall check the compliance with the requirements prescribed for each crop.

2.2.1   OECD Labels

After harvesting and conditioning, the seed lot must be sampled to test the quality of the seed. All activities of sampling, fastening, and labelling must be conducted by the designated authority or by authorized persons. The sample size must be enough (see chapter "Testing Seed for Quality" for details) to carry out the tests by the laboratory. The tests shall include analytical purity and germination. The designated authority shall store a sample for 1 year under appropriate conditions that ensure the maintenance of the seed quality. Based on the results of field inspections and seed testing, the designated authority will issue the OECD certificates.

In the case of certified seeds, tests of post-control are conducted to verify the purity and identity of the variety. This is a comparison between plants grown from the seed lot produced and those grown from the standard sample. This post-control test will help the designated authority check the efficiency of the seed certification process for the verification of the maintenance of varietal purity and identity. The post-control test shall be conducted by the designated authority or under their supervision who will define the percentage of post-control of the certified seed. Generally, that level is between 5 and 10%. OECD has developed guidelines for control plots and field inspection of seed crops (Guidelines-control plot and field inspection, OECD Seed Scheme 2019). The guidelines include methods and techniques that help to determine varietal purity and identity at different stages of seed production.

2.3   International Seed Testing Association (ISTA)

ISTA was founded in 1924 to develop, adopt, and publish standard procedures for sampling and seed testing with the primary purpose of promoting uniform use and application of the methods for testing of seed that is moving in the international trade. ISTA is an independent non-profit organization supported by the cooperation of seed scientists and analysts.

The ISTA's vision is "uniformity in seed quality evaluation worldwide". ISTA plays an important role also in seed testing at national and regional levels by publishing the ISTA Rules that are globally available, annually updated, and harmonized with uniform seed testing methods. It promotes the application of uniform procedures for the evaluation of seeds intended for the market. Presently it has 226 member laboratories in 82 countries of which 136 are ISTA-accredited laboratories (International Seed Testing Association, Website 2021). This has been accomplished through the publication of the International Rules for Seed Testing, a laboratory accreditation system, the ISTA international certificates (orange and blue), and the knowledge of science and technology. Following ISTA rules for seed testing at the national and international level facilitates seed trade, ensures the quality of seed available to the farmers, and contributes to food security.

Twenty technical committees of ISTA work on seed testing issues and are integrated by more than 200 technologists and scientists around the globe. The technical committees are:

Advanced Technologies Committee; Bulking and Sampling Committee; Flower Seed Testing Committee; Forest Tree and Shrub Seed Committee; Germination Committee; GMO Committee; Moisture Committee; Nomenclature Committee; Proficiency Test Committee; Purity Committee; Rules Committee; Seed Health Committee; Statistics Committee; Seed Storage Committee; Tetrazolium Committee; Variety Committee; Vigor Committee; Editorial Board (SST); Seed Science Advisory Group; and Wild Species Working Group.

The ISTA Rules describe the principles and definitions of the standard methodologies, techniques, and procedures for seed sampling, testing, and reporting of results. The quality tests included are heterogeneity, physical purity, other seed determination, germination, moisture content, seed viability, vigour, seed health, varietal/cultivar purity, and detection of genetically modified organisms in the seed. The ISTA rules are developed and supported by experts that develop and validate methods for each component of seed testing. ISTA also publishes different handbooks on specific aspects of seed testing such as Germination; Seedling Evaluation; Seed Health Testing; Flower Seed Testing; Forest Tree and Shrub Seed Testing; Moisture; Nomenclature; Purity; Laboratory Equipment; Statistics; Tetrazolium; Variety Testing, Vigor, and Tolerances. These are widely used by researchers and seed analysts globally.

ISTA has an accreditation programme with a quality management system that includes the quality documentation developed by the laboratory based on the ISTA accreditation standards. When the applicant provides the quality manual to the technical department, ISTA auditors conduct the on-site assessment regarding staff, facilities, seed sampling, and seed testing. If the auditors identify any non-conformity, it must be addressed with a formal corrective action procedure. If the audit approves the quality system, then the accreditation approval procedure is initiated through the Executive Committee. Once it is approved, the lab will receive a certificate of accreditation and authorization to issue international certificates. The proficiency tests are compulsory and the Quality Management System is audited every 2 years to maintain its accreditation.

Laboratories that are accredited by ISTA are entitled to issue the "international certificates", called "Orange Certificate", provided by ISTA as a seed quality passport for international seed trade. ISTA accreditation is a formal recognition of the laboratory's technical capacity to carry out seed tests that are repeatable and reproducible in any laboratory around the world. Therefore, countries with laboratories accredited by ISTA ensure their capability to issue seed certificates for the international movement of seeds.

ISTA also supports advances in seed research by publishing original papers and articles on various aspects of seed science and technology, namely seed quality, physiology, production, harvest, processing/conditioning, sampling, testing, storage, packaging, treatment, genetic conservation, habitat regeneration, distribution, etc. in "Seed Science and Technology" (SST), an international journal; and articles on advancements of seed testing in Seed Testing International (STI), a news bulletin.

2.4   International Plant Protection Convention (IPPC)

Phytosanitary measures are handled nationally by the National Plant Protection Organization (NPPO) of each country. The International Plant Protection Convention (IPPC) is a multilateral treaty commissioned in 1951 that promotes effective actions to prevent and control the introduction and spread of pests of plants and plant products (International Plant Protection Convention, IPPC, Website FAO 2021).

IPPC allows countries to evaluate the risks to their national plant resources and to use science-based measures for the safety of their cultivated and wild plants. Under this principle, IPPC is protecting the farmers, biodiversity, ecosystems, industry, and consumers from pests and diseases. The active participation of member states as parties of the IPPC is very important as the states are part of developing international standards that help to protect the movement/import and export of commodities including seeds. The effective implementation of the Convention is based on exchanging technical and official phytosanitary information among the member states.

Specific information, such as the NPPO contact information and description; phytosanitary restrictions and legislation; entry/exit points; list of regulated/objectionable weeds, pests, diseases, organisms, and objects; emergency actions; non-compliance; pest status; the rationale for phytosanitary requirements; and pest-free areas are published by the member states.

As defined in the International Standards for Phytosanitary Measures, in the context of the international movement of seeds (ISPM38), a seed-borne pest is carried by seeds externally or internally that may or may not be transmitted to plants growing from the seeds, causing their infection/infestation. Whereas, a seed-transmitted pest is a seed-borne pest that is transmitted via seeds directly to plants grown from these seeds, causing their infection/ infestation. Seed can be a pathway for the introduction and dissemination of pests in a new geographic area through seed trade. Safeguarding seed health is critical in avoiding the spread of pests. It is of crucial importance to provide national plant protection organizations (NPPOs) with an updated, scientifically-evidenced list of seed-associated pests.

The International Standards for Phytosanitary Measures (ISPM) are standards adopted by the Commission on Phytosanitary Measures to protect sustainable agriculture, facilitate trade development, protect the environment, and enhance food security. The ISPM on Seed, ISPM #38, was adopted by the Commission in 2017 and provides guidance to assist NPPOs in identifying, assessing, and managing the pest risk associated with the international movement of seeds. ISPM #38 also provides guidance on the procedures to establish phytosanitary requirements to facilitate the seed movement internationally. Its guidelines also include inspections, sampling, and testing of seeds and the phytosanitary certification of seeds for export and re-export of commercial seed, as well as the seed used for research, breeding, and multiplication.

Pest Risk Analysis (PRA) is a tool promoted by IPPC to evaluate the risk of pest introduction and dissemination. For seed imports, the PRA will depend on the purpose or scope of the seed importation. In the case of seed imported to conduct a laboratory test, it may not be necessary to conduct a PRA if the seed is to be destroyed during or after the laboratory test. In the case of imported seed for planting under restrictive conditions such as germplasm or seed to be used as breeding material, an NPPO may require relevant phytosanitary measures. If the seed is to be imported for planting in the field, the NPPO in the importing country may require a pest risk assessment depending on the country/zone of production. In the case of importing seeds of a mixture of different species or varieties, testing and inspections may be done on the components or the mixture or blend to be certified. One advantage of ISPM #38 is the establishment of a system approach as a preventive control or risk reduction in the entire seed production process, from breeding materials to commercial sales. Some of these seed chain components require risk analysis, and others are measures for mitigation of that risk.

Another element to keep in mind is the new technologies applied to breeding programmes where seed varieties with pest resistance are being developed. This may allow, for example, the NPPO of the importing country to consider the use of resistant varieties as an appropriate phytosanitary measure.

Seed treatments with pesticides or disinfectants and physical or biological treatments also can be applied to eliminate any infection/infestation. For imported seed, the importing country may require the "post-entry quarantine" in cases where the pest is difficult to detect when the expression of symptoms takes time, or there are no other phytosanitary measures.

In addition, specific requirements for inspection, sampling, and testing of seeds are established for phytosanitary seed certification. Inspections may be conducted on the seed consignment or in the field production or both. However, sometimes if the visual examination is not enough, it must be combined with laboratory seed health testing. In the case of using laboratory testing for pest detection, NPPO should ensure the use of internationally validated tests that follow diagnostic protocols for regulated pests.

The adoption of ISPM#38 provides a new systematic approach to establishing phytosanitary requirements for the import, inspection, sampling, and testing of seed that facilitates the international movement of seed. The advantage of a system approach is that preventive controls or risk reduction measures are put in place during the entire seed supply chain process.

At ISTA, the Seed Health Committee (SHC) is focusing on seed-borne pests (bacteria, fungi, oomycetes, viruses, nematodes) in more than 40 non-vegetable species from 21 botanical families of spermatophytes (seed plants), including cereals, legumes, oleaginous crops, forest trees, and fruit trees. It also includes a list of pests that were included in the last version of the Annotated list of seed-borne diseases, but that lack evidence of seed-borne status in the scientific literature. These pests will be regularly checked for information on their ability to not only be seed-borne but also to transmit to the progeny. The ISTA Reference Pest List will be updated regularly each time a list will be finalized.

The International Seed Federation (ISF) has prepared a training manual that provides information on the elements of the ISPM #38. It also describes the needs of the NPPO to address and implement this phytosanitary measure. At the same time,

ISF has developed a regulated pest risk database based on scientific evidence (International Seed Federation, ISF Regulated Pest List. PESTLIST. WORLDSEED.ORG 2021). For more information, see the ISF website (International Seed Federation, ISF, Movement of Seed, APSA Congress, Bangkok, June 2017). Finally, it is necessary that the countries design and use the system approaches for phytosanitary certification of seeds for the international movement.

Thus, an appropriate system of quality assurance, either through certification or labelling, is an essential instrument to ensure quality seed in the production chain both for the domestic and international markets.

Acknowledgements We acknowledge valuable inputs received from Dr. V. Sankaran, Former GM (QC), National Seed Corporation, India, and Mr. B.S. Gupta, former Director, Rajasthan State Seed & Organic Certification Agency, India.

References

International Seed Federation (ISF). https://worldseed.org/

Integrated Seed Sector Development (ISSD) (2017) Africa synthesis paper-effective seed quality assurance, Kit Working Papers. https://www.wur.nl/en/project/issd_uganda.htm

International Movement of Seed, ISPM #38, IPPC, FAO, July 2021. https://www.ippc.int/en/core-activities/standards-setting/ispms

International Plant Protection Convention. https://www.ippc.int/en/

ISF Regulated Pest List (2021). pestlist.worldseed.org

Koladya DE, Awal MA (2018) Seed industry and seed policy reforms in Bangladesh: impacts and implications. Int Food Agribus Manage Rev 21(7):989. https://doi.org/10.22434/IFAMR2017.0061

Mabaya M, Miti F, Nwale W, Mugoya M (2017) Zambia Brief 2017. The African Seed Index (TASAI), Sept 2017. https://tasai.org/

Prasad SR, Chauhan JS, Sripathy KV (2017) An overview of national and international seed quality assurance systems and strategies for energizing seed production chain in India. Indian J Agric Sci 87(3):287–300

Singh SP, Jain N (2014) Harmonisation of seed certification processes in Bangladesh and India. Discussion paper. Cuts International, pp 1–12

Trivedi RK, Gunasekaran M (2015) OECD varietal certification in India. National Designated Authority, OECD Seed Schemes, Government of India, Ministry of Agriculture, Department of Agriculture & Cooperation, New Delhi, p 452

The International Union for the Protection of New Varieties of Plants (UPOV). https://www.upov.int/overview/en/variety.html

Yadava DK, Chowdhury PR (2021) Indian seed industry: current trends and future perspectives. Lecture delivered at International Capacity Building Training Programme on Seed Production, under the aegis of Indo-German Cooperation on Seed Sector Development, 12 Nov 2021

Testing Seed for Quality

S. Rajendra Prasad

Abstract

The value of seed depends on its quality. Hence, evaluation of seed quality is of critical value in any seed production system. It is, therefore, desired that seed quality is tested for all essential parameters, following the standard procedures and performing the tests in such a manner that the results are consistent and are reproducible, within the permissible limits of tolerance. As the submitted sample is only a minute fraction of the whole seed lot, reproducibility and reliability of results will greatly depend on the precision of sampling. Hence, the accuracy in sampling and precision in testing the vital parameters of seed quality form the essential components of quality assurance process. Though the seed testing procedures are essentially based on the ISTA Rules, each country may modify these to suit its requirement, crop- and variety-specific characteristics, and the available resources without compromising on the accuracy in seed quality evaluation. The essential components of seed quality and standard procedures to evaluate the same are discussed.

Keywords

Seed testing · Sampling · Purity analysis · Germination testing · Seed moisture testing · TZ testing

S. Rajendra Prasad (✉)
Formerly at University of Agricultural Sciences, Bangalore, India

M. Dadlani, D. K. Yadava (eds.), *Seed Science and Technology*,
https://doi.org/10.1007/978-981-19-5888-5_13

1   Introduction

Seed production is a complex process that requires best agronomic practices, technical competence of the seed grower, regular supervision and rigorous monitoring of the seed crop following the crop-specific norms, and care to avoid admixtures and impurity during harvesting and processing. The seed thus produced, therefore, needs to be tested reliably and precisely for all such components that determine its planting value upon sowing.

Seed quality is determined or influenced by many components (Fougereux 2000). They are as follows:

- Genetic quality—the seed should be true to its type because the genetic potential of a superior genotype can only be realized from a uniform, homogeneous population.
- Physiological quality—high germination and vigour ensure successful seedling emergence and crop establishment in the field.
- Physical quality—good-quality seed that is free from contamination by other crop seeds; inert matter; common, noxious, or parasitic weed seeds; mechanical damage; discolouration; and undersize and underweight seeds results in good agronomic performance.
- Seed health status—the absence of infection/infestation with seed-borne pests (fungi, bacteria, viruses, nematodes, insects, etc.) helps in getting a healthy crop.

Therefore, to ensure seed quality, a set of procedures are followed to test its various components, which aims to achieve 'uniformity in seed quality evaluation worldwide'. The International Seed Testing Association (ISTA) develops, adopts, and publishes internationally agreed standard procedures (Rules) for sampling and testing the quality of seeds with the vision of attaining uniformity in seed testing internationally. This basic objective is apparent in the ISTA's monogram which depicts a balance with the motto 'Uniformity in Seed Testing'. Adoption of these procedures for sampling and testing promotes a uniform application of these for the evaluation of seed quality moving in international trade. Hence, the national seed quality assurance system in most countries adopts ISTA procedures for seed sampling and testing, with little or no change. All the procedures discussed here are, therefore, based on ISTA procedures.

2   Seed Sampling

An accurately performed seed testing is of little value unless the actual sample examined is a true representative of the seed lot as a whole. As per ISTA (2022), the objective of seed sampling is to obtain a truly representative sample of a size suitable for tests, in which the probability of constituents being present is similar to that of the seed lot. The probability of a constituent being present is determined only by its level of occurrence in the seed lot. Homogeneity of the seed lot is of great significance in ensuring uniform distribution of the constituents, thereby resulting in true representative samples. Therefore, achieving and maintaining seed lot homogeneity is an important step in seed lot handling.

Seed lot—The term seed lot represents/refers to the quantity of seeds, which is physically identifiable. For seed quality testing, the maximum size of a lot will be 10 tonnes (10,000 kg) for small seeds, e.g. rapeseed mustard; 20 tonnes (20,000 kg) to 30 tonnes (30,000 kg) for medium seeds, e.g. wheat, barley, and paddy; and 40 tonnes for bigger/bold seeds, e.g. maize. The importance of sampling can be recognized from the fact that a seed lot of 10,000 kg of a small-seeded species, e.g. *Lolium multiflorum*, with a thousand seed weight of ~2.0 g may contain 5,000,000,000 seeds, out of which, as per OECD seed schemes, only 60 g or 30,000 seeds are examined for other seed counts (1:167,000); 6 g or 3000 seeds are examined for purity analysis (1:1670,000); and only 400 seeds are examined for germination testing (1:12,500,000). Therefore, the objective of sampling should be to minimize the chances of error to the best extent possible and to draw a sample that is a true representative of the lot. In cases of doubt, a heterogeneity test is to be performed.

Primary sample—It refers to a small portion taken from the seed lot during one single sampling action.

Composite sample—The primary samples drawn from different points are combined to form a composite sample of the lot. This sample is generally much larger than the size actually required for analysis and forms the base for submitted sample.

Submitted sample—This is the sample submitted to the seed testing laboratory and is derived from the composite sample by reducing it to the size prescribed. The size of this sample is specific to each crop.

Duplicate sample—An identical sample, as the submitted sample, obtained from the same composite sample and marked as 'duplicate sample'; usually retained in the location where the lot was sampled.

Working sample—This is the whole of the submitted sample or sample derived from the submitted sample that is used for analysis in the seed testing laboratory. The size of this sample is specific to each crop. Minimum weights of working samples are based on the principle that a working sample for purity analysis by weight basis should be of such that it contains at least 2500 seeds; and for the analysis by count of weed seeds and other crops seeds by number, it should be about ten times the weight of the working sample for verifying physical purity by weight. However, in both cases, the maximum weight of the sample will not be more than 1000 g.

2.1   Seed Sampling Procedure

- Under the seed law enforcement programmes, only trained and experienced officials are authorized to undertake sampling and, after giving prior notice, should draw three representative samples:
  - One sample to be delivered to the person from whom it has been taken.
  - Second sample to be sent for analysis to the notified/authorized/accredited seed testing lab (STL).
  - The third sample to be retained for any legal proceedings.
- Samples are drawn using (a) triers, (b) pneumatic seed samplers, or (c) automatic seed samplers or simply with hands.
- In most countries following the ISTA procedures, seeds are sampled from unopened containers or the seed stream, either before or when it enters the containers. The seed lot must be so arranged that each part of the seed lot is conveniently accessible. All the primary samples drawn randomly from the seed lot should approximately be of equal size (Tables 1 and 2). Seed mat/seed tape sampling should be done by taking packets or pieces. All sampling apparatus must be kept clean to avoid cross-contamination (ISTA 2022).
- If the primary samples appear to be uniform, they are combined to form the composite sample. If found to be obviously heterogenous, sampling must be refused or stopped. The seed lot has to be made homogenous by reprocessing so that the constituents get uniformly distributed with in the lot.
- The composite sample may be submitted directly to the STL for the test to be conducted if it is of appropriate size or if it is difficult to mix and reduce the composite sample properly under warehouse conditions.

Table 1 Minimum sampling intensity for seed lots in containers holding up to and including 100 kg seed (as per ISTA 2022)

Number of containers	Minimum number of primary sample
1–4	3 primary samples from each container
5–8	2 primary samples from each container
9–15	1 primary sample from each container
16–30	15 primary samples, 1 each from 15 different containers
31–59	20 primary samples, 1 each from 20 different containers
60 or more	30 primary samples, 1 each from 30 different containers

Table 2 Minimum number of primary samples to be taken from seed lots in containers of >100 kg seed

Seed lot size	Minimum number of primary sample
Up to 500 kg	At least 5 primary samples
501–3000 kg	One primary sample for each 300 kg, but not less than 5
3001–20,000 kg	One primary sample for each 500 kg, but not less than 10
20,001 kg and above	One primary sample for each 700 kg, but not less than 40

- To prevent any damage during transit and handling, submitted samples must be packed properly. Samples submitted for moisture testing must be packed with care in moisture-impervious containers.

3 Seed Testing

3.1 Physical Purity

An important criterion for quality seed is to be free from undesirable substances. Purity is an expression of how 'clean' the seed lot is, which not only improves the marketability but also gives better planting value to the cultivator. It refers to the different components of purity, viz. pure seeds, other crop seeds, weed seeds [both normal and objectionable/obnoxious], other distinguishable varieties/variants, and inert matter.

3.1.1 Components of Physical Purity Analysis
1. *Pure seed:* Refers to:
    (a) Seeds of the kind/species stated by the sender or found to be predominant in the sample; it includes intact seeds of the stated species as well as dead, shrivelled, diseased, immature, and pre-germinated seeds.
    (b) Achenes and similar fruits, such as samara with or without perianth, regardless of whether they contain a true seed unless it is apparent that no seed is contained.
    (c) Fractions of broken seeds, achenes, mechanically/insect damaged seeds, etc., which are more than half of the original size.
2. *Other crop seeds*: Refer to seeds of plants that are of crops, other than the main crop.
3. *Inert matter:* Includes seed units and all other matter (including dust, soil particles, plant parts pieces, etc.) and structures not defined as pure seed, weed seed, or other crop seed.
4. *Weed seeds:* Seeds of a species that are recognized as weeds by law/general usage.
5. *Other distinguishable varieties/variant*: Seeds of the main crop species, which could be distinguishable as other variant (ODV).

3.1.2 Objectives of Physical Purity Analysis
- To determine the percentage composition of the components by weight of the sample being tested and by inference to know the composition of the seed lot.
- To identify seeds of various species and inert particles in the sample.

3.1.3 Procedure of the Purity Test
Purity analysis will be performed on the working sample. The working sample shall be of a weight estimated to contain at least 2500 seed units or not less than the weight indicated by the national rules or ISTA Rules.

1. Before starting the separation, clean the working board and purity dishes, and examine the working sample to decide about the use of any particular aid such as a sieves or blower for making the separation.
2. After preliminary separation with the aid of blower or sieves, place and spread the retained or heavier portion on the purity work board. With the help of a forceps or spatula, drag the working sample into a thin row of seeds, and examine each particle individually. The criteria used are the external appearance of the seed in terms of size, shape, gloss, colour, structure, surface texture, etc. and/or appearance under transmitter light.
3. Impurities such as other crop seeds, weed seeds, and inert matter are removed and placed separately in purity dishes, leaving only the pure seed on the purity board.
4. Seeds enclosed in fruits, other than those that do not fit in the definition of pure seeds, should be separated, and the detached empty fruit/appendages are classified as inert matter.
5. Pure seeds should be collected in the sample pan, place lighter portion on the work board, and examine under magnification for further separating into the respective classes (weed seed, other crop seed, and inert matter). After separation, identify the weed seed and other crop seed to the best extent possible, and record their names on the analysis card. The kind of inert matter present in the sample should also be identified and recorded. The weight of each component, viz. pure seed, weed seed, other crop seed, and inert matter, should be recorded in grams to the number of decimal places as given below, and ODVs are counted and reported in number per kg in analysis card:

Weight of the working sample (g)	No. of decimal places for recording the weight of each component	Example
1	2	3
Less than 1	4	5.6789
1–9.990	3	56.789
10–99.99	2	567.89
100–999.99	1	5678.9
1000	0	5678

Note: The values given in column 3 above do not have any relation with the values given in column 1. They are given just to show the decimal pattern using the five digits 56,789

6. During purity analysis, each 'pure' seed fraction from the working sample is separated from the inert matter and other seeds. Purity (%) is calculated as shown below:

$$\% \text{ purity component} = \frac{\text{Weight of the purity component}}{\text{Total weight of the working sample}} \times 100$$

Different components are separated and weighed; and their percentage by weight is calculated and reported.

7. As per the Indian system of certification, the ODVs should be separated, counted, and recorded as number/kg.

3.1.4 Reporting of Results

The results of the purity analysis should be recorded and reported as percentages by weight up to a single decimal point. The percentages of all components should add up to 100%. Components less than 0.05% are reported as 'trace'. If the result for a particular component is nil, that must be reported as 0.0. The results should be reported in the analysis certificate in the space provided for the particular components. Some components such as the seeds of noxious weeds, seeds infected with designated seed-borne diseases, etc. are also to be reported by number. In some national seed rules, e.g. India, the number of ODVs per kg is to be reported separately in some crops, e.g. paddy, soybean, etc.

3.1.5 Advanced Equipment for Seed Purity Testing

Ergonomic Vision System for Purity Testing—Purity analysis is a tedious and time-consuming activity requiring long hours of accurate examination of the seed samples and differentiating each fraction precisely. Purity separations have traditionally been performed using diaphanoscopes and wooden work boards equipped with a drawer or pan in front and armrests on either side. Now some ergonomically designed purity apparatus such as ERGO has been developed that helps in purity analysis by reducing physical drudgery. Such purity testing boards enable analysts to work comfortably increasing the accuracy and productivity. The basic station is supplied with a contoured front, independently adjustable rest for both right and left arms, easy access to controls, two storage drawers, white laminated plywood work surface, substantial plywood base with rigid stainless risers, and non-slip, vibration damping, closed-cell rubber base pad (Fig. 1).

Videometer: As the purity analysis relies solely on the inferences made based on visual examination and identification, some seed equipment manufacturers have come out with purity instruments using imaging and AI. Imaging techniques, NIR spectroscopy, or precise remote sensors could be combined with conventional

Fig. 1 ERGO Vision System for purity testing (Source: seedburo.com)

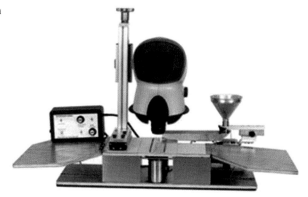

Fig. 2 Videometer

Fig. 3 Example of images generated using the Videometer Lab 2 system for (**a**) 100% *T. durum* wheat grains; (**b**) 100% *T. aestivum* wheat grains; and (**c**) 10% adulteration of *T. durum* wheat grains with *T. aestivum* wheat grains

methods for better results. A combination of spectral imaging, thermal imaging, fluorescence imaging, X-ray imaging, and magnetic resonance imaging offers reliable alternatives to the traditional methods (Li et al. 2014). One such instrument, Videometer (Fig. 2, (Source: https://analytik.co.uk/)), could be used in purity analysis to find out ODVs (Wilkes et al. 2016), inert matter, and weed seeds (Fig. 3).

These instruments show the potential to improve seed purity analysis by increasing the accuracy and reducing the examination time and drudgery, but are yet to be validated and recommended by ISTA.

4   Seed Moisture Determination

The moisture content, which is the amount of water in the seed and is usually expressed in percentage on a wet weight basis, is associated with almost every aspect of seeds and their physiological functions, including maturity, longevity and vigour, and injury due to heat, insects, and pathogens (Elias et al. 2012).

Factors that influence the determination of moisture content during seed quality analysis are (a) seed constituents, particularly volatile components; (b) seed size, which mainly affects the speed of moisture exchange; and (c) seed coat permeability, thickness, and constitution, which might hamper the exchange of moisture. The objective is to measure moisture at the time of sampling in the seed; hence, for moisture testing, seed should be packed in moisture-impervious container. Care should be taken while testing to see that seed does not absorb or exude moisture.

4.1   Methods of Moisture Determination

The principle used in moisture determination is to ensure the removal of seed moisture as much as possible giving no scope for oxidation, the loss of other volatile substances, or decomposition. Tang et al. (2000) determined the moisture of corn seed using three methods. Out of DICKEY-John moisture meter, microwave (intact seed as well as ground seeds), and oven-drying (intact seed as well as ground seeds) methods, the oven-drying method using ground seed at 130 °C for 4 h was the most accurate, whereas the DICKEY-John moisture meter readings were more inconsistent. The microwave (intact) method showed a more variable relationship (lower r^2 values) between weight loss and seed moisture, whereas the microwave (ground) method revealed that it can provide a fast ($\sim$60 min) estimate for accurately measuring high-moisture corn seed.

Two types of methods can be employed to determine seed moisture. These are:
Destructive/hot air oven method
Low constant temperature method (103 °C for 17 h).
High constant temperature method (130 °C for 2–3 h).

Non-destructive/quick method
Moisture meters are based on indirect quantification of water content, commonly
    based on electrical resistance/conductance of the moisture available in the seed.

4.1.1   Destructive/Hot Air Oven Method
The procedure recommended by the ISTA is followed by most seed testing labs globally. The seed moisture is removed by drying the seed sample at a specified temperature for a specified duration depending upon the crop (Table 3). The instruments required are:

- Constant temperature electrical oven.
- Analytical balance.
- Grinder.
- Weighing bottles.

The oven must be electrically heated and should be capable of being maintained in such a way that the temperature of the oven set at 103/130 °C is regained in less than 30 min of opening the door and placing samples inside. The temperature inside should be uniform and not fluctuate during operation. The balance must be capable to weigh with an accuracy of at least ±0.001 g. The containers/weighing bottles

Table 3 Description of seed moisture determination of some important crop species by hot air oven method (ISTA 2019)

Species	Common name	Grinding	Drying [h]	Oven temperature	Pre-drying to bring the seed moisture content to
1	2	3	4	5	6
Abelmoschus esculentus	Okra	NR	3	High	–
Allium spp.	Onion, garlic, shallot, leek	NR	17	Low	–
Arachis hypogaea	Groundnut	Cut	17	Low	17% or less
Avena sativa	Oat	Fine	2	High	17% or less
Beta vulgaris	Beetroot	NR	1	High	–
Brassica spp.	Cabbages	NR	17	Low	–
Capsicum spp.	Sweet pepper, chilli	NR	17	Low	–
Cicer arietinum	Chickpea	Coarse	1	High	17% or less
Citrullus	Watermelon	NR	1	High	–
Cucumis spp.	Cucumber	NR	1	High	–
Cucurbita spp.	Squash, pumpkin, zucchini, gourds	NR	1	High	–
Dactylis glomerata	Orchardgrass	NR	1	High	–
Daucus carota	Carrot	NR	1	High	–
Glycine max	Soybeans	Coarse	17	Low	12% or less
Gossypium spp.	Cotton	Fine	17	Low	17% or less
Helianthus annuus	Sunflower	NR	17	Low	–
Hordeum vulgare	Barley	Fine	2	High	17% or less
Lactuca sativa	Lettuce	NR	1	High	–
Lycopersicon esculentum	Tomato	NR	1	High	–
Oryza sativa	Paddy	Fine	2	High	13% or less
Phaseolus spp.	Beans	Coarse	1	High	17% or less
Pisum sativum	Pea	Coarse	1	High	17% or less
Ricinus communis	Castor	Cut	17	Low	–
Vigna spp.	Black gram, cowpea, green gram	Coarse	1	High	17% or less
Zea mays	Maize	Fine	4	High	17% or less

NR not recommended

Fig. 4 Instruments for seed moisture determination (Source: A- theinstrumentguru.com/moisture-measurement; B- directindustry.com)

should be made of glass or non-corrosive metal to give a mass per unit area of not more than 0.3 g/cm^2. Wire sieves with meshes of 0.50, 1.00, 2.00, and 4.00 mm and a desiccator are also required (Fig. 4).

Procedure

- The submitted sample shall be accepted for moisture determination only if it is sealed in a vapour-proof container with minimum air space. Two replicates/ duplicates or two independently drawn working samples should be used for moisture estimations. The samples are weighed based on the diameter of the containers used; if the diameter is >5 cm and <8 cm, then a 4.5 ± 0.5 g sample would be taken, whereas, if the diameter is >8 cm, then 10.0 ± 1.0 g sample is used.

- *Grinding*: Large seeds and seeds with seed coats that impede water loss from the seeds must be ground before drying unless their high oil content makes them difficult to grind or (particularly in seed such as Linum with oil of a high iodine number) liable to gain in weight through oxidation of the ground material. If grinding is not possible, splitting/cutting is also permitted as per the ISTA Rules. For the species which require fine grinding, at least 50% of the ground material should pass through the sieve of 0.50 mm, and not more than 10% should remain on meshes of 1.0 mm. For species requiring coarse grinding, at least 50% of the ground material should pass through the sieve of 4.0 mm, and not more than 55% should pass through a sieve of 2.0 mm. The grinding process must not exceed more than 2 min.

- *Pre-drying:* In species where grinding is required and moisture is high (see Table 3), pre-drying is recommended.

- Containers are weighed with their lids before (M1) and after filling (M2) with the prescribed quantity of seed. After weighing, containers are covered with the lids to prevent loss of sample or possible contamination. Ovens can be preheated and set at the desired temperature, or the drying period is set to begin at the time the oven reaches the required temperature.

- The container and its cover are weighed before and after seed filling. During drying in the oven, the lid is removed to facilitate moisture exit. After the prescribed period of drying, in a desiccator, the container is covered and placed to cool at ambient temperature. The container with its cover and contents is weighed after cooling (M3).

4.1.2 Calculation and Expression of Results

The moisture content as percentage by weight is calculated to three decimal places by using the formula below:

$$\text{Percent moisture content(Wb)} \frac{\text{Loss of weight}}{\text{Initial weight}} \times 100 = \frac{M2 - M3}{M2 - M1} \times 100$$

where:

M1 = weight of the container and its lid in grams to a minimum of three decimals.
M2 = weight of the container, its lid, and seeds before drying (in grams to a minimum of three decimals).
M3 = weight of the container, its lid, and seeds after drying in (grams to a minimum of three decimals).

The moisture content is calculated from the results obtained for the first (pre-drying) and second stages of the procedure if the material is pre-dried. The moisture lost in the first stage S1, and the moisture lost in the second stage S2, each is calculated as above and expressed as a percentage, then the original/actual moisture content is calculated by using the formula

$$\{S1 + S2\} - \frac{\{S1 \times S2\}}{100}$$

The result is the arithmetic mean of the replicate/duplicate seed moisture content for a given seed sample. As per the ISTA Rules, the maximum difference of 0.2% is accepted between the two replicates. If the difference between two replicates exceeds 0.2%, the seed moisture determination has to be repeated. As it is difficult to meet the replicate difference of seed moisture up to 0.2% in tree or shrub species, the maximal limit of 0.3–2.5% is accepted between two replicates for seed moisture in these species.

4.1.3 Determination of Moisture Content by Non-destructive/Quick Method

These estimations are not accurate, but convenient and quick in use. These are frequently used in seed processing plants, stores, etc. for on-the-spot information. These deduce the seed water content by measuring other physical parameters like electrical resistance or electrical conductivity of the moisture present in the seed. These values are transformed into seed moisture content with the help of species-specific charts, calibrated against the standard air-oven method or basic reference method. Some of the moisture meters in use are capacitance meters and conductance meters. All moisture meters need to be periodically calibrated and cross-checked before use with the values obtained by the oven-drying method for the specific species.

4.2   Reporting of Results

The seed moisture content test should be reported in the analysis certificate/report to the nearest 0.1% as per the ISTA Rules, which are followed by most countries. If the seed moisture content is determined using any moisture meter, the brand name and type of the equipment have to be mentioned on the analysis certificate, along with the range for which the moisture meter is calibrated.

The standard procedures of seed moisture determination for some important field crops and vegetables as recommended by ISTA (2019) and followed in most countries are presented in Table 3.

5   Germination

Germination is the sum of physiological processes culminating in the emergence of the embryo from its enclosing coverings, including the endosperm, perisperm, testa, or pericarp (Bewley et al. 2013). The metabolic processes in the non-dormant and viable seed are activated with the absorption of water, subsequently leading to expansion of the embryo and protrusion of the radicle. Seed germination is a complex process that involves several biochemical, physiological, and morphological changes (see chapter "Seed Dormancy and Regulation of Germination" for more details). Three conditions must be fulfilled for germination to be initiated. These are as follows: a) the seed should be viable, i.e. the embryo should be alive and have the capacity to germinate; b) the seed must be non-dormant; and c) favourable conditions in terms of moisture, temperature, air (O_2), and light [in some species] must be available in required amounts. When the above conditions are met, the quiescent embryo resumes growth, commencing the process of germination. The activation of metabolic machinery of seed embryos is the first and foremost step initiating the seed germination process. However, from the seed technology perspective, the development of healthy and normal seedlings is critical, in determining the planting value of a seed lot.

5.1   Seed Germination Stages

The process of germination involves several successive and overlapping events. These are i) imbibition/absorption of water; ii) cell enlargement and cell division initiation; iii) enhanced enzymatic activity; iv) food translocation from cotyledons/ endosperm to the growing embryo; v) increase in respiration and assimilation; vi) increase in cell division and enlargement; and vii) cell differentiation into tissue and organs resulting into a seedling.

5.2   Types of Seed Germination

Radicle emerges from the base of the embryo axis and is the growing point of the root, and the plumule, the growing point of shoot, is at the upper end of the embryo axis, above the cotyledons. Above the cotyledons, a section of seedling stem is called the epicotyl and below the cotyledons is called the hypocotyl. In cultivated plants, two types of germination are commonly found, viz. epigeal and hypogeal germination.

5.2.1   Epigeal Germination

Germination in dicots in which the cotyledons emerge above the soil surface. Here, the cotyledons rise above the ground surface due to the elongation of the hypocotyl; it is called *epigeous or epigeal germination* (Fig. 5). This type of germination is very common in beans, gourds, castor, tamarind, onion, etc.

5.2.2   Hypogeal Germination

Seed germination in dicots where the cotyledons remain below the soil surface. Here, the epicotyl elongates, and the hypocotyl does not raise the cotyledons above

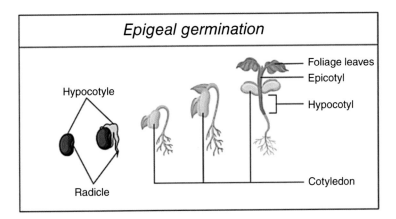

Fig. 5 Epigeal germination of beans seed (Source: https://www.vedantu.com/)

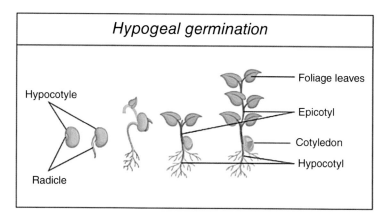

Fig. 6 Hypogeal germination of pea seed (Source: https://www.vedantu.com/)

the ground, which is called hypogeous or hypogeal germination (Fig. 6). It is commonly seen in mango, custard apple, pea, gram, lotus, maize, etc.

5.3 Germination Testing

The main objective of seed germination testing in laboratory is to evaluate the seed quality and to predict the seedling performance in the field. The ultimate aim of testing the germination in the seed testing laboratory is to determine the planting value of the seed sample and by inference the quality of the seed lot. Germination tests shall be conducted with the pure seed fraction. A minimum of 400 seeds are required in 4 replicates of 100 seeds each or 8 replicates of 50 seeds each or 16 replicates of 25 seeds each depending on the size of the seed and size of the containers of the substrata. The test is conducted under optimum conditions of moisture, prescribed temperature, suitable substratum, and light, if necessary. Pre-treatment to the seed, mostly to release some degree of dormancy, is given for those crops as recommended by ISTA.

5.3.1 Essential Equipment Required to Conduct Germination Test

(a) Seed Germinator

Seed germinators are useful for germination tests for maintaining the specific conditions of temperature, relative humidity, aeration, and light. The seed germinators are generally of two types: (1) cabinet germinator (single and double chamber) and (2) walk-in germinator. Like all seed testing equipment, it must be calibrated regularly to ensure that the temperatures are precisely maintained.

The cabinet-type seed germinators are preferred in a laboratory where a comparatively smaller number of samples of different kinds of seeds, requiring different conditions, are being tested at a time. The number of germinators required by the seed testing laboratory will depend mainly on the number of seed samples and the species being analysed. Whereas, laboratories that handle a large number of seed samples and maintain fewer (two to three) sets of temperature conditions, the walk-in germinators are preferred. For conducting germination tests in sand media, occupying more space, such germinators are more useful.

Double chamber germinator   Walk in germinator

Source: Dr. S. R. Prasad, STR unit, NSP, UAS, Bangalore. Unpublished

(b) Counting Devices

As germination tests are conducted on a fixed number of seeds per replication (e.g. 25, 50, 100), counting devices are used for quick and precise drawing of seeds. The counting devices include the automatic seed counter, seed counting boards, and vacuum seed counter. These devices not only increase the efficiency of testing by minimizing the time spent on counting and planting/putting the seeds but also provide proper spacing of the seed on the germination substratum. Counting boards are suitable for medium- and bold-sized seeds, while a vacuum counter can be used for small-sized seeds. In the absence of counting devices, the work may be accomplished manually.

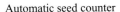

Automatic seed counter   Seed counting boards   Vacuum seed counter

Source: Dr. S. R. Prasad, STR unit, NSP, UAS, Bangalore. Unpublished

New computer-based devices are being developed for accurate visualization of germinating seeds and rapid evaluation of the germination ability of a seed lot using X-ray techniques, nuclear magnetic resonance (NMR), micro-imaging, multispectral analysis, image analysis, etc. These techniques include capturing seed digital images and their processing; discerning and measuring shape, size, and colour components; and establishing their modification and connectivity over germination time. If such techniques are employed, periodical cross-check with the traditional method is worth. However, these techniques are yet to be validated, and though many seed companies use such methods for their internal quality control, these are not yet accepted for official quality assurance and certification purposes.

5.3.2 Substrate/Media for Seed Germination

The basic supplies required for conducting germination tests include germination paper (Crepe Kraft paper or towel paper, sunlit filter paper, and blotter paper) and sand. In some cases, soil may be used in place of sand.

(a) Paper

In case of tests which require top-of-paper (TP), between-paper (BP), or pleated-paper (PP) methods, paper is used as the substratum. Paper towel method (rolled towel test—BP method) is commonly used for seeds of medium and bold sized, whereas the pleated-paper method is more common in the case of small seeds. Germination paper must be cellulosic with excellent moisture absorption/retention capacities and capillary rise potential [i.e. -30 mm/min], have neutral pH and good bursting strength [i.e. 2.0 kg/cm^2], and must be free from phytotoxicity. The paper substratum used once in the germination test is not reusable. Organic substrates and fibres such as peat, coconut, or wood fibres, with a recommended size of less than 5 mm, can also be used as germination media for certain spp. (see the ISTA Rules for details).

(b) Sand Substratum

Sand as the substrate for germination has the advantage of being reusable and less expensive. Especially in the case of seed lots that are heavily treated with chemicals, the results obtained in sand media are more accurate and reproducible compared to 'roll towel' tests. The sand (normally quartz/white sand is used) should be uniform and free from very large and small particles. It should be free from toxic substances, and the pH should be within the range of 6.0–7.5. To achieve this, the sand is washed, sterilized, and graded with a sieve size of 0.8 mm diameter (upper sieve) and 0.05 mm diameter (bottom sieve). Only the sand retained on the bottom sieve should be used.

Testing of Substratum: Before accepting the supplies in the laboratory, the germination paper substratum should be tested for various properties, viz. phytotoxicity, capillary rise, moisture-holding capacity, pH, and bursting strength. Periodic checks of the quality of the germination medium should also be conducted. To assess the phytotoxicity of the substrate, bioassay method can be adopted by germinating seeds of brassica, onion, chillies, or berseem which are relatively more sensitive to toxic substances resulting in seedling abnormalities.

5.3.3 Germination Test Conditions

(a) Moisture and aeration: The moisture requirements of the seed will vary according to its kind. Small-seeded species with thin seed coats require less water than the large-seeded species, with thick seed coats. Throughout the germination period, the substratum must be kept moist. However, care needs to be taken that the substratum should not be too wet as excessive moisture restricts aeration and may cause rotting of the seedlings or the development of watery seedlings.

(b) Temperature: The temperature required for germination is specific to the kind of crop or species. This varies with the age of seeds and within the species. The prescribed temperatures provided in the Rules for Seed Testing (ISTA 2022) for agricultural and horticultural seed germination can be broadly classified into two groups:

  • Constant temperatures (specific temperature during the entire germination period).
  • Alternate temperatures (in some crops, seeds are maintained at lower temperatures for 12 h [during 6 pm to 6 am] and at higher temperatures for 12 h [from 6 am to 6 pm]).

(c) Light: Seeds of most species can germinate in both light and darkness, but it is always better to illuminate the tests for the proper growth of the seedlings. For the germination of seeds that require light, cool fluorescent light is illuminated for at least 8 h in every 24 h cycle.

Due to physiological dormancy, hard seededness [mechanical/physical dormancy], or the presence of inhibitory substances, several hard seeds or fresh (ungerminated) seeds may remain at the end of the germination test. In such cases, retesting may be done with one or a combination of dormancy-breaking treatments. Dormancy-breaking treatments, e.g. mechanical scarification—scrubbing, KNO_3, GA_3, etc., may be applied as pre-germination treatments or used during germination.

5.3.4 Evaluation of Germination Test

The evaluation of the germination tests needs to be done on the day of the final count, which varies according to the kind of seed. Of course, if/when needed, 'first count' can be taken before the 'final count' on the day specified for 'first count'. This helps not only in concluding the test ahead of the 'final count day' but also in providing an idea of the 'vigour status' of the seed. The vigorous seeds will show up as normal seedlings in the first count itself. The seed analyst may terminate the germination test before the prescribed final count day or extend the test beyond the period, depending on the actual seedling growth. Usually, the first and second counts are taken in the case of top-of-paper (TP) and between-paper (BP) media. Only normal and dead seeds (which are a source of infection) are removed and recorded at the first and subsequent counts. While evaluating the germination test, the seedlings and seeds are categorized into normal seedlings, abnormal seedlings, dead seeds, fresh-ungerminated seeds, and hard seeds. In some species, it may also be necessary to remove the seed coat and separate the cotyledons to examine the plumule where essential structures are still enclosed at the end of the test (Table 4).

(a) *Normal Seedlings*: Though in a strict botanical context a sprouted seed is considered to have germinated, for seed testing, germination (%) is calculated based on normal seedlings only. To achieve uniformity in evaluating normal seedlings, one of the following definitions must be followed:
   - Seedlings that show the capacity to develop into a normal plant when grown in good-quality soil and under favourable conditions of water supply, temperature, and light.
   - Seedlings that possess all the essential structures mentioned below when tested on artificial substrata:
     – A well-developed root system (including a primary root) except in the case of plants (e.g. in species of Gramineae) which normally produce seminal roots, of which there still are at least two.
     – A well-developed and intact hypocotyl with no damage to the conducting tissues.
     – An intact plumule with a well-developed green leaf, which is within or emerging through the coleoptile or an intact epicotyl with a normal plumular bud.
     – In monocotyledons, one cotyledon for seedlings and in dicotyledons two cotyledons for seedlings.
   - Seedlings with slight defects provided they show an overall vigorous and balanced development of other essential structures as mentioned below:
     – Seedlings of crops such as *Pisum, Vicia, Phaseolus, Lupinus, Vigna, Glycine, Gossypium, Zea*, and all species of Cucurbitaceae, with primary root damage, but with several sufficient length and vigorous secondary roots to support the seedlings in soil.
     – Superficial damage to the seedlings or hypocotyl, epicotyl, or cotyledons decay limited to one area and no effect to the conducting tissues.
     – Dicotyledons seedlings with only one cotyledon.
   - Epigeal germination in seedlings of tree species when the radicle is four times the length of the seed provided all structures that have developed appear normal.
   - Seedlings damaged by bacteria or fungi, but only when it is apparent that the seed parent is not the source of infection and it can be determined that all the essential structures were present.

Table 4 Germination methods as recommended by ISTA (2022) for some important field and horticulture crop seeds

Species	Substrate	Temperature (°C)	First count [days]	Final count [days]	Recommendations for breaking dormancy
1	2	3	4	5	6
Abelmoschus esculentus (okra)	TP; BP; S	20–30	4	21	–
Allium cepa (onion)	TP; BP; S	20–15	6	12	Prechill
Arachis hypogaea (groundnut)	BP; S	20–30; 25	5	10	Remove shells; preheat at 40 ± 2 °C
Avena sativa (oats)	BP; S	20	5	10	Preheat at 30–35 °C; prechill
Beta vulgaris (beet root)	TP; BP; S	20–30; 15–25; 20	4	14	Prewash (multigerm, 2 h; genetic monogerm, 4 h); dry at max. 25 °C
Brassica juncea (mustard)	TP	20–30; 20	5	7	KNO_3; prechill
Brassica napus (rape seed)	BP; TP	20–30; 20	5	7	KNO_3; prechill
Brassica oleracea (cabbage)	BP; TP	20–30; 20	5	10	KNO_3; prechill
Cajanus cajan (red gram)	BP; S	20–30; 25	4	10	–
Capsicum spp. (chilli)	TP; BP; S	20–30	7	14	KNO_3
Cicer arietinum (Bengal gram)	BP; S	20–30; 20	5	8	–
Corchorus spp. (jute)	TP; BP	30	3	5	–
Crotalaria juncea (sunhemp)	BP; S	20–30	4	10	–
Cucumis melo (muskmelon)	BP; S	20–30; 25	4	8	–
Cucumis sativus (cucumber)	TP; BP; S	20–30; 25	4	8	–
Cucurbita maxima (winter squash)	BP; S	20–30; 25	4	8	–
Cucurbita moschata (pumpkin)	BP; S	20–30; 25	4	8	–
Cucurbita pepo (summer squash)	BP; S	20–30; 25	4	8	–
Daucus carota (carrot)	TP; BP	20–30; 20	7	14	–
Glycine max (soybean)	BP; S	20–30; 25	5	8	–
Gossypium spp. (cotton)	BP; S	20–30; 25	4	12	–
Helianthus annuus (sunflower)	BP; S	20–30; 25	4	10	Preheat; prechill

(continued)

Table 4 (continued)

Species	Substrate	Temperature (°C)	First count [days]	Final count [days]	Recommendations for breaking dormancy
Hordeum vulgare (barley)	BP; S	20	4	7	Preheat (30–35 °C); prechill; GA$_3$
Lactuca sativa (lettuce)	TP; BP	20	5	10	Prechill
Lens culinaris (lentil)	BP; S	20	5	10	Prechill
Linum usitatissimum (linseed)	TP; BP	20–30; 20	3	7	Prechill
Lycopersicon lycopersicum (tomato)	TP; BP	20–30	5	14	KNO$_3$
Medicago sativa (alfalfa)	TP; BP	20	4	10	Prechill
Nicotiana tabacum (tobacco)	TP	20–30	7	16	KNO$_3$
Oryza sativa (paddy)	TP; BP; S	20–30; 25	5	14	Preheat; soak in H$_2$O or KNO$_3$
Pennisetum typhoides (pearl millet)	TP; BP	20–30	3	7	–
Pisum sativum (pea)	BP; S	20	5	8	–
Ricinus communis (castor)	BP; S	20–30	7	14	–
Secale cereale (rye)	TP; BP; S	20	4	7	Prechill; GA$_3$
Sesamum indicum (sesame)	TP	20–30	3	6	–
Solanum melongena (brinjal)	TP; BP	20–30	7	14	–
Sorghum vulgare (jowar)	TP; BP	20–30; 25	4	10	Prechill
Triticum aestivum (wheat)	TP; BP; S	20	4	8	Preheat (30–35 °C); prechill; GA$_3$
Triticum durum (wheat)	TP; BP; S	20	4	8	Preheat (30–35 °C); prechill; GA$_3$
Vicia faba (broad bean)	BP; S	20	4	14	Prechill
Vigna mungo (black gram)	BP; S	20–30; 25	4	7	–
Vigna radiata (green gram)	BP; S	20–30; 25	5	7	–
Vigna unguiculata (cowpea)	BP; S	20–30; 25	5	8	–
Zea mays (maize)	BP; S	20–30; 25	4	7	–

Note: Substrate: *TP* top of paper, *BP* between paper, *S* sand. In temperature, for some species, two options are indicated, e.g. 20–30 [in alternating cycle of 12 h each] or 20 or 25 constant. Any one of the options chosen for a sample should be adopted for the entire test duration of that sample. Ideal is to adopt the same option for all the samples in that crop for the production/testing season

(b) *Abnormal Seedlings:* The seedlings, when grown in good-quality soil and under favourable conditions of water supply, temperature, and light, which do not exhibit the capacity for continued development into normal plants are abnormal seedlings.

Abnormal seedlings with the following defects shall be classified as follows:

- *Damaged seedlings*: Damaged seedlings are those with no cotyledons; seedlings with constrictions, splits, cracks, or lesions that affect the conducting tissues of the epicotyl, hypocotyl, or root; and species where primary root is an essential structure but without a primary root, except for *Pisum, Vicia, Lupinus, Vigna, Glycine, Arachis, Gossypium, Zea,* and all species of Cucurbitaceae, where several vigorous secondary roots have developed to support the seedlings in soil.
- *Deformed seedlings*: Seedlings where essential structures are weak or with ill-balanced development such as spirally twisted or stunted plumules, hypocotyls, or epicotyls; swollen shoots and stunted roots; split plumules or coleoptiles without a green leaf; watery and glassy seedlings; or no further development after the emergence of the cotyledons.
- *Decayed seedlings*: Any of the essential structures of the seedlings, so diseased or decayed that normal development is prevented, except when there is clear evidence to show that the cause of infection is not the seed itself.

(c) *Hard Seeds:* Seeds that remain hard at the end of the test period because they have not absorbed water due to an impermeable seed coat, e.g. Leguminosae, Malvaceae [*Gossypium* and *Hibiscus*], etc.

(d) *Fresh-Ungerminated Seeds*: Seeds which neither are hard nor have germinated but remain clean and firm and apparently viable at the end of the test period.

(e) *Dead Seeds:* Seeds which at the end of the test period are neither hard nor fresh and not have produced any part of a seedling. They are collapsed, and milky paste comes out when pressed at the end of the test.

5.3.5 Evaluation of Seedlings

Seedling evaluation and separating normal seedlings from the abnormal ones in all economically important plant species are performed as per the ISTA Rules. For a detailed description of normal and abnormal seedlings of important plant species, see *ISTA Handbook on Seedling Evaluation* (2018).

Representative examples of epigeal and hypogeal germination in monocot and dicot seeds (as per ISTA) are described here.

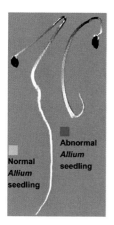 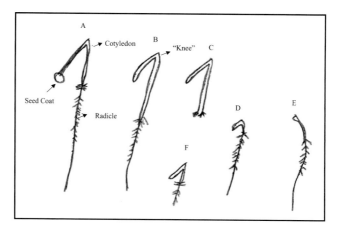

Fig. 7 Seedlings of onion (*Allium* sp.). (*A, B*) Normal seedling (cotyledon bent, long; root well developed). (*C*) Abnormal seedling (no radicle). (*D*) Abnormal seedling (cotyledon short, weak). (*E*) Abnormal seedling (indefinite knee). (*F*) Abnormal seedling (weak radicle). (Source: https://scholarsjunction.msstate.edu/seedtechpapers/92)

1. *Monocotyledons with epigeal germination:* Representative genus: *Allium* (Fig. 7).

Normal and abnormal seedling differentiation

Normal seedlings	Abnormal seedlings
All essential structures of the seedling present as a whole	The seedling is called abnormal if:
The primary root should be intact and show only the acceptable defects like:	1. Deformed and fractured
1. Necrotic or discoloured spots	2. From the seed coat, the cotyledon is released before the primary root
2. Healed splits and cracks	3. Consists of fused twin seedlings
3. Superficial splits and cracks	4. The seedling is white or yellow and is glassy and spindly
The shoot system should be intact with a definite 'knee' and show only acceptable defects like:	If the primary root is stunted, is retarded, is missing, is broken, is split from the tip, is trapped in the seed coat, shows negative geotropism, is constricted, is spindly, is glassy, and as a result of primary infection is decayed
1. Discoloured or necrotic spots	If the cotyledon is short and thick, is broken, is bent over or forming a loop, is forming a spiral, does not show a definite 'knee', is constricted, is spindly glassy, and as a result of primary infection is decayed
2. Loose twists	

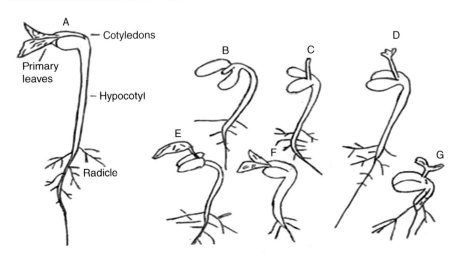

Fig. 8 Seedlings of beans (*Phaseolus* sp.). (*A*) Normal seedling (epicotyl, hypocotyl, and radicle well developed). (*B*) Abnormal seedling (bold head, no terminal bud; no primary leaves). (*C*) Abnormal seedling (terminal bud present but no primary leaves). (*D*) Abnormal seedling (primary leaves chlorotic, stunted). (*E*) Normal seedling (one primary leaf present). (*F*) Abnormal seedling (hypocotyl thickened; no radicle). (*G*) Abnormal seedling (hypocotyl weak; no radicle). (Source: https://scholarsjunction.msstate.edu/seedtechpapers/92)

2. *Dicotyledons with epigeal germination*: Representative genus: *Phaseolus* except *P. coccineus* (Fig. 8).

Normal and abnormal seedling differentiation

Normal seedlings	Abnormal seedlings
The primary root should be intact and show only acceptable defects like:	The seedling is categorized as abnormal if:
1. Necrotic or discoloured spots	1. Deformed and fractured
2. Healed splits and cracks	2. Cotyledons emerging before the primary root
3. Superficial splits and cracks	3. Consists of fused twin seedlings
	4. The seedling is white or yellow and is glassy and spindly and as a result of primary infection it is decayed
• In the shoot system, hypocotyl and epicotyl are intact and show acceptable defects like:	If the primary root is stunted, is retarded, is missing, is broken, is split from the tip, is trapped in the seed coat, shows negative geotropism, is constricted, is spindly, is glassy, and as a result of primary infection is decayed
– Necrotic or discoloured spots	
– Healed splits and cracks	
– Superficial splits and cracks	
– Loose twists	

(continued)

Normal seedlings	Abnormal seedlings
• Intact terminal bud	If hypocotyl and epicotyl show the following defects, viz. are short and thick, are cracked deeply, are split right through, are missing, are bent over or forming a loop, are forming a spiral, are tightly twisted, are constricted, are spindly glassy, and as a result of primary infection are decayed
• Intact cotyledons and show acceptable defects, viz.:	
– Up to 50% of tissue not functioning normally	
– Only one cotyledon intact	
– Three cotyledons	
The primary leaves should be intact and show acceptable defects like:	
1. Up to 50% of tissue not functioning normally	
2. Only one intact cotyledon	
3. Three primary leaves	
4. Normal shape but retarded growth	

3. *Monocotyledon with hypogeal germination:* Representative genus: *Oryza* (Fig. 9).

Normal and abnormal seedling differentiation

Normal seedlings	Abnormal seedlings
The primary root should be intact and shows acceptable defects like:	Abnormal seedling if:
1. Necrotic or discoloured spots	1. Deformed and fractured
2. Healed splits and cracks	2. Cotyledons emerging before the primary root
3. Superficial splits and cracks	3. Consists of fused twin seedlings
	4. The seedling is white or yellow and is glassy and spindly and as a result of primary infection it is decayed
	5. Scutellum detached from the endosperm
• In the shoot system, if the mesocotyl is developed, it should be intact and show only acceptable defects like:	If the primary root is stunted, is retarded, is missing, is broken, is split from the tip, is trapped in the seed coat, shows negative geotropism, is constricted, is spindly, is glassy, and as a result of primary infection is decayed
– Discoloured or necrotic spots	
– Superficial cracks and splits	
– Loose twists	
• The coleoptile is intact and shows acceptable defects like:	• The the mesocotyl is defective like cracked or broken, forming a loop or spiral, tightly twisted, and decayed as a result of primary infection
– Discoloured or necrotic spots	
– Loose twists	
– A split of one third or less from the tip	
• The primary leaf is intact and shows acceptable defects like:	• Coleoptile is deformed, broken, or missing, has a damaged or missing tip, and is strongly bent over, forming a loop or spiral, tightly twisted, split for more than one third from the tip, spindly, and decayed as a result of primary infection

(continued)

Normal seedlings	Abnormal seedlings
– Discoloured or necrotic spots	• The primary leaf is defective if it extends less than half the length of the coleoptile; is missing, shredded, or otherwise deformed; is yellow or white; and is decayed as a result of primary infection
– Slightly retarded growth	

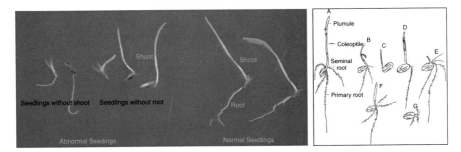

Fig. 9 Seedlings of paddy (*Oryza sativa*). (*A*) Normal seedling (plumule, root well developed). (*B*) Abnormal seedling (weak plumule and root). (*C*) Abnormal seedling (no root). (*D*) Abnormal seedling (only one root). (*E*) Normal seedling (weak plumule-coleoptile empty). (*F*) Abnormal seedling (no plumule). (*G*) Abnormal seedling (weak plumule and root development). (Source: https://scholarsjunction.msstate.edu/seedtechpapers/92)

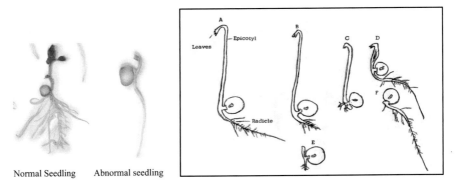

Fig. 10 Seedlings of peas (*Pisum sativum*). (*A*) Normal seedling (good epicotyl and radical development). (*B*) Abnormal seedling (stunted radicle). (*C*) Abnormal seedling (no radicle development). (*D*) Abnormal seedling (split, swollen epicotyl). (*E*) Normal seedling (weak epicotyl and radicle). (*F*) Abnormal seedling (no epicotyl). (Source: https://scholarsjunction.msstate.edu/seedtechpapers/92)

4. *Dicotyledon with hypogeal germination:* Representative genus: *Pisum* sp. (Fig. 10).

Normal and abnormal seedling differentiation

Normal seedlings	Abnormal seedlings
The primary root should be intact and shows acceptable defects like:	Abnormal seedling if:
1. Necrotic or discoloured spots	1. Deformed and fractured
2. Healed splits and cracks	2. Consists of fused twin seedlings
3. Superficial splits and cracks	3. The seedling is white or yellow and is glassy and spindly and as a result of primary infection it is decayed
• In the shoot system, if cotyledons are intact and show acceptable defects like:	If the primary root is stunted, is retarded, is missing, is broken, is split from the tip, is trapped in the seed coat, shows negative geotropism, is constricted, is spindly, is glassy, and as a result of primary infection is decayed
– Up to 50% of tissue not functioning normally	
– Only one intact cotyledon	
– Three cotyledons	
• The epicotyl is intact and shows acceptable defects like:	• Cotyledons are defective to such an extent that only less than 50% of the original or estimated tissue is functioning normally
– Necrotic or discoloured spots	• Deformed, broken, or missing
– Healed splits and cracks	• Discoloured or necrotic and is decayed as a result of primary infection
– Superficial splits and cracks	• The epicotyl is defective if it is too short, thick, deeply cracked or broken, split right through, missing, bent over or forming a loop or spiral, tightly twisted, constricted, spindly, and as a result of primary infection is decayed
– Loose twists	• The primary leaves are defective to such an extent that only less than 50% of the original leaf area is functioning normally, deformed, damaged, separated or missing, discoloured, necrotic, and as a result of primary infection decayed

5.3.6 Calculation and Expression of the Result

Results are expressed as a percentage by number.

$$\text{Germination}(\%) = \frac{\text{Number of seeds germinated}}{\text{Number of seeds put for test}} \times 100$$

When hundred seeds of four replicates of a test are within the maximum tolerated range, the average represents the percentage germination to be reported on the analysis certificate. The average percentage is calculated to the nearest whole number. The total % of all the category of seeds (normal, abnormal, dead hard, fresh-ungerminated) should be 100.

Germination rate may be deduced as the average number of seeds that germinate over the first and final count period.

5.3.7 Retesting

The results of a test are considered unsatisfactory under the following circumstances.

1. Replicates performance is out of tolerance.
2. Results being inaccurate due to wrong evaluation of seedlings or counting or errors in test conditions.
3. Dormancy persistence or phytotoxicity or spread of fungi or bacteria.

In such a situation, the result will not be reported, and a second test will be made by the same method or by an alternative method. The average of the two tests shall be reported.

6   Seed Viability

Seed viability indicates the capacity of a seed to germinate and produce a normal seedling. However, a seed may contain both live and dead tissues and may or may not have the capacity to germinate indicating the viability of the tissues as well as the whole seed (Copeland and McDonald 2001). Hence, the viability test may be performed for rapid assessment to determine the germination potential of the seed lots to emerge under the field condition and also for deciding the marketing and storage by the seed companies.

6.1   Seed Viability Test

Seed viability test is a technique used to determine whether individual seeds within a sample appear to be alive or dead. It reveals the proportion of live seeds in a population to be estimated (Pradhan et al. 2022). Estimating the germination potential of a seed lot by actually germinating the representative sample is often time-consuming, specially in tree seeds. For this reason, a quick viability test is needed.

The objectives of quick viability tests:
- To determine quickly the viability of seeds in a species that normally germinate slowly or show deep dormancy under normal germination methods.
- To determine the viability of samples which at the end of the germination test reveal a high percentage of fresh-ungerminated or hard seeds.

6.2   Tetrazolium Test

Tetrazolium test is also known as TZ test, is a reliable and widely used quick test for seed viability, introduced by the German scientist Lakon in 1942. It is based on topographic evaluation of a biochemical test which demonstrates that specific embryo structures have to be alive for the seed to germinate.

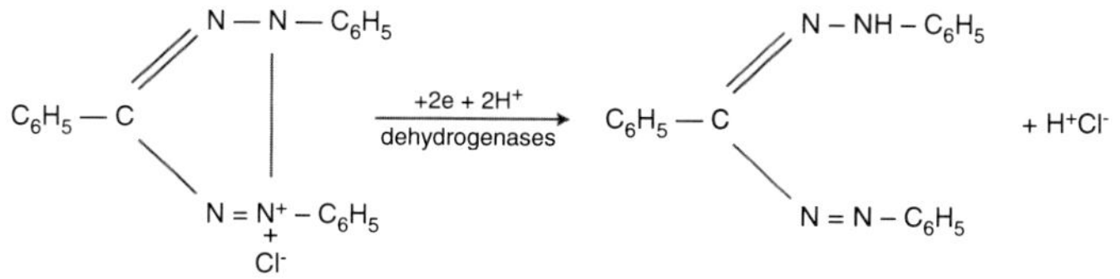

Fig. 11 Reduction reaction catalysed by enzyme dehydrogenases in live seed tissues in darkness

Principle and mechanism: The test is established on the fact that all respiring tissues are capable of reducing 2,3,5-triphenyltetrazolium bromide or chloride, a colourless chemical, by H+ transfer reactions catalysed by the dehydrogenase enzyme into a red-coloured compound, formazan (Fig. 11), in darkness. Thus, the non-diffusible formazan stains the living tissues red. Respiring tissues can be found within the embryo; in cotyledons, radicle, and scutellar tissue; in some nutritive endosperm tissues; in female gametophyte tissues in gymnosperms; and in the aleurone cell layer inside the pericarp of grasses (França-Neto and Krzyzanowski 2019). The analyst must have a sound knowledge of the seed structure like the location and shape of the embryo, storage tissue type, properties of the seed coat, etc. for the successful conduct of the test. Thus, the viability pattern of the seed can be estimated by both the staining intensity and the staining pattern to make viability determinations. Once the principle and procedure are established for one species, it is relatively easy to test other species even of unfamiliar ones with necessary modifications (Elias et al. 2012).

Merits/advantages of tetrazolium test
– Alternative to standard germination test.
– Can determine seed viability before harvesting.
– Helps evaluate the seed's physical and physiological condition.
– Can help detect structural abnormalities and the degree of insect damage and mechanical damage during harvesting and post-harvest handling.
– The test is simple and requires inexpensive equipment.
– Determines viability of dormant/hard/fresh-ungerminated seeds.
– The test is not influenced by environmental factors.

6.2.1 Procedure

Preparation of seed: The TZ test is conducted on the pure seed fraction of the representative sample. Two to four replicates of 100 seeds should be randomly taken for each test as specified by the ISTA (2022). Many species have hard and impermeable seed coats that need some kind of mechanical abrasion for moisture to penetrate the interior tissues of the seed. The seed is, therefore, hydrated to soften the seed coat and activate the dehydrogenase enzyme as the respiration rate increases, which facilitates pink-red staining of the viable tissues. Moistening also facilitates the cutting, piercing, or removal of the seed coat or other structures during preparation for staining. Softening by soaking overnight is followed with exceptions in some species to avoid imbibitional injury to the seeds. Fully imbibed seeds are easily sectioned and uniformly stained. The interpretation of staining patterns on a clean-cut surface of the embryo is more reliable.

Preparation for staining: Fully imbibed seed is cut, pierced, or punctured to remove the seed coat for easy entry and absorption of TZ solution. Preparation of seeds for staining includes:

1. With a razor blade, bisect longitudinally or transversely.
2. Seed coat punctured with a sharp needle.
3. With a razor blade or scalpel, make seed coat incisions.
4. Remove the seed coat.
5. Excise the embryo.

After the seed coat removal, proper sectioning procedures need to be followed for rapid and uniform staining of the seed internal structure. In general, seeds of cereals and large-seeded grasses are sectioned longitudinally, whereas in dicot seeds, the testa is removed exposing the cotyledons for direct contact with TZ solution.

Preparation of staining solution: 0.5–1.0% solution (w/v) of 2,3,5-triphenyltetrazolium bromide or chloride is used for seeds of grasses, legumes, vegetables, flowers, and tree species that are not bisected through the embryo, and 0.1–0.5% solution is used for seeds of grasses and cereals bisected through the embryo. To prepare a 1.0% solution, 1 gram of tetrazolium salt is dissolved in distilled or tap water or phosphate buffer to make 100 ml. The pH of the solution should be around 7.0. The phosphate buffer solution is prepared by mixing 400 ml of solution A (9.078 g of potassium dihydrogen phosphate dissolved in 1000 ml water) and 600 ml of solution B (11.876 g of disodium hydrogen phosphate in 1000 ml water). 1.0% TZ solution of pH 7.0 is obtained by dissolving 10 g of tetrazolium salt in 1 lt of phosphate buffer solution. The TZ solution should be kept only in dark-coloured glass bottles to prevent the effect of light on the solution.

Staining with TZ solution: Place the seeds in the recommended concentration of 2,3,5-triphenyltetrazolium chloride in a Petri plate or beaker in such a manner that all the parts of the seed are in touch with the TZ solution. Temperatures between 25 and 35 °C are generally preferred. The staining time depends on the species—ranging from 4 to 8 h. Usually, the seeds are soaked overnight in darkness. After the staining period, the seeds are removed from the tetrazolium solution and rinsed thoroughly with water. The seeds are then evaluated for staining patterns.

6.2.2 Evaluation of the Staining Pattern

Large seeds can be examined under magnifying lenses, whereas small seeds can be observed under the microscope. Interpretation of the staining pattern requires special attention to the internal seed structures and its capacity to develop a normal root and shoot system. Bisected embryos of the seeds may be examined directly, while others may require a bit of manipulation for a precise evaluation, e.g. from the surrounding seed tissues, the embryo may be removed, opaque seed coats may be removed, or in grasses the outer coverings may be treated with lactic acid [85% V/V]. Evaluation of embryo staining pattern is comparable to that of normal and abnormal seedling in case of germination test. The embryos may show a range of staining patterns, stained completely to completely unstained. The embryos that are stained uniformly are

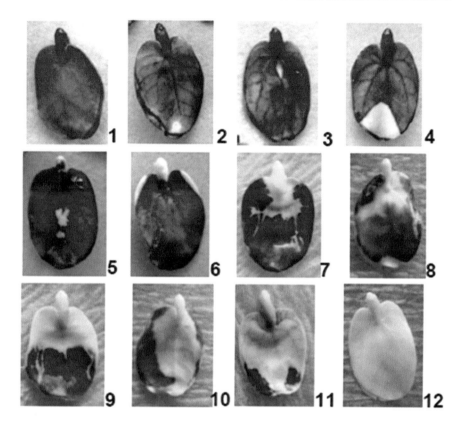

Fig. 12 Tetrazolium staining patterns and their interpretation for *Jatropha curcas* seeds (Parreño-de Guzman et al. 2011). *1*: Germinable—embryo completely stained. *2–4*: Germinable—minor unstained areas in plumule. *5*: Non-germinable—more than extreme tip of radicle unstained. *6*: Non-germinable—more than half of radicle unstained. *7, 8*: Non-germinable—whole radicle unstained plus juncture of plumule and radicle axis. *9, 10*: Non-germinable—whole radicle and half of plumule unstained. *11*: Non-germinable—radicle and more than half of plumule unstained and greenish in colour. *12*: Non-germinable—embryo completely unstained

considered viable, while the unstained are non-viable. However, some may be stained partially and partially unstained, pointing that they are neither completely alive nor completely dead, which must be evaluated thoroughly to determine their potential to produce normal seedlings (Fig. 12). A complete examination of seed structures, viz. embryo, radicle, plumule, cotyledons, scutellum, aleurone layer, coleorhiza, and coleoptile, is essential in concluding the result. Usually, endosperm is made up of non-living tissues and hence does not stain.

6.2.3   Tetrazolium Testing Method for Various Species

For detailed descriptions and explanations of TZ tests of different plant species, please see the *ISTA Handbook on TZ Testing* (2003). Following the general principle and procedures of the tetrazolium test, it is not difficult to suitably modify test methods for several familiar species (Elias et al. 2012) (Table 5). With good

Table 5 Tetrazolium testing method for various familiar crop species (Elias et al. 2012)

Family	Crop	Moistening duration (h)	Preparation for staining	Staining time (h)	Staining pattern evaluation
1	2	3	4	5	6
Poaceae	*Zea mays* (large-sized cereal grain)	18	Seeds are cut through midsection longitudinally to almost full depth, and cut surfaces are spread slightly apart	6–24	Viable—Embryo completely stained, and less than one third of the scutellum unstained
					Non-viable—Both plumule and radicle are not completely stained
	Triticum aestivum (medium-sized cereal grain)	6–18	(a) Seeds are cut through midsection of the embryonic axis longitudinally to almost full depth and into the nutritive tissue within the basal half	6–24	Viable—Embryo stained entirely, and some area of the scutellum unstained. Endosperm comprising of non-living cells does not stain
			(b) Narrow tip of a blade or a lance needle is thrust through the seed coat to remove the intact embryonic axis and the scutellum and then into the endosperm immediately above the germ. Seed basal end is		Non-viable—Radicle completely unstained, and the plumule or more than one third of scutellum are unstained

(continued)

Table 5 (continued)

Family	Crop	Moistening duration (h)	Preparation for staining	Staining time (h)	Staining pattern evaluation
			split, while the embryo is lifted out		
	Lolium spp. (small-sized grasses)	6–18	(a) Seeds are cut through midsection of the embryonic axis longitudinally to almost full depth and into the nutritive tissue within the basal half	6–24	Viable— Embryo evenly stained, and part of scutellum unstained
			(b) Seeds are cut laterally, full depth from the midsection outward to one side slightly above the embryo		Non-viable— Essential parts of the embryonic axis unstained, and entire embryo unstained
			(c) Distal end of the seed is removed, and one fourth to one third of length is discarded		
Fabaceae	*Trifolium* spp. (small-seeded legume)	18	(a) Seeds are cut through the seed coat longitudinally and into the nutritive tissue near the midsection of the distal half	6–24	Viable— Entire embryo is stained, and small areas of the cotyledons and radicle or hypocotyl unstained
			(b) Distal end of the radicle including a fragment of the nutritive tissue is		Non-viable— Extended areas of the embryonic axis and

(continued)

Table 5 (continued)

Family	Crop	Moistening duration (h)	Preparation for staining	Staining time (h)	Staining pattern evaluation
			removed and discarded		more than half of cotyledons (living cells) unstained
	Phaseolus vulgaris (large-seeded legumes)	18–24	Seeds are cut longitudinally through the seed coat and into the nutritive tissue along the entire length near the midsection	6–24	Viable— Entire embryo stained, and some unstained or dark red areas in the radicle and cotyledons
					Non-viable— Extended areas of the embryonic axis and more than half of cotyledons are unstained or darkly stained
Chenopodiaceae	Beta vulgaris	18	(a) One side or the back of the seed, including a thin slice of the embryo, is removed	24–48	Viable— Entirely stained embryo and green embryo with firm tissues
			(b) Seed coat near the border of the embryo and the nutritive tissue is punctured		Non-viable— Unstained radicle or plumule and green embryo with soft tissues

(continued)

Table 5 (continued)

Family	Crop	Moistening duration (h)	Preparation for staining	Staining time (h)	Staining pattern evaluation
Solanaceae	*Lycopersicon esculentum*	18	(a) Puncture or cut through the seed coat near the Centre	18–24	Viable— Embryo stained entirely and a bit unstained areas of the endosperm (living cells)
			(b) Seeds are cut longitudinally starting in the midsection of the curved back their entire length and almost full depth and cut toward the radicle and cotyledon tips		
			(c) Seeds are cut laterally full depth from the Centre of the seed outward between the radicle and the cotyledons		Non-viable— Extended areas of the radicle are unstained; areas of the cotyledons and the shoot apex are unstained
			(d) Basal end of the seed is removed, including a tip of the nutritive tissue, if present		

knowledge of seed anatomy and morphology, tetrazolium testing for unfamiliar species can be developed with guidelines from AOSA and ISTA tetrazolium testing handbooks.

Besides the above-mentioned quality testing, seed health is the other vital parameter of seed quality which has been described in the chapter Seed Health: Testing and Management. Seed vigour, another important attribute of seed quality, is discussed in the chapter Seed Vigour and Invigoration.

References

Bewley JD, Bradford KJ, Hilhorst HWM, Nonogaki H (2013) Germination. In: Seeds. Springer, New York. https://doi.org/10.1007/978-1-4614-4693-4_4

Copeland LO, McDonald MB (2001) Seed viability and viability testing. In: Principles of seed science and technology. Springer, Boston, pp 124–139

Elias SG, Copeland LO, McDonald MB, Baalbaki RZ (2012) Seed testing: principles and practices. Michigan State University Press, East Lansing

Fougereux J (2000) Germination quality and seed certification in grain legume. Special Report. Grain Legumes 27:14–16

França-Neto JB, Krzyzanowski FC (2019) Tetrazolium: an important test for physiological seed quality evaluation. J Seed Sci 41(3):359–366

ISTA (2003) Working sheets on tetrazolium testing. In: Leist N, Krämer S (eds) Agricultural, vegetable and horticultural species, vol I. ISBN: 978-3-906549-40-8

ISTA (2018) In: Don R, Ducournau S (eds) Handbook on seedling evaluation. International Seed Testing Association, Zurich. ISBN 978-3-906549-39-2

ISTA (2019). International rules for seed testing, Full Issue. i–19-8 (300). ISTA, Zurich. https://doi.org/10.15258/istarules.2019

ISTA (2022) International rules for seed testing, chap 2, i–2-44 (52). https://doi.org/10.15258/istarules.2022.02

Li L, Zhang Q, Huang D (2014) A review of imaging techniques for plant phenotyping. Sensors 14:20078–20111

Parreño-de Guzman LE, Zamora OB, Borromeo TH, Sta Cruz PC, Mendoza TC (2011) Seed viability and vigor testing of *Jatropha curcas* L. Philip J Crop Sci 36(3):10–18

Pradhan N, Fan X, Martini F, Chen H, Liu H, Gao J, Goodale UM (2022) Seed viability testing for research and conservation of epiphytic and terrestrial orchids. Bot Stud 63(1):1–14

Tang S, TeKrony DM, Collins M, McKenna C (2000) Determination of high seed moisture in maize. Seed Technol: 22(1):43–55

Wilkes T, Nixon G, Bushell C, Waltho A, Alroichdi A, Burns M (2016) Feasibility study for applying spectral imaging for wheat grain authenticity testing in pasta. Food Nutr Sci 7:355–361

Seed Health: Testing and Management

Karuna Vishunavat, Kuppusami Prabakar, and Theerthagiri Anand

Abstract

Healthy seeds play an important role in growing a healthy crop. Seed health testing is performed by detecting the presence or absence of insect infestation and seed-borne diseases caused by fungi, bacteria, and viruses. The most detrimental effect of seed-borne pathogens is the contamination of previously disease-free areas and the spread of new diseases. Sowing contaminated or infected seeds not only spreads pathogens but can also reduce yields significantly by 15–90%. Some of the major seed-borne diseases affecting yield in cereals, oilseeds, legumes, and vegetables, particularly in the warm and humid conditions prevailing in the tropical and sub-tropical regions, are blast and brown spot of rice, white tip nematode and ear-cockle in wheat, bacterial leaf blight of rice, downy mildews, smuts, head mould, seedling rots, anthracnose, halo blight, and a number of viral diseases. Hence, detection of seed-borne pathogens, such as fungi (anthracnose, bunt, smut, galls, fungal blights), bacteria (bacterial blights, fruit rots, cankers), viruses (crinkle, mottle, mosaic), and nematodes (galls and white tip), which transmit through infected seed to the main crop, is an important step in the management strategies for seed-borne diseases. Thus, seed health testing forms an essential part of seed certification, phytosanitary certification, and quarantine programmes at national and international levels. Detection of seed-borne/ transmitted pathogens is also vital in ensuring the health of the basic stock used for seed production and in maintaining the plant germplasm for future research and product development. Besides the precise and reproducible testing methods,

K. Vishunavat (✉)
Department of Plant Pathology, College of Agriculture, Govind Ballabh Pant University of Agriculture and Technology, Pantnagar, Uttarakhand, India

K. Prabakar · T. Anand
Department of Plant Pathology, Tamil Nadu Agriculture University, Coimbatore, Tamil Nadu, India

© The Author(s) 2023
M. Dadlani, D. K. Yadava (eds.), *Seed Science and Technology*,
https://doi.org/10.1007/978-981-19-5888-5_14

335

appropriate practices during seed production and post-harvest handling, including seed treatment and storage, are important components of seed health management and sustainable crop protection.

Keywords

Seed health testing methods · Seed-borne diseases · Management of SBDs · Seed treatment

1   Significance of Seed Health Testing

Seed health is an essential component of seed quality. It is estimated that 30% of seed-borne diseases can be controlled by using disease-free seeds. For many of these diseases, fungicides are not available or registered, and resistant cultivars are not available, necessitating the use of disease-free seed as the only means of crop protection. Hence, research and development priorities to facilitate and improve the scope of seed health testing need special attention.

Seeds are regularly moved internationally, in small or large quantities, for trade and research purposes. These are often produced in one or more countries and distributed from those to several other countries. Seed health issues are becoming increasingly important in the international seed trade. With the advent of free trade, many countries are redefining their phytosanitary requirements to prevent the introduction of new and harmful pathogens into their countries (McGee 1997). To provide scientific answers to the problems encountered in the worldwide movement of seeds, an internationally accepted programme is needed to standardize seed health tests and inspection practices.

Development of seed health testing methods needs to be viewed in light of the general evolution of the seed sector. The importance of seed health can be established by the extent of losses attributed to seed-borne pathogens (Mathur et al. 1988; Mekonnen Gebeyaw 2020), which could be as high as 15–90%. Hence, predictive relationships need to be established between the seed-borne pathogens which cause significant yield loss (Hajihasani et al. 2012), and reliable, effective, inexpensive, and rapid detection methods need to be standardized for detecting the same.

Irrespective of the detection methodology, the specificity, sensitivity, reliability, and efficiency of the assay and pathogen tolerance in the seed lot also need to be understood before a technique is considered acceptable for seed health test as a tool for disease management and can be routinely used in seed quality assessment. These considerations also help develop national policies and methods for conducting seed health tests as part of crop protection to increase crop yields.

The purpose of seed health testing may be any of the following:

- Testing for seed certification schemes.
- Testing to make accurate decisions regarding the appropriate use of seed treatment.
- Testing for quarantine purposes to avoid the spread of disease to new regions and to issue a phytosanitary certificate.
- Testing for the evaluation of planting value in the field.
- Testing for treated seeds.
- Testing for storage quality.
- Testing to assess the prevalence of seed-borne infection or the importance of a seed-borne disease in a research programme.
- Testing seeds for resistance of cultivars.

2 Seed Sampling

Seed sampling is the key to obtain accurate seed health test results. The low frequencies of the many important seed-borne pathogens in and on the seed and also the low to very low seed infection thresholds may cause disease outbreaks leading to huge economic losses. Therefore, seed sampling for seed health testing needs special attention. A seed sample should be the representative of the entire seed lot. Seed health tests are nearly always performed on a sample drawn from the seed lot which may be as large as 10,000–30,000 kg. It is, therefore, critical that the samples used for testing are reliable representatives of the seed lot, and this requires standardized sampling procedures. Seed testing organizations such as the International Seed Testing Association (ISTA) and the Association of Official Seed Analysts (AOSA) have developed specific rules and procedures for seed sampling for the evaluation of seed quality traits like germination and physical purity and seed health testing.

The objective of seed sampling, in particular reference to seed health testing, as per the ISTA Rules is 'to obtain a sample of a size suitable for tests, through which the probability of an infection being present is established only by its level of occurrence within the seed lot'. The two basic considerations of seed sampling are firstly to obtain a test sample that accurately represents the composition of the seed lot as a whole and secondly to keep in mind that regardless of how accurately an analysis is performed, the results represent only the standard of the sample submitted for analysis.

The general procedure for the seed sampling method is shown in Fig. 1 and is in accordance with AOSA and ISTA procedures. Primary samples of equal size are taken randomly from the whole seed lot and combined and blended to make a homogeneous composite sample. The submitted sample, which is distributed to the testing facility, is drawn from the composite sample. The working sample, which is used for performing the tests, is obtained from the submitted sample after prescribed blending and dividing to ensure sample homogeneity. In some cases, the working sample may be the entire submitted sample or a composite sample. These

Primary Sample
Small samples of equal size taken from the seed lot
↓
Composite Sample
Primary samples bulked and blended for homogeneity
↓
Submitted Sample
All or a part of the composite sample submitted for testing
↓
Working Sample
All or a part of the submitted sample on which the test is performed

Fig. 1 Procedure for seed sampling for the evaluation of seed quality, including seed health testing

sampling schemes are designed to reduce variability and ensure uniformity of working samples.

3   Genesis of Seed Health Testing

It was Frederick Nobbe, under whose leadership the first seed testing laboratory was established in 1869 in Tharandt, Germany, and this was followed by the second lab in 1871, in Copenhagen, Denmark, under the guidance of E. Moller Holst. Subsequently, seed testing spread rapidly in Europe during the next 20–30 years. By the beginning of the twentieth century, around 130 seed testing stations were operating in Europe. In the United States, the first seed testing laboratory was opened in 1876. The first seed health testing laboratory was established in 1918 at the Government Seed Testing Laboratory in Wageningen, the Netherlands. In India, currently notified Central Seed Testing Laboratory (CSTL), at the Indian Agricultural Research Institute (IARI) New Delhi, is the first seed testing laboratory, established in 1961 (Jha 1993).

Dorph-Petersen (1921) who later became the ISTA president, in a report entitled 'Remarks on the Investigations of the Purity of Strain and Freedom from Disease', described the field trials for cultivar purity and detection of stripe (*Fusarium*) and smut diseases. The International Seed Testing Association was formed at the 1924 Congress, and one of its first technical committees was aimed at *Investigations of Genuineness of Variety and of Plant Diseases*, the forerunner of the *Plant Disease Committee* (PDC). The first *International Rules for Seed Testing* was published by ISTA in 1928, which contained a special section on *Sanitary Condition* with special mention of *Claviceps purpurea, Fusarium, Tilletia,* and *Ustilago hordei* on cereals; *Ascochyta pisi* on peas; *Colletotrichum lindemuthianum* on beans; and *Botrytis, Colletotrichum linicola,* and *Aureobasidium lini* on flax (Wold 1983).

The Plant Disease Committee of ISTA in 1957 established a comparative seed health testing programme aimed at standardizing techniques for the detection of seed-borne pathogens (Mathur and Jorgensen 2002). In 1981, the referee groups were re-organized into crop groups with working groups conducting comparative tests on seeds of beet, crucifers, legumes, temperate cereals and grasses, tropical and sub-tropical crops, and viruses.

The Seed Health Committee (formerly the Plant Disease Committee) started the development of Guidelines for Comparative Testing of Methods for the Detection of Seed-borne Pathogens in 1993. This resulted in a complete revision of the method validation process for seed health testing as described in the Handbook of Method Validation. The methods have been validated and included in the Annexe to Chap. 7 of the International Rules for Seed Testing. The aim of the Seed Health Committee has been to develop and publish the validated procedures for seed health testing and to promote uniform application of these procedures for the evaluation of seeds moving in international trade (Hampton 2005, 2007).

In 1993, the International Seed Health Initiative-Vegetables (ISHI-Veg) was started to give stimulus to the vegetable seed industry to put more emphasis on seed health for quarantine pests and their impact on the international seed trade. Later, in the year 2000, the International Seed Federation (ISF) took over the secretariat and financial administration of ISHI-Veg. Nevertheless, in order to remain flexible and efficient, ISHI-Veg has maintained a special structure within ISF with separate funding by the participating countries. Further, ISF started two more ISHIs, viz. ISHI for herbage crops in 1997 and ISHI for field crops in 1999 (ISF 2022).

In 1995, during the second Seed Health Symposium of Plant Disease Committee, ISHI-Veg and ISTA jointly produced the Guidelines for Comparative Testing Methods for Detection of Seed-borne Pathogens (Sheppard and Wesseling 1998). These generalized test methods were particularly beneficial to the seed companies to make risk analysis specific to the conditions under which they operate, taking also other factors such as resistance in their varieties, seed production region, and the region where the seeds will be sold into consideration.

3.1 Advancements in Seed Health Testing Methodologies

Since the establishment of comparative seed health testing programme by the Plant Disease Committee of ISTA in 1957, seed health testing has undergone many changes. The seed health testing was primarily focused on the detection of seed-borne fungi and mostly relied upon incubation methods, morphological identification, or grow-out tests for the detection of these pathogens on seed. Presently, due to advances in technology, the seed-borne pathogens are detected both by conventional methods and using immunodiagnostic and molecular methods.

3.1.1 Conventional Methods

Direct Examination

This refers to the detection of such seed-borne pathogens which cause discoloration of the seed or cause change in the shape and size of the seed and hence are visibly detectable. Visual examination also helps for detecting fungal structures present on or with seed, such as the sclerotia, galls, smut balls, discolouration, malformation, resting hypha, fruiting bodies of fungi (oospores, pycnidia, perithecia), and bacterial masses (Rao and Bramel 2000). Some examples of visibly detectable diseases are maize seeds infected with *Nigrospora*, which have white streaks with black spore masses near the tips; sorghum seeds infected with *Acremonium* wilt which are completely deformed; and soybean seeds infected with *Cercospora kikuchii* showing symptoms of purple seed stain. The fungus *Claviceps*, the cause of ergot of sorghum and pearl millet, often is mixed with seed as sclerotia. Yellow ear rot or tundu disease-infected wheat seeds can also be detected by visual inspection of black galls caused by the nematode *Anguina tritici* (Agarwal and Sinclair 1997) (Plate 1).

Seed Washing Test

The washing test is helpful for detecting surface-borne fungal spores causing smuts, bunts, rusts, downy mildews, powdery mildews, etc. and of bacterial crusts on seed surface (Maddox 1998). Spores of fungi or bacterial cells are washed from seeds with water, and then the suspension is centrifuged. The supernatant is discarded, and the pellet is re-suspended in sterile distilled water. This spore suspension is then examined under the microscope for the presence of fungal spores. The spore load per seed can be estimated using a haemocytometer.

NaOH Seed Soak Method

This method is used for the easy detection of *Tilletia indica* and *T. barclayana*, causing Karnal bunt in wheat and paddy kernel smut/bunt, respectively (Agarwal and Srivastava 1981). In this method, a sample of minimum 2000 seeds is soaked in sodium hydroxide solution (NaOH 2%) for 20 h at 20 °C. The infected seeds exhibit loose spores of shiny jet black in colour of the fungus at the infected portions in contrast to the pale yellow healthy seed. Upon rupturing the black seeds in a drop of water, a mass of teliospores is released (Plate 2). The sodium hydroxide treatment increases the colour contrast between diseased and healthy wheat and paddy seeds (Agarwal and Mathur 1992), making the detection easier.

Embryo Count Method

This technique was developed exclusively for the detection of loose smut fungi (*Ustilago tritici*) in wheat (Rennie 1982). The fungal mycelium is localized in the embryo of the seed. To separate the intact embryo, from the rest of the seed, for microscopic observation is not possible. Therefore, the embryos are released from the wheat seeds by soaking in a solution of NaOH (5%) + trypan blue (0.03%) for 12 h. The embryos thus released are processed, dehydrated with sprit, and cleared in a solution of glycerine + lactic acid (1:1) and examined under stereo binocular microscope. The test is performed with a sample size of minimum 2000 seeds (Cappelli and Covarelli 2005) (Plate 3).

Plate 1 (**a**) Karnal bunt (*Nevossia indica*)-infected wheat seeds. (**b**) Nematode galls (*Anguina tritici*) mixed with wheat seed. (**c**) Seeds infected with paddy bunt (*Tilletia barclayana*). (**d**) Common bunt-infected wheat seed (*Tilletia laevis*). (**e**) Soybean mosaic virus (SMV) seed discolouration. (**f**) Purple seed stain (*Cercospora kikuchii*) of soybean

Incubation Test

The incubation test is the most common and widely used method for detecting a large number of seed-borne pathogens. In this, the seeds are incubated on a substrate, under specific environmental conditions for specified times to allow pathogens to grow on the seed. At the end of the incubation period, seeds are examined for the fungal growth of the pathogen on each seed using a stereomicroscope or compound

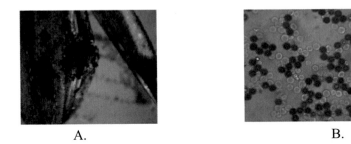

A.   B.

Plate 2 Detection of rice bunt by NaOH seed soak method. (**a**) Paddy seed infected with bunt. (**b**) Teliospores of *Tilletia barclayana*

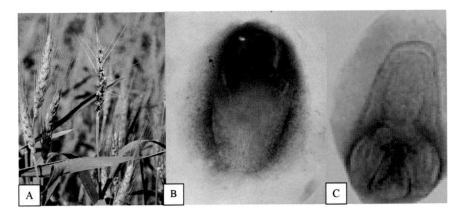

Plate 3 (**a**) Loose smut infected ear of wheat. (**b**) Wheat embryo showing loose smut infection by embryo count method. (**c**) Healthy non-infected embryo

microscope. Fungi are identified on the basis of their morphological characters, such as growth of the mycelium its septation, size and shape of the fruiting bodies, spores, their size and shapes, arrangement of conidia on conidiophores, etc. (Warham et al. 1990). Standard blotter, 2,4-D blotter, deep freezing blotter, and agar plate methods are the commonly used incubation methods for the detection of various seed-borne pathogens (Rao and Bramel 2000; Tsedaley 2015).

Doyer, in 1938, developed the standard blotter method, the most widely practised seed health testing incubation method, which was later included in the International Seed Testing Application Rules of 1966. Many laboratories still use this method as the first screening test for health condition of a seed lot. The blotter method is widely used for detecting fungi which are able to produce mycelial growth and fruiting structures under the incubation conditions (Plate 4). The method is good in testing seeds for fungi such as *Alternaria, Ascochyta, Bipolaris, Botryodiplodia, Botrytis, Cercospora, Cladosporium, Colletotrichum, Curvularia, Drechslera, Fusarium, Macrophomina, Myrothecium, Phoma, Phomopsis, Rhizoctonia, Sclerotinia*, etc.

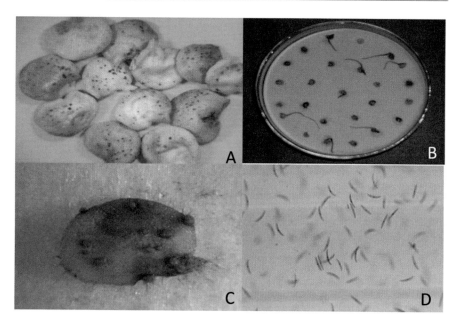

Plate 4 Blotter method for the detection of chilli anthracnose pathogen. (**a**) Infected seeds showing acervuli of the fungus. (**b**) Blotter test showing the growth of the fungus on seed surface. (**c**) Enlarged view of a single seed showing acervuli on seed surface. (**d**) Spores of *Colletotrichum capsici*

All kinds of cereals, vegetables, legumes, ornamentals, and forest seeds are tested by this method.

For routine seed examination, usually a seed sample of 400 seeds is used. In the standard blotter method, until and unless not specified otherwise, seeds are placed in 9 mm Petri dishes containing three-layered water-soaked blotter sheets as substrate (water-holding capacity of 40 cc). Seeds, sterilized with NaOCl solution (1.0%) and subsequently washed at least three times with sterilized water, are placed at 25 seeds/plate in small seeds and 10 seeds/plate in large seeds. These plates are incubated for 7 days at 20 ± 2 °C, under white fluorescent light and alternate cycle of 12 h light and 12 h darkness examined for the growth of the fungi under stereo binocular microscope. The major limitation of this method is that certain crop seeds germinate quite fast and obstruct the observations. To overcome this problem, the 2,4-D blotter method is used where the blotters, instead of soaking in ordinary water, are treated with a solution of 200 ppm of 2,4-D solution, as it either kills the seed or retards the seed germination and thus facilitates the easy observation of the seed samples. The rest of the procedure, incubation conditions, and period for incubation are the same as in the standard blotter method. The method was first used by Hagborg et al. (1950) for testing bean seeds for the presence of *Colletotrichum lindemuthianum*. However, this method too stances certain limitations because the 2,4-D reduces the recovery of certain fungi, and in that case, the deep freezing blotter method (Limonard 1966)

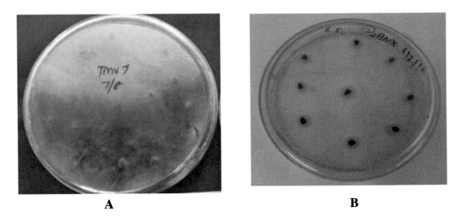

Plate 5 Agar plate method for the detection of seed-borne pathogens of sesame and tomato. (**a**) Sesame seeds showing *Macrophomina phaseolina* growth on PDA medium. (**b**) Tomato seeds showing *Clavibacter michiganensis* subsp. *michiganensis* growth on D2ANX medium

may be used. In this method, after plating the seed, as mentioned in the blotter method, the Petri dishes are incubated initially for 24 h at 20 ± °C and then transferred to a freezer at −20 °C for 24 h followed by a 5-day incubation at 20 ± 2 °C under white fluorescent light in alternate cycle of 12 h light and 12 h darkness.

In the agar plate method, the substrate used is the culture medium. Commonly, used media are either potato dextrose agar or Czapek's Dox agar or malt agar on which most of the fungi display their growth. Sterilized culture medium is poured in sterilized Petri plates of 9 mm diameter at 15 ml/plate. After solidification of the medium, seeds are placed in culture plates in the same way as in the blotter test. These plates are incubated under similar conditions as in the standard blotter test for the same period of time. However, certain pathogen requires selective or semi-selective media for their recovery or growth. For example, *Clavibacter michiganensis* subsp. *michiganensis* gives better recovery on D2ANX medium (Tripathi et al. 2018) (Plate 5).

Seedling Grow-Out Test

Certain seed-borne pathogens exhibit characteristic symptoms on developing seedlings, and thus, seedling grow-out test can be used as a direct method to assess the seed-borne pathogens on their living host and their transmission through seed. Under controlled greenhouse conditions, seedling symptom test reveals seed viability and helps detect the presence or absence of seed-borne pathogens on host plants. To perform this assay, seed samples are planted under controlled greenhouse conditions, conducive to disease development, and seedlings thus raised are examined for the appearance of symptoms (Plate 6). Seedling grow-out test is one of the most applicable and widely used assays for the detection of seed-borne pathogens in the living host (Lee et al. 1990; Yang et al. 1997). However, this test has certain

Plate 6 Grow-out test for the detection of urdbean leaf crinkle disease. (**a**) Naturally infected plant. (**b**) Screening of plants for seed transmission studies. (**c**) Plants exhibiting infection on exposure from nethouse

limitations as it may not always reveal the infection on the seedlings or the symptoms produced by certain seed-borne pathogens are not very distinct and conspicuous. In such cases, pathogens need to be isolated from suspected seedlings for confirmation. These additional steps further increase the time required to complete the seedling grow-out assay. Another limitation is that such test requires a large sample size to statistically confirm the infection percentage in seed. Besides, the test requires controlled conditions for the growth of seedlings and expression of symptoms failing, of which the symptoms are either obscure or ambiguous.

Conventional methods for seed health testing are mostly based on visual symptoms, culturing, and laboratory identification of the pathogens. These methods are performed at different levels (multi-stage) and time- and labour-intensive and in addition require extensive taxonomic expertise. They are not reliable at times and are difficult to apply in those cases where the symptoms are ambiguous or not expressed. Thus, within the last 25 years, advanced techniques for the accurate and feasible detection of the many seed-borne pathogens are developed. These include various immunoassays and nucleic acid-based techniques.

3.1.2 Immunoassay Methods

These methods are usually applied to detect many bacterial and viral pathogens. Among immunoassays, enzyme-linked immunosorbent assay (ELISA) is widely used to detect seed-derived fungi, bacteria, and viruses. Chang and Yu (1997) used DAS-ELISA for the detection of moulds, viz. *Aspergillus parasiticus*, *Penicillium citrinum*, and *Fusarium oxysporum*, in rice and corn seeds. The seed immunoblot binding assay (SIBA) has been an effective method in detecting *Tilletia indica*, in wheat seed. Immunoassays are popularly used in detecting mycotoxins produced by fungi such as *Aspergillus*, *Claviceps*, and *Fusarium* spp. The consistent use of ELISA for the detection of *Erwinia stewartii* in corn seeds has been demonstrated (Lamka et al. 1991). Detection and identification of *Xanthomonas oryzae* pv. *oryzae* in rice seed, using pathovar-specific monoclonal antibodies, could be performed using ELISA (Gnanamanickam et al. 1994). Likewise, ELISA has been an effective method for the detection of multiple viruses present in seeds (Fegla et al. 2000; Forster et al. 2001; Gillaspie Jr et al. 2001; Faris-Mukhayyish and Makkouk 2008; Ojuederie et al. 2009; Chalam et al. 2017; Torre et al. 2019).

The dot-immunobinding assay (DIBA) or dot-ELISA is similar to ELISA, except replacing the microtiter plate with a nitrocellulose or nylon-based membrane. The cut surface of the pre-soaked test seed is brought in-tuned with the membrane. The free protein binding sites present within the membranes are blocked using bovine serum albumin (BSA) or non-fat powdered milk, followed by the application of virus-specific antisera. The positive reaction is indicated with coloured dots. The presence of barley stripe mosaic virus and bean common mosaic virus in a single seed of French bean could be detected by the DIBA technique (Lange and Heide 1986). DIBA is optimized and used successfully for the rapid detection of 15 known soybean viruses as well (Ali 2017).

Flow cytometry (FCM) is a technique that enables fast multi-parameter analysis and quantification of the inoculum such as bacterial cells. The analysis is based on size and granularity and may be based on fluorescence after staining with a fluorescent dye. FCM has already been used in combination with antibody staining (immuno-FCM) to detect *Clavibacter michiganensis* subsp. *michiganensis* in tomato seed extracts and *Xanthomonas campestris* pv. *campestris* in cabbage seed extracts (Chitarra et al. 2002, 2006).

Immunoassays are suitable for the detection of seed-borne bacteria and viruses; however, the lack of species-specific antibodies remains a major constraint of their use for the detection of seed-borne fungal pathogens. Additionally, serology-based assays can also detect non-viable propagules which can lead to ambiguity in results (Mancini et al. 2016). Furthermore, these methods cannot detect all strains of pathogens and thus are limited in their applicability.

3.1.3 Molecular/Nucleic Acid-Based Diagnostic Methods

The detection, quantification, and characterization of seed-borne pathogens using multitude molecular marker are steadily increasing a routine practice for seed health testing. These molecular methods are now available for the detection of a number of seed-borne pathogens. In contrast to conventional seed health tests, DNA-based molecular techniques often have the advantage of being specific to the species level, sensitive, and rapid with the potential of being automated. Rapid detection of a specific pathogen in host tissue itself may be achieved using polymerase chain reaction (PCR). PCR-based assays have high sensitivity and specificity and often require as little as 24 or 48 h to complete. They are applicable to a wide range of pathogens and can be used to separate closely related species (see Chap. 15 for more).

Molecular methods have also been worked out for a number of seed-borne fungal pathogens of significance such as *C. purpurea* (ergot) (Correia et al. 2003), *Microdochium nivale* (foot rot) (Scherm et al. 2013), and *T. tritici* (bunt) (Majumder et al. 2013; Anil et al. 2008) in wheat and *Plenodomus lingam* (black leg) in cabbage (Mancini et al. 2016). PCR-based assay using internal transcribed spacer (ITS) primers specific to the regions of the ribosomal repeat (rDNA) was developed for the identification of the three *Alternaria* species on carrot seeds. The primers were highly specific, sensitive, and capable of differentiating the fungal pathogens (Konstantinova et al. 2002). Black spot disease of crucifers caused by *Alternaria* sp. is an important seed-borne disease. PCR-based diagnostic procedure involving the use of specific primers designed from DNA sequence in the ITS region of nuclear rDNA was employed to detect and differentiate *Alternaria brassicae*, *A. brassicicola*, and *A. japonica*, causal agents of black spot of crucifers. These pathogens were detected in the DNA extracted from seed macerates (Iacomi-Vasilescu et al. 2002).

By employing two different sets of primers, the conventional and real-time PCR, *A. brassicae* was specifically detected in the DNA extracted from seed. The presence of seed-borne pathogens such as *A. brassicicola* and *A. japonica* in radish; *A. alternata* in radish and cabbage; *Stemphylium botryosum*, *Penicillium* sp., and *Aspergillus* sp. in cabbage; and *Verticillium* sp. in tomato seeds was detected by the quantitative real-time PCR (Guillemette et al. 2004).

The BIO-PCR technique involves a combination of biological and enzymatic amplification of DNA sequences of target bacteria. Using this method, it is possible to detect *P. syringae* pv. *phaseolicola*, even if only 1 seed in a lot of 400 to 600 seeds is infected (Mosqueda-Cano and Herrera-Estrella 1997).

Incorporation of an immunological step before PCR assay significantly improves the sensitivity of detection of target bacteria present in seeds. Immunomagnetic separation (IMS) was performed before PCR assay to concentrate *A. avenae* subsp. *citrulli* present in watermelon seeds. A significant increase in sensitivity (100-fold) of detection by IMS-PCR was observed in comparison to an on-the-spot PCR assay, the detection limit being 10 CFU/ml. As low as 0.1% of seed infection was revealed by IMS-PCR assay (Walcott and Gitaitis 2000).

However, these methods are applied to detect viruses in seeds in mere some cases. In the case of RNA plant viruses, reverse transcriptase is employed to provide a complementary cDNA before PCR assay for the detection of the virus concerned within the seeds. *Clavibacter michiganensis* subsp. *michiganensis* in tomato seed extracts (Tripathi et al. 2022) and pea seed-borne mosaic virus (Kohnen et al. 1992) and cucumber mosaic virus in lupin seeds (Wylie et al. 1993) were detected by RT-PCR assay. Similarly, CABMV may be detected in samples consisting of 1 infected and 99 healthy leaves, indicating the sensitivity of the RT-PCR assay (Gillaspie Jr et al. 2001). RT-PCR was found rapid and sensitive in detecting viruses in seeds of vegetables (Gumus and Paylan 2013). Nonetheless, for the detection of several seed-borne pathogens, with real-time PCR, it's crucial to choose the appropriate target DNA fragments to design the primers and probes with adequate specificity and sensitivity and comparable amplification (Mancini et al. 2016).

A sensitive multiplex RT-PCR-based method could simultaneously assay the presence of five seed-borne legume viruses such as alfalfa mosaic virus (AMV), bean yellow mosaic virus (BYMV), clover yellow vein virus (CYVV), cucumber mosaic virus (CMV), and subterranean clover mottle virus (SCMoV) (Bariana et al. 1994). Primers are so designed that the size of RT-PCR product was indicative of the virus amplified and the sequence of more than one stain of virus was available.

Recently, a simple, rapid, and cost-effective method for nucleic acid amplification termed loop-mediated isothermal amplification (LAMP) has been developed for the detection of various plant pathogens. A sensitive reverse transcription loop-mediated isothermal amplification (RT-LAMP) method is developed for the rapid detection of BSMV isolates. The RT-LAMP assay can be used for the routine diagnostics of BSMV in seed and plant material (Zarzyńska-Nowak et al. 2018).

4 Management of Seed-Borne Diseases

4.1 Management of Seed-Borne Diseases Through Crop Production Practices

4.1.1 Identification of Disease-Free Areas for Healthy Seed Production

An important consideration for seed production is to select the site and the field where the climatic conditions are favourable for the plant to grow, flower, and set seed. Besides, it is also important to know about the past history of the field such as previous crops cultivated, weed populations, predominant diseases, and other intrinsic factors for successful disease-free seed production. For example, chickpeas should not be planted on the land on which lucerne was previously grown, as *Phytophthora* root rot affects both the crops (Ogle and Dale 1997).

Some plant pathogens are more pronounced in certain geographical areas, and hence, to escape the diseases, such areas need to be avoided for seed production. In general, most fungal and bacterial diseases are more pronounced in wet areas than in dry areas, e.g. ergot and smut of pearl millet. Healthy seed production is, therefore, recommended in those areas and seasons which are not predisposed for disease development. In addition, the same crop must not be cultivated in the same field year after year.

4.1.2 Time of Sowing

Altering the sowing or planting date can help to reduce the disease outbreaks by avoiding the weather conditions favourable for pathogens to grow, multiply, and infect the crop. Prolonged wet weather favours the occurrence of many diseases, e.g. downy mildew requires prolonged high moisture to cause infection. Chickpea (*Cicer arietinum*) root rot caused by *Rhizoctonia* spp. intensifies if the crop is sown immediately after rainfall. Early maturing varieties of wheat and pea are able to escape infection by *Puccinia graminis tritici* and *Erysiphe polygoni*, respectively. Similarly, in pearl millet, early sowing reduces the incidence of ergot disease (Gupta and Kumar 2020). Early planting also helps to avoid a pronounced bacterial black rot disease in crucifers (*Xanthomonas campestris*) because the environmental conditions are usually dry and not conducive for the development and spread of the pathogen.

4.1.3 Cultural Practices

The aim of the good cultural practices is to create favourable environmental conditions for crop development, promote good plant health, and limit the spread of plant pathogens, thereby minimizing the disease outbreaks. Some common practices include tillage, removal, and destruction of diseased crop residues or debris, eradication of alternate and collateral hosts, cultivation of non-host crops, selection of disease-free seeds, maintenance of appropriate isolation distances, timely rouging of infected plants and weeds, adequate irrigation, and balanced fertilization (Gupta and Kumar 2020).

A minimum isolation distance between seed production plots and commercial plots must be maintained to reduce the incidence of loose smut in wheat and barley seed crops. However, the distance recommended between seed plots and commercial plots may vary from one region to another depending upon the prevailing weather conditions. In countries, such as Germany and the United Kingdom, the distance between seed production plot and commercial cultivation plots for wheat and barley is maintained 50 m, whereas in the Netherlands, for barley, a minimum distance of 100 m between seed production plots and commercial field is secured. However, in India, for loose smut disease, an isolation distance of at least 150 m is recommended between seed plots and commercial field in the case of wheat, barley, oat, and rye (IMSCS 2013).

There are a number of the cultural practices which influence the incidence and severity of diseases such as spacing between the rows and between the plants in a row, time and methods of sowing/planting, depth of sowing, time and number of irrigation, quantity and composition of fertilizers/organic manures, cropping patterns, etc. In the nursery, overcrowding can lead to seedling damping off resulting in seedling's death. The incidence of bunt and smut in wheat is higher in deeply sown crops. Deep ploughing of the soils can effectively reduce the inoculum of *Phytophthora infestans*. Sowing of trap crops stimulates the dormant pathogen, and thus the host crop gets protected from pathogen attacks. Similarly, mixed cropping may reduce or increase the disease incidences. For instance, when soybean and maize are grown together, soybean rust (*Phakopsora pachyrhizi*) is more pronounced, and when maize and cowpea are grown together, anthracnose (*Colletotrichum lindemuthianum*) is more severe on the cowpea than when a single crop is grown (Ogle and Dale 1997). Similarly, less incidence of *Macrophomina* stem and root rot was observed in sesame, grown as a mixed crop or intercropped with green gram (Rajpurohit 2002). Crop rotation and burning of stubble also helps to reduce the build-up of inoculum affecting seed setting. Good control of *Cephalosporium gramineum* and *Pseudocercosporella* spp. in cereal crops has been obtained following stubble burning (Ogle and Dale 1997). Crop rotation also helps to reduce the build-up of pathogens causing seed rot (Francl et al. 1988).

The most effective disease control strategy is to use, as far as possible, the disease-resistant or disease-tolerant varieties and disease-free seeds for cultivation and adopt a cropping pattern which does not aggravate a disease.

4.2   Seed Certification for the Management of Seed-Borne Diseases

Seed certification is a regulatory management practice for disease-free seed production and to reduce the seed-borne infection. Seed certification ensures that the seed has certain set standards and the quality and history of each seed lot are evident and may be traced. Seed certification procedures are established to maintain the standards of purity and permissible level of infection in both field and laboratory testing. Certain legislation and standard have been recommended by the seed certification board/committee for seed testing including seed health; however, any organization/country/state may enforce even higher voluntary standards (HVS) for varietal and physical purity, for weed seed contents, and for seed health in a seed lot.

Seed certification includes testing of seed lots before sowing as well as after harvesting and crop inspection in the field in compliance with standards set forth for isolation and freedom from diseases. Implementation of seed certification procedures helps to regulate and control the distribution and spread of certain seed-borne pathogens to uninfected newer areas. Field inspection during seed production helps to reject the seed lots with high incidence of seed-transmitted pathogens there in the field itself. Nevertheless, certain seed-borne pathogens are carried through seed asymptomatically. For that matter, seed samples are tested under laboratory conditions for notified seed-borne diseases under seed certification programme, and those seed lots exhibiting higher level of seed infection than prescribed are rejected. Despite the fact that seed certification standards (tolerance limit) for seed-borne diseases are difficult to establish in different categories of seed (basic, foundation, or certified seed), inoculum threshold has been established with supportive correlation data for a few seed-borne pathogens such as *Phoma lingam*, *Xanthomonas campestris* pv. *campestris*, *Pseudomonas syringae* pv. *phaseolicola*, and *Diaporthe phaseolorum* var. *sojae* (Kuan 1988).

There are many examples of successful seed health management through seed certification. In Australia, certification of bean seed production (*Phaseolus vulgaris* L.) began in the 1930s in New South Wales and Victoria as a measure against the devastating diseases bacterial blight and anthracnose (Persley et al. 2010). Successful management of barley stripe mosaic virus (BSMV) following seed certification prevented yield losses in Montana (Carroll 1983) and North Dakota, USA. At both the places, the programme followed field inspection, supplemented with serological assays (Carroll et al. 1979) of the foundation seed lot. In North Dakota, no specific field inspection is done, but the samples of foundation seed, having 1000 seeds in each seed lot, are assayed by ELISA. In either case, only the virus-free seeds are certified. Similarly, schemes based on rigorous field inspections were a complete success for the management of seed-borne fungal diseases (Gabrielson 1988). A standard of no infection of *Plenodomus lingam* in brassica, *Septoria apiicola* and *Phoma betae* on celery, *Colletotrichum lindemuthianum* and *Pseudomonas syringae* pv. *phaseolicola* on beans, and *Ascochyta fabae* in broad bean for seed multiplication is proposed (Hewett 1979a, b). Besides, inoculum thresholds for seed-borne fungi (Gabrielson 1988), bacteria (Schaad 1988), and viruses (Stace-Smith and Hamilton 1988) are used as part of disease management. In India, Karnal bunt of wheat was the first disease for which tolerance limit was fixed for seed-borne inoculum as early as in 1970, and strict field inspection, followed by seed testing, could effectively control this disease.

In India, in 1971, the seed certification standards were formulated and published in the manual of 'Indian Minimum Seed Certification Standards', which contained general seed certification standards applicable to all major crops. These minimum seed certification standards are further revised and upgraded in 1988 and in 2013 based on information generated through scientific studies.

4.2.1   Designated Seed-Borne Diseases

Diseases specified by national regulatory authority for the certification of seeds and for which certification standards must always be complied with are known as designated diseases. These diseases would cause contamination if present in the seed field or within the specified isolation distance. In order to produce disease-free seeds, some diseases are designated, and standards have been fixed for those diseases in concerned crops in India. The permissible limits of seed-borne designated diseases of important agricultural and horticultural seed crops are listed in Table 1 and exemplified in Plate 7. During field inspection, the guidelines given in the Indian Minimum Seed Certification Standards should be carefully followed by checking the symptoms of these designated diseases at different plant stages. Simultaneously, compliance for seed health testing under laboratory conditions as per the set standards should be followed.

Also, use of tissue culture and micro-propagation techniques helps in the multi-plication of pathogen-free planting stock which can be grown in greenhouse to create a barrier for disease incidents and their spread (IMSCS 2013).

4.3   Management of Seed-Borne Diseases Through Quarantine Regulations

Plant quarantine is a legislative procedure to exclude the plant pathogens from invading into an area where they do not exist, by monitoring the import and export of plant, seeds, grafts, planting material, or equipment to prevent the spread of diseases and pests through these sources.

Plant quarantine may, therefore, be defined as 'Rules and Regulations' proclaimed by the government(s) to regulate the introduction of plants, planting materials, plant products, soil, living organisms, etc. from one region to another with a view to prevent unintended introduction of exotic pests, weeds, and pathogens harmful to the agriculture or to the environment of a region or country and, in case introduced, prevent their establishment and further spread without adversely affect-ing the trade (Gupta and Khetarpal 2004). Quarantine, thus, aims to prevent the introduction of dangerous pathogens, but not to stop the movement of other biological material.

Initially, it was France where a Quarantine Act was enacted in 1660 for the eradication of barberry plants. Subsequently, Germany in 1873, Britain in 1866, and the United States in 1912 passed Acts and legislations to prohibit the entry of plants or planting material carrying pest and diseases harmful to agriculture.

Table 1 (A, B, and C) Designated seed-borne diseases and permissible limits for important agricultural and horticultural seed crops in India (IMSCS 2013)

A. Designated diseases and their maximum permissible limits in seed

Crop	Disease/causal organism	Seed standards (%)	
		Foundation	Certified
Wheat and hybrids	Karnal bunt (*Tilletia tritici*)	0.05	0.25
Paddy and hybrids	Bunt (*Tilletia barclayana*)	0.10	0.50
Sorghum and hybrids	Ergot (*Claviceps sorghi*) (teleomorph) (*Sphacelia sorghi*) (anamorph) Sclerotia, seed entirely or partially modified as sclerotia, broken sclerotia, or ergotted seed	0.02	0.04
Pearl millet and hybrids	Ergot (*Claviceps sorghi*) (teleomorph) (*Sphacelia sorghi*) (anamorph) Sclerotia, seed entirely or partially modified as sclerotia, broken sclerotia, or ergotted seed	0.02	0.04
Triticale	Karnal bunt (*T. indica* = *Tilletia tritici*)	0.05	0.25
Forage sorghum including Sudan grass	Ergot (*Claviceps* spp. = *Sphacelia sorghi*) Sclerotia, seed entirely or partially modified as sclerotia, broken sclerotia, or ergotted seed	0.02	0.04
Sweet potato	Storage rots	None	None
	Scurf (*Monilochaetes infuscans*)	None	1.0
	Wilt (*Fusarium oxysporum* f.sp. *batatas*)	None	1.0
	Block rot (*Ceratostomella fimbriata*)	None	1.0
	Internal cork	5.0	5.0
	Nematode	None	1.0
Ginger	Dry rot	1.0	5.0
	Phyllosticta	5.0	10.0
Turmeric	Dry rot	1.0	5.0

B. Designated diseases and their maximum permissible limits in field

Crop group	Disease/causal organism	Field standards (%)	
		Foundation	Certified
Cereals			
Wheat and hybrids	Loose smut (*Ustilago tritici*)	0.10	0.50
Sorghum and hybrids	Kernel smut (*Sphacelotheca sorghi* = *Sporisorium sorghi*) Head smut (*Sphacelotheca reiliana* = *Sporisorium reiliana*)	0.050	0.10
Barley and hybrids	Loose smut (*Ustilago nuda* = *Ustilago segetum* var. *nuda*)	0.10	0.50
Oat	Loose smut (*Ustilago nuda* = *Ustilago segetum* var. *avenae*)	0.10	0.50
Pearl millet and hybrids	Grain smut (*Tolyposporium penicillariae*) Downy mildew/green ear (*Sclerospora graminicola*) Ergot (*Claviceps microcephala* = *C. fusiformis*)	0.050 0.050 0.020	0.10 0.10 0.040
Triticale	Ergot (*Claviceps purpurea*)	0.020	0.040

(continued)

Table 1 (continued)

B. Designated diseases and their maximum permissible limits in field

Crop group	Disease/causal organism	Field standards (%)	
		Foundation	Certified
Oilseeds			
Sesame	Leaf spot (*Cercospora sesami*)	0.50	1.0
Sunflower and hybrids	Downy mildew (*Plasmopara halstedii*)	0.050	0.50
Pulses			
Green gram	Halo blight (*Pseudomonas phaseoli* = *P. savastanoi* pv. *phaseolicola*)	0.10	0.20
Bean legume vegetable			
Cowpea	• Ashy stem (*Macrophomina phaseolina*) • Anthracnose (*Colletotrichum lindemuthianum*) • Blight (*Ascochyta* spp.) (for hill areas only) • Cowpea mosaic	0.10	0.50
French and Indian bean	• Bacterial blight (*Xanthomonas* spp.) • Anthracnose (*Colletotrichum lindemuthianum*) • Blight (*Ascochyta* spp.) (for hill areas only) • Bean mosaic	0.10	0.20
Cluster bean	• Bacterial blight (*Xanthomonas campestris* pv. *cyamopsidis*) • Anthracnose (*Colletotrichum lindemuthianum*) • Blight (*Ascochyta* spp.) (for hill areas only)	0.10	0.20
Cucurbits			
Musk melon and hybrids	• Cucumber mosaic virus (CMV)	0.10	0.20
Summer squash	• CMV • Watermelon mosaic virus (WMV)	0.10	0.50
Solanaceous vegetables			
Eggplant	• *Phomopsis* blight (*Phomopsis vexans*)	0.10	0.20
Eggplant hybrids		0.10	0.50
Chilli	• Leaf blight (*Alternaria solani*) • Anthracnose (*Colletotrichum capsici*)	0.10	0.50
Tomato and hybrids	• Early blight (*Alternaria solani*) • Leaf spot (*Stemphylium solani*) • Tobacco mosaic virus	0.10	0.50
Leafy vegetables			
Celery	• Leaf blight (*Septoria apiicola*) • Root rot (*Phoma apiicola*)	0.10	0.50
Lettuce	• Lettuce mosaic virus	0.10	0.50
Parsley	• Leaf spot (*Septoria petroselini*)	0.10	0.50

(continued)

Table 1 (continued)

B. Designated diseases and their maximum permissible limits in field

| Crop group | Disease/causal organism | Field standards (%) | |
		Foundation	Certified
Cole crops			
Cabbage, cauliflower, broccoli, turnip, radish	• Black leg (*Leptosphaeria maculans*) • Black rot (*Xanthomonas campestris* pv. *campestris*) • Soft rot (*Erwinia carotovora*)	0.10	0.50
Fibre crops			
Jute	• Jute chlorosis	1.00	2.00
Forage crops			
Forage sorghum including Sudan grass	Kernel smut (*S. sorghi*) Head smut (*S. reiliana*)	0.050	0.10
Others			
Sweet potato	• Black rot (*Ceratocystis fimbriata*)	None	None
Plant bed	• Wilt (*Fusarium oxysporum* f.sp. *batatas*) • Scurf (*M. infuscans*)	None None	None None
Seed bed	• Wilt (*Fusarium oxysporum*) • Mosaic	0.050	0.10
Taro (Arvi)	• *Phytophthora* rot (*Phytophthora colocasiae*) • Dasheen mosaic	None 0.50	None 0.10
Tapioca	• Mosaic	0.10	0.50
Flower crops			
Annual carnation	• Mosaic (streak mosaic virus)	0.10	0.20
Chrysanthemum spp.	Grey mould (*Botrytis cinerea*) • Blotch (*Septoria chrysanthemella*)	0.10	0.20
Marigold	Leaf spot (*Alternaria tagetica*) Flower bud rot (*A. alternata*) Collar rot (*Rhizoctonia solani*)	0.10	0.20
Ornamental sunflower (*Helianthus* spp.)	• Downy mildew (*P. halstedii*)	0.050	0.050
Petunia/hybrid	Leaf blotch (*Cercospora petuniae*) Leaf spot (*Ascochyta petuniae*) *Phyllosticta* leaf spot (*Phyllosticta petunia*) Leaf blight (*Alternaria alternata*) Crown rot (*Phytophthora parasitica*) Tobacco mosaic virus (TMV) Cucumber mosaic virus (CMV)	0.10	0.20
Snapdragon/hybrid	• Anthracnose (*Colletotrichum antirrhini* and *C. fuscum*) • Blight (*Phyllosticta antirrhini*)	0.10	0.20

C. Certification for micro-propagation/tissue culture recommended for disease-free propagation

Crop	Diseases
Apple	Apple mosaic virus, apple chlorotic leaf spot virus
Bamboos	Bamboo mosaic virus (BaMV)

(continued)

Table 1 (continued)

C. Certification for micro-propagation/tissue culture recommended for disease-free propagation	
Crop	Diseases
Banana	Bunchy top virus, cucumber mosaic virus, banana bract mosaic virus, banana streak virus
Black pepper	CMV and *Badnavirus*
Citrus	Indian citrus ringspot virus (ICRSV), citrus tristeza virus (CTV), and citrus yellow mosaic virus (CYMV)
Potato	Potato virus A, potato virus S, potato virus M, potato virus Y, potato virus X, potato leafroll virus (PLRV), potato apical leaf curl virus (PALCV), potato spindle tuber viroid (PSTVd) Wart (*Synchytrium endobioticum*) Cyst-forming nematodes Brown rot (*Pseudomonas solanacearum*) Common scab (*Streptomyces scabies*)
Sugarcane	Sugarcane mosaic virus Yellow leaf and *Luteovirus* Grassy shoot (*Candidatus Phytoplasma sacchari*) Red rot (*Colletotrichum falcatum*) Smut (*Sporisorium scitamineum*)
Vanilla-tissue culture	Vanilla mosaic potyvirus Vanilla necrosis potyvirus Cymbidium mosaic potexvirus Odontoglossum ringspot tobamovirus Uncharacterized potyvirus/rhabdovirus

In India, the Destructive Insects and Pests Act, 1914, provides regulatory measures for the protection of plants and planting material. India has a well-established network to offer plant quarantine services at both national and state level. The Directorate of Plant Protection, Quarantine and Storage, Ministry of Agriculture and Farmers Welfare, Government of India, monitors the plant quarantine services at national level. Plant quarantine services are rendered through 73 plant quarantine stations in the country at different airports, seaports, and land frontiers with headquarter at Faridabad. In India, domestic quarantine under DIP act is in place to restrict the movement of invasive pests, viz. flute scale, San Jose scale, coffee berry borer, codling moth, banana bunchy top, mosaic viruses, potato cyst nematode, potato wart, and apple scab. Many plant pathogens, for instance, downy mildew (*Sclerospora graminicola*) spores and ergot (*Claviceps fusiformis*) sclerotia in pearl millet, contaminate the seeds during threshing and may disseminate the inoculum with the infected seed to other un-infested areas. The ICAR-National Bureau of Plant Genetic Resources (NBPGR) is the nodal agency authorized to undertake quarantine processing of all the plant genetic material including transgenics and research material in the country and for issuance of import permit under PQ order, 2003 (Regulations of Import into India). NBPGR has intercepted and detected several pathogens, not prevalent in India, in the imported plant

| Wheat loose smut | Sorghum kernel smut | Pearl millet green ear | Pearl millet ergot |
| Pearl millet smut | Cowpea mosaic | Tomato early blight | Chilli anthracnose |

Plate 7 Important designated diseases in different crops

germplasm using conventional, serological, and molecular techniques (Chalam et al. 2017; Bhalla et al. 2018).

4.4   Management of Seed-Borne Diseases Through Seed Treatment

Seed treatment refers to an application of physical, chemical, biological, or organic matter(s) to the seed to provide protection against pests and pathogens to germinating seed and seedlings and to improve the establishment of healthy crops. Seed treatment improves the seed quality and manages the seed-borne pathogens. Seed treatment not only benefits in seed disinfestations by cleaning the spores, mycelia, or propagules of microorganisms on the seed surface but also supports seed disinfection by eliminating the pathogen that has penetrated deep into the living cells of seed (e.g. smut or bunt) and gives protection to germinating seedlings from soil-borne pathogens. Various methods are applied for seed treatment.

4.4.1   Physical Seed Treatment

Physical seed treatments consist of heat treatments, most common being hot water and hot air treatments. The oldest practice of physical seed treatments is hot water treatment, where the seeds are immersed in hot water at a precise temperature for a certain period. Hot water treatment was generally practised to sterilize contaminated cereal seeds (Gilbert et al. 2005), though Nega et al. (2003) reported the successful management of seed-borne pathogens, viz. *Alternaria* spp., *Phoma* spp., *Peronospora valerianellae*, *Septoria* spp., and *Xanthomonas* spp., in carrots, celery, parsley, and lettuce by hot water treatment. Koch et al. (2010) observed hot air treatment of carrot seeds as effective as chemical treatment for the management of seed-borne infection of *Alternaria dauci* and *A. radicina*.

Aerated steam treatment has been highly effective in multiple host pathogen systems (Forsberg et al. 2002; Tinivella et al. 2009; Schaerer 2012) and exhibited an effective management of seed-borne infection of *Septoria apiicola* in celery, *F. moniliforme* in sweet corn, *C. michiganensis* subsp. *michiganensis* in tomato, and *X. campestris* in cauliflower (Groot et al. 2006, 2008). Precisely managed aerated steam seed treatment kills the pathogen leaving seeds intact without adversely affecting the seed vigour and viability.

Electronic seed treatment is a new seed treatment method which may help to destroy the DNA of harmful organisms present on seed surface. However, there is a need for further investigation to reach an inference whether this method of seed treatments can be used to eradicate pathogens on seed surface (Schmitt et al. 2006).

4.4.2 Chemical Seed Treatment

Application of fungicidal seed treatments improves seed emergence and plant vigour and avoids the transmission of seed-borne pathogens from seed to seedlings. In addition, chemical seed treatments protect the emerging seed and seedlings from soil-borne pathogens as well. Two organic fungicides captan (dicarboximide fungicide) and concurrently thiram (dithiocarbamate), when introduced, were widely applied for seed treatment in various crops. Systemic fungicides, carboxin and thiabendazole, introduced in the early 1970s also got wide acceptability for seed treatment in a number of crops. Systemic fungicide is an important disease control strategy for several agricultural and horticultural crops worldwide. In India, benzimidazole fungicides are registered for use in 18 crops including rice, wheat, barley, peanuts, cotton, jute, mango, apples, grapes, beans, eggplants, cucurbits, peas, sugar beets, tapioca, and roses. Seed treatment with systemic fungicides is considered as an economically viable disease management strategy for several agricultural and horticultural crops worldwide (Bhushan et al. 2013; Lamichhane et al. 2020). Some other classes of fungicides, such as phenylpyrroles, phenylamides, strobilurins, and triazoles, are also used for the management of a number of seed-borne diseases through seed treatment (Zeun et al. 2013).

Fungicides introduced in recent past are the formulations of mixtures of multiple active ingredients with different mechanisms of action to enhance control of a variety of pathogens, e.g. strobilurins (e.g. azoxystrobin, pyraclostrobin, and trifloxystrobin), triazoles (e.g. difenoconazole, tebuconazole, and prothioconazole), pyrazole, and carboxamides (e.g. sedaxane). These systemic fungicides are used as seed dressing fungicides.

Seed treatments are commonly applied as seed dressing, seed coating, or seed pelleting before sowing (Pedrini et al. 2017). Seed dressing is the most common seed treatment method and involves dressing the seed with a dry or wet formulation of a fungicide. Seed dressing with fenfuram, triadimefon, triadimenol, tebuconazole, and hexaconazole enabled the effective control of wheat flag smut caused by *Urocystis agropyri* (Singh and Singh 2011; Shekhawat and Majumdar 2013; Kumar et al. 2019).

Seed coating and pelleting require a special binder to be used with the formulation to improve the adherence of the product to seed surface. Anwar and Shafi Bhat

(2005) evaluated several fungicides, isoprothiolane, tricyclazole, hexaconazole, and mancozeb, as seed coatings in two different doses and found isoprothiolane and tricyclazole to be the most effective in controlling the nursery blast disease. Seed pelleting requires specialized machinery and application techniques; in this case, fungicides can be segregated at different layers of the coating (Ahmed and Kumar 2020), to make it more effective.

4.4.3 Biological Seed Treatment

Application of beneficial antagonistic microbes to seed for managing seed- and soil-borne pathogens is a classical delivery system which reduces water pollution from chemicals, enriches soil microbiota, and is safe for the environment. Seed treatment with microbial antagonists protects seeds and seedlings from various pathogens. Majority of microbial antagonists are bacteria (84%) and fungi (16%). In the last decades, various bacterial biocontrol agents have been identified and used for the management of seed-borne diseases. Various bacteria which are mainly investigated and successfully commercialized as the biological control agents are *Pseudomonas fluorescens*, *Bacillus subtilis*, *Serratia marcescens*, *Streptomyces griseoviridis*, and *Burkholderia cepacia* (Singh 2014; Bisen et al. 2015; Keswani et al. 2016; Gouda et al. 2018). Species of *Pseudomonas* and *Bacillus* are the most commonly used bacterial biocontrol agents in controlling various phytopathogens including seed-borne pathogens (Abhilash et al. 2016; Bhat et al. 2019; Khan et al. 2020). Similarly, various fungi have also been studied for their effect as biocontrol agents on seed-borne pathogens. The important fungal biocontrol agents are the species of *Phomopsis*, *Ectomycorrhizae*, *Trichoderma*, *Cladosporium*, *Gliocladium*, etc. Among various fungal biocontrol agents, *Trichoderma* spp. are widely studied and globally used biopesticide (Singh 2006; Keswani et al. 2013; Bisen et al. 2016; Singh et al. 2016). *Trichoderma* spp. are potential plant symbionts and reported for their antagonistic activity against a wide range of seed- and soil-borne fungi. More than 60% of the global biopesticide market is based on *Trichoderma* formulation (Keswani et al. 2013).

Extracts from several plant species are also known to contain natural antimicrobial compounds which can be used for seed disinfection as an alternative to fungicidal treatments, singly or in combination (Begum et al. 2010). These extracts include essential oils, showing virtuous antifungal activities in in vitro trials using tea, clove, peppermint, rosemary, laurel, oregano, and thyme oils. Such oils have been reported to be active against pathogens like *Drechslera*, *Ascochyta*, and *Alternaria* spp. (Alice and Rao 1986; Riccioni et al. 2013). Of these, thyme oil which contains thymol and other antifungal compounds has shown greater efficacy in both in vitro and field tests against seed-borne fungi and bacteria (Van der Wolf et al. 2008). Other effective natural antifungal compounds have been extracted from plants that belong to the genus *Allium*, which produce various sulphur-containing compounds with antimicrobial effects (Nelson 2004; Lanzotti 2006).

5 Conclusions

Seed-borne pathogens pose a serious threat to crop establishment and yield. Seeds also serve as a way of dispersal and survival of plant pathogens. Therefore testing of seed lots for the presence of the pathogen is the most efficient way to avoid spread of diseases. Seed health testing and detection are the first line of approach in managing seed-borne diseases of plants. This can most effectively be accomplished by exclusion, using seed detection assays to identify contaminated seed lots that can be discarded or treated. In comparison to conventional and serological techniques, PCR assays have much higher sensitivity and specificity and often require very short time to detect the pathogens associated with seed. These are applicable to a wide range of pathogens and can be used to separate closely related species. Different modified PCR techniques including real-time PCR, BIO-PCR, IMS-PCR, RT-PCR, and IC-RT-PCR and LAMP assay hold great potential for enhancing pathogen detection in seeds, because it embodies some of the key characteristics, which include specificity, sensitivity, rapidity, ease of implementation and interpretation, and applicability (see chapter on Molecular Techniques for Testing Genetic Purity and Seed Health for more details).

Many techniques, measures, strategies, and procedures are applied in the management of seed-borne diseases in both field and horticultural crops. These techniques include (1) agronomic practices, viz. selection of disease-free areas for seed production, use of disease-resistant crop varieties and disease-free seeds for cultivation, adjustment of the time of sowing, removal and destruction of pathogen-infected crop residues, alternate and collateral weed hosts, timely rouging of designated disease-infected plant and/or plant parts in the seed production plot, proper isolation distance, crop rotation, and balanced nutrient management, (2) seed certification, (3) quarantine regulations, and (4) seed treatment methods, viz. physical, chemical, and biological methods. The treatments of seeds with fungicides or biocontrol agents represent good methods for their protection, disinfestation, or disinfection from seed-borne pests and pathogens. The success of such treatments relies upon the pathogen localization on the seed level and can provide improved crop stand in the field and increased yields.

References

Abhilash PC, Dubey RK, Tripathi V, Gupta VK, Singh HB (2016) Plant growth-promoting microorganisms for environmental sustainability. Trends Biotechnol 34:847–850

Agarwal VK, Mathur SB (1992) Detection of *Tilletia indica* (Karnal bunt) wheat seed samples treated with fungicides. FAO Plant Prot Bull 40:148

Agarwal VK, Sinclair JB (1997) Principles of seed pathology, 2nd edn. CRC Press, Boca Raton, p 539

Agarwal VK, Srivastava AK (1981) A simpler technique for routine examination of rice seed lots for rice bunt. Seed Technol News 11(3):1–2

Ahmed S, Kumar S (2020) Seed coating with fungicides and various treatments for protection of crops: a review. Int J Agric Env Sustain 2(1):6–13

Ali A (2017) Rapid detection of fifteen known soybean viruses by dot-immunobinding assay. J Virol Methods 249:126–129

Alice D, Rao AV (1986) Management of seed borne *Drechslera oryzae* of rice with plant extracts. Int Rice Res Newslett 11:19

Anil A, Singh US, Kumar J, Garg GK (2008) Application of molecular and immunodiagnostic tools for detection, surveillance and quarantine regulation of Karnal bunt (Tilletia indica) of wheat. Food Agric Immunol 19(4):293–231

Anwar A, Shafi Bhat M (2005) Efficacy of fungicides as seed treatment in the management of blast disease of rice in nursery bed. Agric Sci Digest 25(4):293–295

Bariana HS, Shannon AL, Chu PWG, Waterhouse PM (1994) Detection of five seed-borne legume viruses in one sensitive multiplex polymerase chain reaction test. Phytopathology 84:1201–1205

Begum J, Anwar MN, Akhter N, Nazrul Islam Bhuiyan MD, Hoque MN (2010) Efficacy of essential oils as jute seed protectant. Chittagong Univ J Biol Sci 5:1–7

Bhalla S, Chalam VC, Singh B, Gupta K, Dubey SC (2018) Biosecuring of plant genetic resources in India: role of plant quarantine. ICAR – National Bureau of Plant Genetic Resources, New Delhi, p 216

Bhat MA, Rasool R, Ramzan S (2019) Plant growth promoting rhizobacteria (PGPR) for sustainable and eco-friendly agriculture. Acta Sci Agric 3:23–25

Bhushan C, Bhardwaj A, Misra SS (2013) State of pesticide regulations in India. Report of Centre for Science and Environment, New Delhi

Bisen K, Keswani C, Mishra S, Saxena A (2015) Unrealized potential of seed biopriming for versatile agriculture. In: Rakshit A, Singh HB, Sen A (eds) Nutrient use efficiency: from basics to advances. Springer, New Delhi, pp 193–206

Bisen K, Keswani C, Patel JS, Sarma BK, Singh HB (2016) *Trichoderma* spp.: efficient inducers of systemic resistance in plants. In: Chaoudhary DK, Varma A (eds) Microbial-mediated induced systemic resistance in plants. Springer, Singapore, pp 185–195

Cappelli C, Covarelli L (2005) Methods used in seed pathology and their current improvements. Phytopathol Pol 35:11–18

Carroll TW (1983) Certification schemes against barley stripe mosaic. Seed Sci Technol 11:1033–1042

Carroll TW, Gossel PL, Batchelor DL (1979) Use of sodium dodecyl sulphate in serodiagnosis of barley stripe mosaic in embryo and leaves. Phytopathology 69:12–14

Chalam VC, Parakh DB, Maurya AK (2017) Role of viral diagnostics in quarantine for plant genetic resources and preparedness. Indian J Plant Genet Resour 30:271–285

Chang GH, Yu RC (1997) Rapid immunoassay of fungal mycelia in rice and corn. J Chin Chem Soc 35:533–539

Chitarra LG, Langerak CJ, Bergervoet JHW, van den Bulk RW (2002) Detection of the plant pathogenic bacterium *Xanthomonas campestris* pv. *campestris* in seed extracts of *Brassica* sp. applying fluorescent antibodies and flow cytometry. Cytometry 47(2):118–126

Chitarra LG, Breeuwer P, Abee T, Van Den Bulk RW (2006) The use of fluorescent probe to assess viability of the plant pathogenic bacterium *Clavibacter michiganensis* subsp. *michiganensis* by flow cytometry. Fitopatol Bras 31:349–356

Correia T, Grammel N, Ortel I, Keller U, Tudzynski P (2003) Molecular cloning and analysis of the ergopeptine assembly system in the ergot fungus *Claviceps purpurea*. Chem Biol 10(12):1281–1292

Dorph-Petersen K (1921) Stats-Frokontrollen Dansk Frokontrol, Frederiksberg Bogtr Publisher, p 160

Faris-Mukhayyish S, Makkouk K (2008) Detection of four seed-borne plant viruses by the enzyme-linked immunosorbent assay (ELISA). J Phytopathol 106(2):108–114

Fegla GI, El-Samra IA, Younes HA, Abd El-Aziz MH (2000) Optimization of dot immunobinding assay (DIA) for detection of tomato mosaic virus (ToMV). Adv Agric Res 5:1495–1506

Forsberg G, Andersson S, Johnsson L (2002) Evaluation of hot, humid air seed treatment in thin layers and fluidized beds for pathogen sanitation. J Plant Dis Prot 109:357–370

Forster RL, Seifers DL, Strausbaugh CA, Jensen SG, Ball EM, Harvey TL (2001) Seed transmission of the high plains virus in sweet corn. Plant Dis 85:696–699

Francl LJ, Wyllie TD, Rosenbrock SM (1988) Influence of crop rotation on population density of *Macrophomina phaseolina* in soil infested with *Heterodera* glycines. Plant Dis 72:760–764

Gabrielson RL (1988) Inoculum thresholds of seed-borne pathogens: fungi. Phytopathology 78: 868–887

Gebeyaw M (2020) Review on: impact of seed-borne pathogens on seed quality. Am J Plant Biol 5(4):7–81

Gilbert J, Woods SM, Turkington TK, Tekauz A (2005) Effect of heat treatment to control *Fusarium graminearum* in wheat seed. Can J Plant Pathol 27:448–452

Gillaspie AG Jr, Pio-Ribeiro G, Andrade GP et al (2001) RT-PCR detection of seed-borne cowpea aphidborne mosaic virus in peanut. Plant Dis 85:1181–1182

Gnanamanickam SS, Shigaki T, Medalla ES, Mew TW, Alvarez AM (1994) Problems in detection of *Xanthomonas oryzae* pv. *oryzae* in rice seeds and potential for improvement using monoclonal antibodies. Plant Dis 78:173–171

Gouda S, Kerry RG, Das G, Paramithiotis S, Shin HS, Patra JK (2018) Revitalization of plant growth promoting rhizobacteria for sustainable development in agriculture. Microbiol Res 206: 131–140

Groot SPC, Birnbaum Y, Rop N, Jalink H (2006) Effect of seed maturity on sensitivity of seeds towards physical sanitation treatments. Seed Sci Technol 34:403–413

Groot SPC, Birnbaum Y, Kromphardt C, Forsberg G, Rop N, Werner S (2008) Effect of the activation of germination processes on the sensitivity of seeds towards physical sanitation treatments. Seed Sci Technol 36:609–620

Guillemette T, Iacomi-Vasilescu B, Simoneau P (2004) Conventional and real-time PCR-based assay for detecting pathogenic *Alternaria brassicae* in cruciferous seed. Plant Dis 88:490–496

Gumus M, Paylan IC (2013) Detection of viruses in seeds of some vegetables by reverse transcriptase polymerase chain reaction (RT-PCR). Afr J Biotechnol 12(25):3891–3897

Gupta K, Khetarpal RK (2004) Concept of regulation pests their risk analysis and Indian scenario. Annu Rev Plant Pathol 4:409–441

Gupta A, Kumar R (2020) Management of seed-borne diseases: an integrated approach. In: Kumar R, Gupta A (eds) Seed-borne diseases of agricultural crops: detection, diagnosis & management. Springer Nature, Singapore, pp 717–745

Hagborg WAF, Warner GM, Phillips NA (1950) Use of 2,4-D as an inhibitor of germination in routine examinations of beans for seed-borne infection. Science 111:91

Hajihasani M, Hajihassani A, Khaghani S (2012) Incidence and distribution of seed-borne fungi associated with wheat in Markazi Province, Iran. Afr J Biotechnol 11:629–629

Hampton J (2005) ISTA method validation. Seed Test Int 130:22–23

Hampton J (2007) ISTA method validation. Seed Test Int 133:39

Hewett PD (1979a) Regulating seed borne diseases by certification. The scientific basis for administrative control of plant diseases and Pests. In: Ebbels DL, King JE (eds) Plant health. Blackwell Scientific, Oxford, p 163

Hewett PD (1979b) Seed standards for disease in certification. J Nat Agric Bot 15:373–384

Iacomi-Vasilescu B, Blancard D, Guénard M, Molinero-Demilly V, Laurent E, Simoneaui P (2002) Development of a PCR-based diagnostic assay for detecting pathogenic *Alternaria* species in cruciferous seeds. Seed Sci Technol 30:87–95

Indian Minimum Seed Certification Standards (2013) The Central Seed Certification Board, Department of Agriculture & Co-operation Ministry of Agriculture, Government of India

ISF (2022) The International Seed Health Initiative (ISHI) https://worldseed.org/. Accessed 5 Mar 2022

Jha DK (1993) A textbook on seed pathology. Vikas Publishing House Pvt. Ltd, New Delhi, p 132

Keswani C, Singh SP, Singh HB (2013) A superstar in biocontrol enterprise: *Trichoderma* spp. Biotech Today 3:27–30

Keswani C, Bisen K, Singh V, Sarma BK, Singh HB (2016) Formulation technology of biocontrol agents: present status and future prospects. In: Arora NK, Mehnaz S, Balestrini R (eds) Bioformulations for sustainable agriculture. Springer, India, pp 35–52

Khan N, Bano A, Ali S, Ali Babar M (2020) Crosstalk amongst phytohormones from planta and PGPR under biotic and abiotic stresses. Plant Growth Regul 90:189–203

Koch E, Schmitt A, Stephan D, Kromphardt C (2010) Evaluation of non-chemical seed treatment methods for the control of *Alternaria dauci* and *A. radicina* on carrot seeds. Eur J Plant Pathol 127:99–112

Kohnen PD, Dougherty WG, Hampton RO (1992) Detection of pea seed-borne potyvirus by sequence-specific enzymatic amplification. J Virol Methods 37:253–258

Konstantinova P, Bonants P, Gent-Pelzer MV, Zouwen PVD, Bulk RVD (2002) Development of specific primers for detection and identification of *Alternaria* spp. in carrot material by PCR and comparison with blotter and plating assays. Mycol Res 106:23–33

Kuan TL (1988) Inoculum threshold for seed borne pathogens. An overview. Am Phytopathol Soc 78(6):868–872

Kumar S, Kashyap PL, Saharan MS, Singh I, Jasrotia P, Singh DP, Singh GP (2019) Difenoconazole: a new seed dressing molecule for effective management of flag smut (*Urocystis agropyri*) of wheat. J Cereal Res 11(1):37–40

Lamichhane JR, You MP, Laudinot V, Barbetti MJ, Aubertot JN (2020) Revisiting sustainability of fungicide seed treatments for field crops. Plant Dis 104:610–623

Lamka GL, Hill JH, McGee DC, Braun EJ (1991) Development of an immunosorbent assay for seed-borne *Erwinia stewartii* in corn seeds. Phytopathology 81:839–846

Lange L, Heide M (1986) Dot immunobinding (DIB) for detection of virus in seed. Can J Plant Pathol 8:373–379

Lanzotti V (2006) The analysis of onion and garlic. J Chromatogr A 1112:3–22

Lee KW, Lee BC, Park HC, Lee YS (1990) Occurrence of green mottle mosaic virus disease of watermelon in Korea. Kor J Plant Pathol 6:250–255

Limonard T (1966) A modified blotter test for seed health. Neth J Plant Pathol 72:319–321

Maddox DA (1998) Implications of new technologies for seed health testing and the worldwide movement of seed. Seed Sci Res 8:277–284

Majumder D, Rajesh T, Suting EG, Debbarma A (2013) Detection of seed borne pathogens in wheat: recent trends. Aust J Crop Sci 7(4):500–507

Mancini V, Murolo S, Romanazzi G (2016) Diagnostic methods for detecting fungal pathogens on vegetable seeds. Plant Pathol 65(5):691–703

Mathur SB, Jorgensen J (2002) A review of the activities of the plant disease committee of ISTA through its 75 years of existence, 1924–1999. ISTA Hist Papers 1:1–34

Mathur SB, Haware MP, Hampton RO (1988) Identification, significance and transmission of seed borne pathogens. In: Summerfield RJ (ed) World crops: cool season food legumes. Kluwer Academic Publishers, pp 351–365

McGee DC (1997) World phytosanitary system: problems and solutions. In: McGee DC (ed) Plant pathogens and the worldwide movement of seeds. APS Press, St Paul, MN, pp 67–80

Mosqueda-Cano G, Herrera-Estrella L (1997) A simple and efficient PCR method for the specific detection of *Pseudomonas syringae* pv. *phaseolicola* in the bean seeds. World J Microbiol Biotechnol 113:463–467

Nega E, Ulrich R, Werner S, Jahn M (2003) Hot water treatment of vegetable seed - an alternative seed treatment method to control seed-borne pathogens in organic farming. J Plant Dis Prot 110:220–234

Nelson EB (2004) Microbial dynamics and interactions in the spermosphere. Annu Rev Phytopathol 42:271–309

Ogle H, Dale M (1997) Disease management: cultural practices. In: Brown JF, Ogle HJ (eds) Plant pathogens and plant diseases. Rockvale Publications, Armidale, NSW, pp 390–404

Ojuederie OB, Odu BO, Ilori CO (2009) Serological detection of seed-borne viruses in cowpea regenerated germplasm using protein a Sandwich enzyme linked immunosorbent assay. Afr Crop Sci J 17(3):125–132

Pedrini S, Merritt DJ, Stevens J, Dixon K (2017) Seed coating: science or marketing spin? Trends Plant Sci 22:106–116

Persley D, Cooke T, House S (2010) Diseases of vegetable crops in Australia. CSIRO Publishing, Collingwood, VIC, p 292

Rajpurohit TS (2002) Influence of intercropping and mixed cropping with pearl millet, green gram and mothbean on the incidence of stem and root rot (*Macrophomina phaseolina*) of sesame. Sesame Safflower Newsl 17:40–41

Rao NK Bramel PJ (eds) (2000) Manual of genebank operations and procedures. Technical manual No. 6., ICRISAT, Andhra Pradesh, India

Rennie WJ (1982) Wheat loose smut. ISTA handbook on seed health testing, section 2, working sheet No. 48, International Seed Testing Association, Zürich, Switzerland

Riccioni L, Orzali L, Marinelli E (2013) Seed treatment with essential oils. Proceedings of the international conference 'future IPM in Europe', March 19–21, Riva del Garda, Italy

Schaad NW (1988) Bacteria. Part of inoculum threshold of seed borne pathogens symposium. Phytopathology 78:872–875

Schaerer H (2012) Seed treatments for healthy vegetable seedlings. Cost action FA1105 meeting – organic greenhouse horticulture, state of the art and future trends, 15–17 October, Bucharest, Romania

Scherm B, Balmas V, Spanu F, Pani G, Delogu G, Pasquali M, Migheli Q (2013) *Fusarium culmorum*: causal agent of foot and root rot and head blight on wheat. Mol Plant Pathol 14(4):323–341

Schmitt A, Jahn M, Kromphardt C, Krauthausen HJ, Roberts SJ, Wright SAI, Amein T, Forsberg G, Tinivella F, Gullino ML, Wikström M, van der Wolf J, Groot SPC, Werner S, Koch E (2006) STOVE: seed treatments for organic vegetable production. European Joint Organic Congress, Odense, Denmark

Shekhawat PS, Majumdar VL (2013) Management of flag smut of wheat through seed-cum- soil treatment with *Trichoderma* alone and in combination with fungicides. Plant Dis Res 28(2): 169–170

Sheppard JW, Wesseling JBM (1998) ISTA/ISHI guide for comparative testing of methods for the detection of seed-borne pathogens. Seed Sci Technol 26:237–255

Singh HB (2006) *Trichoderma*: a boon for biopesticides industry. J Mycol Plant Pathol 36:373–384

Singh HB (2014) Management of plant pathogens with microorganisms. Proc Indian National Sci Acad 80:443–454

Singh D, Singh A (2011) Raxil 060 FS – a new seed dressing fungicide formulation for the control of flag smut and loose bunt of wheat. Plant Dis Res 26(2):189

Singh V, Upadhyay RS, Sarma BK, Singh HB (2016) Seed biopriming with *Trichoderma asperellum* effectively modulates plant growth promotion in pea. Int J Agric Environ Biotechnol 9:361–365

Stace-Smith R, Hamilton RI (1988) Viruses. Part of inoculum threshold of seed borne pathogens symposium. Phytopathology 78:875–880

Tinivella F, Hirata LM, Celan MA (2009) Control of seed-borne pathogens on legumes by microbial and other alternative seed treatments. Eur J Plant Pathol 123:139–151

Torre C, Agüero J, Gómez-Aix C, Aranda MA (2019) Comparison of DAS-ELISA and qRT-PCR for the detection of cucurbit viruses in seeds. Ann Appl Biol 176(2):158–169

Tripathi R, Tiwari R, Vishunavat K (2018) Evaluation of different growth media for Clavibacter michiganensis subsp, michiganensis and formation of biofilm like structures. Int J Curr Microbiol Appl Sci 7:207–216

Tripathi R, Vishunavat K, Tiwari R (2022) Morphological and molecular characterization of *Clavibacter michiganensis subsp, michiganensis* causing bacterial canker in tomatoes. Physiol Mol Plant Pathol 119:101833

Tsedaley B (2015) Review on seed health tests and detection methods of seed-borne diseases. J Biol Agric Healthc 5(5):176–184

Van der Wolf JM, Bimbaum Y, van der Zouwen PS, Groot SPC (2008) Disinfection of vegetable seed by treatment with essential oils, organic acids and plant extracts. Seed Sci Technol 36:76–88

Walcott RR, Gitaitis RD (2000) Detection of *Acidovorax avenae* subsp. *citrulli* in watermelon seed using immunomagnetic separation and polymerase chain reaction. Plant Dis 84:470–474

Warham EJ, Butler LD, Sutton BC (1990) Seed testing of maize and wheat: a laboratory guide. CYMMYT, CAB International, UK, p 84

Wold A (1983) Opening address. Seed Sci Technol 11:464

Wylie S, Wilson CR, Jones RAC (1993) A polymerase chain reaction assay for cucumber mosaic virus in lupin seeds. Aus J Agric Res 44:41–51

Yang Y, Kim K, Anderson EJ (1997) Seed transmission of cucumber mosaic virus in spinach. Phytopathology 87:924–931

Zarzyńska-Nowak A, Hasiów-Jaroszewska B, Jeżewska M (2018) Molecular analysis of barley stripe mosaic virus isolates differing in their biological properties and the development of reverse transcription loop-mediated isothermal amplification assays for their detection. Arch Virol 163:1163–1117

Zeun R, Scalliet G, Oostendorp M (2013) Biological activity of sedaxane–a novel broad-spectrum fungicide for seed treatment. Pest Manag Sci 69:527–534

Molecular Techniques for Testing Genetic Purity and Seed Health

Santhy V., Nagamani Sandra, Kundapura V. Ravishankar, and Bhavya Chidambara

Abstract

With the globalization of seed trade and transgenic variety development, the application of molecular technologies for seed quality gained more significance in both the internal and international markets. Besides germination, genetic purity and seed health are the two most important seed quality parameters that determine the planting value of a seed lot. Compared to the conventional methods of testing, molecular marker technologies are more efficient for quality analysis as these are more accurate, robust, abundant, and faster. Among the various markers, simple sequence repeats (SSRs), due to their genome-wide presence, reproducibility, multi-allelic nature, and co-dominant inheritance, have emerged as the best markers, for establishing varietal distinctness, identity, and variety/hybrid seed purity testing. With the advent of the next-generation sequencing (NGS) technology, single nucleotide polymorphic (SNP) markers also became widely popular, and the closest to being an ideal marker besides SSRs, in seed genetic purity testing. With large-scale GM crop cultivation, testing for the adventitious presence and trait purity are two added components of seed quality testing. The methods of GM seed quality testing include DNA-based (conventional and real-time PCR), protein-based (lateral flow test and ELISA), and bioassay-based technologies. DNA-based methods including PCR/real-time PCR assays have been successfully employed to detect the adventitious presence of transgenic seeds in seed trade especially at international level, as well as in the national gene

S. V. (✉)
Division of Crop Improvement (Seed Technology), ICAR-CICR, Nagpur, India

N. Sandra
Division of Seed Science and Technology, ICAR-Indian Agricultural Research Institute, New Delhi, India

K. V. Ravishankar · B. Chidambara
Division of Basic Sciences, ICAR- Indian Institute of Horticultural Research, Bengaluru, India

© The Author(s) 2023
M. Dadlani, D. K. Yadava (eds.), *Seed Science and Technology*,
https://doi.org/10.1007/978-981-19-5888-5_15

365

banks for germplasm conservation. ISTA plays a prominent role in international harmonization and providing universal guidelines on use of different methods to detect GM seeds. The BMT group of UPOV and the Working Group on DNA Methods of the Variety Committee of ISTA, work in tandem to standardize suitable molecular techniques for establishing variety identity and purity testing, respectively. In the area of seed health testing also, molecular detection assays such as, PCR (nested PCR, multiplex PCR, real-time PCR), loop-mediated isothermal amplification (LAMP), and DNA microarray with many advantages over the conventional assays have been proven highly useful. However, there is a need to validate the usefulness of molecular markers through stringent multi-laboratory tests for their reproducibility before recommending them in routine seed purity and health testing.

Keywords

Variety identity · Genetic purity testing · ISTA · UPOV · PVP · Hybrid purity · DNA markers · Trait purity · ELISA · GM detection · PCR · Transgenic crops · Adventitious presence · Seed-borne viruses

1   Introduction

With the globalization of the seed market, competition in the seed trade has increased manifold with more stringent quality standards. High seed quality can only be obtained by a thorough control of the entire seed production process, from planning to final delivery. Among the many parameters of seed quality; genetic purity and seed health are of high importance in determining the authenticity and planting value of a seed lot. While field-based grow-out method is still much in vogue for variety authentication, several DNA-based modern techniques have been developed to test the genetic purity and seed health. Genetic purity of a variety determines the extent of conformity of the submitted seed sample with the claimed variety or the extent of purity of a variety within its seed lot or purity of F1 hybrid in a given hybrid seed lot. Maintaining absolute (100%) varietal purity during production is difficult to achieve in spite of following the recommended steps, allowing some degree of varietal off-types, which occurs inadvertently due to outcrossing, incomplete roguing of the off-types, or physical admixture during harvest, storage, or seed handling (Bradford 2006). Hence, to ascertain the desired levels of purity, seed lots are subjected to post-control grow-out tests or laboratory-based tests for internal quality control by the seed companies/producer/supplier. The scope of seed purity testing got broadened with the introduction of transgenic varieties in the global market, making the tests for GM trait purity and the adventitious presence of GM seeds in non-GM seed lots an essential requirement. Seed health, the other major quality determinant, is a measure of freedom of seeds from incidence of fungi, bacteria, viruses, and (very serious error and need to be removed) pests such as nematodes and insects in the seed lot, many of which pose difficulty in detection due to lack of unambiguous, precise, and rapid methods. Molecular techniques offer high potential for testing variety identity, genetic purity of seed lots, trait purity, and adventitious presence of transgenes as well as ascertaining good seed health.

2 Variety Characterization/Identification and Genetic Purity Testing

With the proliferation of a large number of varieties and narrowing gene pool, characterization and identification of varieties and establishing distinctness among closely related genotypes have become crucial aspects in plant breeding, variety maintenance, seed production, seed trade and protecting the interests of farmers and consumers (Smith and Register 1998). Variety characterization based on genetically inherited markers (morphological, biochemical, or molecular) aids in the development of identification keys and in distinguishing each variety unambiguously. As the number of crop varieties in cultivation is rising every year, the need to identify these based on stable and robust genetic markers is becoming more challenging (Santhy and Meshram 2015). The scope of variety identification using molecular markers is unlimited and can be used to test the uniqueness of a variety using a small sample size. ISTA (International Rules for Seed Testing) (2021) termed 'the combination of alleles determined for a specific set of DNA markers within a sample or variety' as 'allele profile' which helps in identifying varieties. The characterization done at molecular level generating a marker profile is called DNA fingerprinting and can provide additional evidence for the uniqueness of a variety. While the number of bands generated by a polymorphic marker is used to differentiate varieties (Law et al. 1998), the number and genomic distribution of the markers determine the robustness of a DNA fingerprint. In general, for n number of varieties to be distinguished, the minimum number of polymorphic bands scored should be between n and 2n (Singh and Singh 2019). It is pertinent to mention that while one robust marker shall be enough for testing hybrid purity, more numbers or multiple markers with high discrimination are needed for varietal purity testing. The markers required are still higher for variety identification and, the number required may be the highest for variety registration as EDVs (Hwu 2013).

Varietal purity testing is a quality assurance tool for crop producers and suppliers to comply with seed regulations for both commercial marketing and international seed exchanges. Due to their stability, reliability, and abundance, molecular markers can be employed for seed purity-related issues such as (1) determining the genetic identity of a variety or parental lines and verifying if the variety offered for sale is the same or not; (2) testing purity of elite varieties/inbred lines, GM/non-GM seeds, and/or F1 hybrid seeds; (3) trait-specific testing of seed (GM seeds).

3   Molecular Markers for Varietal Identity and Genetic Purity

Molecular markers have several advantages in comparison with conventional markers such as, high polymorphic information content (PIC), insensitivity to environment, stability, developmental stage independence and abundance. These methods are best suited for unequivocal identification of varieties (Singh and Singh 2019) and to determine whether the allele profile of a sample is identical to that of an authentic reference variety (www.seedtest.org). Based on the loci studied, the marker can be multi-locus (RAPD, AFLP) or single locus (SSRs and SNPs), and based on dominance, they can be dominant (RAPD, AFLP) or co-dominant (SSRs, SNPs). Based on the method, they can be hybridization-based (RFLP) or amplification-based (RAPD, SSRs, AFLP). Based on the number of alleles, markers are biallelic (SNP) or multi-allelic (SSR). Markers can be those where prior genome information is required (SSR) or those where genome information is not required (RAPD, AFLP, SNP).

Among the various marker techniques available, a few markers which have been well studied for their potential in seed quality testing such as variety identification, distinctness testing, and hybridity determination are as follows: (1) Restriction Fragment Length Polymorphism (RFLP), (2) Randomly Amplified Polymorphic DNA (RAPD), (3) Amplified Fragment Length polymorphism (AFLP), (4) sequence Characterized Amplified Regions (SCAR), (5) Sequence Tagged Sites (STS), (6) Simple Sequence Repeats (SSRs)/microsatellites, (7) Inter-SSR markers (ISSR), (8) Sequence-Related Amplified Polymorphism (SRAP) and Single Nucleotide Polymorphism (SNP).

3.1   Restricted Fragment Length Polymorphism (RFLP)

RFLP, the first DNA-based technique based on hybridization, produces polymorphisms inherited in a Mendelian fashion. They determine variation among varieties caused due to differences in the restriction sites resulting in the length difference of restricted fragments. The polymorphic bands are identified by Southern hybridization, thus identifying each cultivar and also determining the hybrid purity. The result obtained depends on the number of probes and the restriction enzymes used. RFLP markers have high reproducibility, have co-dominant inheritance, and are locus-specific. Disadvantages of the technique are, they are time-consuming, there is requirement of high quality and quantity of DNA, expensive radioactive probes, there is necessity to perform tedious Southern blotting method and prior sequence information is required for developing radiolabelled probes. The revolutionary invention of the polymorphic chain reactions (PCR) by Kary. B. Mullis in 1983 in making multiple copies of DNA segments led to a significant shift towards the use of markers based on amplification than those based on hybridization.

3.2   Randomly Amplified Polymorphic DNAs (RAPD)

The advent of RAPD markers provided a new tool for the molecular geneticist. RAPD uses low amounts of DNA with no need for high purity and does not require previous knowledge of the host genome. These have been employed extensively to discriminate crop varieties and, for the identification of parental lines and hybrids in, rice (Santhy et al. 2003), cotton (Ali et al. 2008), and vegetables (Kumar et al. 2008). In spite of its advantages, RAPD does not offer enough reliability since it uses short primers (of about ten nucleotides) and low annealing temperatures to amplify random regions in the genome, which makes the strategy non-reproducible across laboratories (Butler 2012). Due to its multiple bands and reproducibility issues across labs, RAPD markers were later replaced by other PCR-based markers such as CAPS, SCAR, AFLP, and SSR/microsatellite markers for plant variety identification and seed purity determination.

3.3   Cleaved Amplified Polymorphic Sequences

The CAPS markers are those in which DNA fragments after digestion with restriction enzymes are amplified with specific primers and separated on an agarose gel. CAPS markers closely linked to the gene of interest are helpful in crop breeding for marker-assisted selection and have been found useful in identifying the true female parents for authentic planting supply in few crops (Babu et al. 2017). The scoring of this type of marker is dependent on the variation in size of fragments following the digestion of the PCR product by a restriction enzyme.

3.4   Amplified Fragment Length Polymorphism (AFLP)

AFLP technology combines RFLP and PCR and is based on the selective amplification of a subset of genome restriction fragments using PCR. AFLP uses restriction enzymes recognizing frequent and rare restriction sites in the genome to generate fragments (that end with frequent or rare sticky ends or a combination of both), some of which are later specifically selected for PCR amplification. The selection is achieved with primers made of double-stranded adaptors linked to sequences complementing those generated by the restriction enzymes and, additionally carrying one to three nucleotides, to reduce the number of fragments for amplification. For higher specificity, this is usually completed in two steps, with a pre-amplification using only one nucleotide and a final amplification using two to three selective nucleotides. The use of AFLP can efficiently reveal multiple polymorphisms in a single reaction and is highly reproducible (Singh and Singh 2019). Their potential in fingerprinting and identification of inbred lines and hybrids has been demonstrated (Grzebelus et al. 2008). Due to the laborious procedure and requirement of high-quality DNA, it is not suited for routine purity testing.

3.5   Sequence Characterized Amplified Regions (SCAR)

To increase the reliability of a PCR-based marker, SCAR have been developed. These are obtained by eluting a selected fragment from RAPD or AFLP gel, cloned, and sequenced at its termini. A pair of primers (forward and reverse) specific for these termini is designed. This primer pair amplifies a single fragment under more restrictive annealing temperatures in PCR bringing higher reliability. Successful primer pairs give rise to SCAR markers. Direct application of these markers for hybrid purity testing in crop seeds has been demonstrated (Jang et al. 2004). Nevertheless, these markers are not adequate for the detection of seed admixtures or mislabelling.

3.6   Simple Sequence Repeats (SSRs)

SSRs or microsatellites are DNA stretches consisting of short, tandem repeats of short nucleotide sequences. SSR markers are also known as sequence-tagged microsatellite markers in which the above repeats are amplified using primers specific to sequences flanking these repeats. The amplification products are size separated by electrophoresis and visualized. The amplicons from different genotypes frequently show length polymorphisms due to allelic variation of the number of repeat motifs in the microsatellite.

SSRs/microsatellite markers became more valuable and reliable owing to their reproducibility, multi-allelic nature, co-dominant inheritance, genome-wide presence, robustness, higher polymorphism, and analytical simplicity (Abd El-Moghny et al. 2017).

SSR allelic profiles are conserved throughout the plant growth stage allowing unambiguous identification of crop varieties (Santhy et al. 2019; Ravishankar and Dinesh (2015). Being locus-specific and co-dominant, SSRs are the most suitable markers for seed hybridity testing as the heterozygosity of the hybrids can be easily determined by the presence of both the parental alleles (Tatiana et al. 2006; Selvakumar et al. 2010; Pallavi et al. 2011). The use of SSR markers for assessing seed purity has been reported in major crops like rice (Bora et al. 2016), maize (Daniel et al. 2012), and vegetable crops (Ravishankar et al. 2018). However, the target DNA region flanking each tandem repeat needs to be sequenced, primers have to be designed for the amplification of the repeat region, and marker screening needs to be done before they can be utilized. Occurrence of null alleles due to poor primer annealing and underestimation of heterozygosity is considered a major drawback associated with SSR markers.

3.7   Multiplex PCR

A greater throughput resolution of multiple markers can be achieved in a single PCR reaction making the technique much faster amplifying two or more loci simultaneously in the same reaction through multiplexing. In multiplex PCR, different segments from the same DNA get amplified simultaneously. We need to ensure that the length of amplified fragments does not overlap and different primer sets have the same melting temperature. Multiplexing of SSRs could discriminate parents of hybrids and prove the hybridity in brinjal and rice (Arun Kumar et al. 2014; Madhuchhanda et al. 2020).

3.8 Expressed Sequence Tags (ESTs)

The availability of enormous data for expressed sequence tags (ESTs) in the public domain made the marker-based studies shift from genomic SSRs to EST-SSRs. EST-SSRs amplify portions of expressed sequences in the genome which may be functionally associated with major component traits as compared to the non-EST-SSRs which may be randomly distributed across the genome. EST-SSRs based polymorphic markers have been identified in major crops (Parthiban et al. 2018) and have been proven useful for hybrid purity testing (Naresh et al. 2009).

3.9 Inter Simple Sequence Markers (ISSR)

Inter simple sequence markers (ISSR) are another set of PCR-based markers which have been widely studied for their utility in confirming F1 hybridity (Khajudparn et al. 2012). It is based on a single primer having a microsatellite sequence and detects variation in the size of the genomic region flanked by microsatellite sequences, generating multi-locus markers. In spite of being highly polymorphic, fast, and inexpensive, ISSR markers are associated with less reproducibility.

3.10 Sequence-Related Amplified Polymorphisms

SRAP markers are appropriate molecular markers for genotype identification since they are based on the amplification of coding regions in the genome utilizing two primers. The primers are 17 or 18 nucleotides long and contain a 13–14-base-long core sequence, followed by the sequences CCGG in the forward primer and AATT in the reverse primer, with the first 10 or 11 bases at the 5′ end being sequences of no specific structure (filler sequences). At the 3′ terminus, three selected nucleotides follow the core. The forward and reverse primers' filler sequences must be distinct from one another. This marker reveals more polymorphic fragments than AFLP markers and is more compatible with genotype morphological variability (Ferriol et al. 2003). These markers were found highly efficient and reproducible for genetic purity testing of cabbage commercial hybrid seeds (Liu et al. 2007).

3.11   Single Nucleotide Polymorphisms (SNPs)

An SNP is a single nucleotide base difference between two DNA sequences or individuals. SNPs can be categorized according to nucleotide substitutions as either transitions (C/T or G/A) or transversions (C/G, A/T, C/A, or T/G) or insertions/deletions in the genome (Jiang 2013). These provide the ultimate/simplest form of molecular markers as a single nucleotide base is the smallest unit of inheritance. They may be present within coding and/or non-coding, intergenic regions of the genome at different frequencies. They are co-dominat markers, often linked to genes, can be easily automated, multiplexing possible making it cost-effective, enabling quick detection of high polymorphisms with lower error rate. The potential SNP markers in detecting varietal polymorphism and in reliable genetic purity testing of Indonesian rice varieties were proven by Utami et al. (2016). Given the availability of whole-genome sequence in many crops (rice, soybean, maize, etc.), SNPs are expected to be the standard marker of choice in the future.

A high-density oligonucleotide SNP array has hundreds to thousands of probes arrayed on a small chip allowing many SNPs to be tested simultaneously. The probes are so designed to have the SNP sites in several locations matching with target DNA as well as not matching to the SNP allele present in the target DNA at several other locations. By comparing the differential amount of hybridization of the target DNA to each of these probes, it is possible to determine specific homozygous and heterozygous alleles. Microarray-based SNP genotyping is more time-consuming, and only hundreds of samples can be genotyped with thousands of SNPs. Compared to SSRs, the information obtained using SNPs are low, and hence, there is need to employ them in large numbers. Some of the high-throughput SNP genotyping methods include KASP (competitive allele-specific PCR) (Peng et al. 2021) and Fluidigm assays (Park et al. 2021).

Target SNP-seq combines the advantages of high-throughput sequencing and multiplex PCR amplification. It uses genome-wide perfect SNPs with conserved flanking sequences captured uniquely in PCR amplification (Zhang et al. 2020). This approach has 1000 times more coverage in a very short time and at a low cost, making it more competitive than the current SNP genotyping methods.

The SNP markers have been reliably employed for variety identification, distinctness testing, fingerprinting, genetic purity testing, assessing the parent-offspring relationship, and diversity analysis (Zhang et al. 2020; Kim et al. 2021; Josia et al. 2021).

An ideal DNA marker should be uniform, have a wide genome distribution, be co-dominant, have multiple alleles, have less DNA requirement, and be simple, easy to execute, less error-prone, reproducible, and amenable to high-throughput automation. SNP markers which fulfil most of these criteria are the closest to being ideal (Jiang 2013) (Fig. 1).

The International Seed Testing Association (ISTA) and the International Union for the Protection of New Varieties of Plants (UPOV) are working in a synchronized manner and sharing information on the use of DNA-based methods in variety testing, though with different objectives. While the BMT working group of UPOV

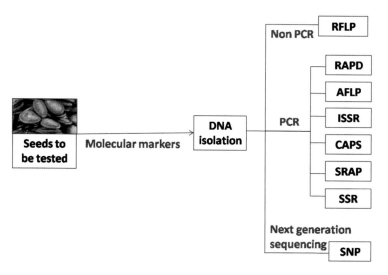

Fig. 1 Types of molecular markers used in seed genetic purity testing

aims the use of molecular markers in variety characterization for registration and protection, the DNA working group of ISTA aims for their use in routine seed testing/certification and protocol development. ISTA has identified a suitable set of SSRs for discriminating varieties in rice, maize, wheat, and soybean (www.upov. int).

4 Use of Molecular Markers for Plant Variety Protection

Plant variety registration and protection has attained critical importance all over the world and emphatically comes under the purview of seed regulation for quality control. Testing for DUS is an essential component of variety registration advocated by the International Union for the Protection of New Varieties of Plants (UPOV). The current plant variety protection system relies on the morphological description of plant varieties, which at times is not sufficient to discriminate a large number of varieties being developed in crops (Jamali et al. 2019). The potential of DNA markers in testing variety distinctness has been tested and proven (Santhy et al. 2000; Shengrui et al. 2020). The UPOV's Working Group on Biochemical and Molecular Techniques (BMT), after thorough research to identify marker-trait associations which are robust, has proposed the use of molecular markers that are directly linked to conventional characteristics. It was proved that molecular markers developed for a subset of DUS traits (genomic DUS) can be robustly used as a tool for determining the distinctness, uniformity, and stability of crop varieties (Yang et al. 2021). BMT-UPOV also proposed to develop/calibrate threshold levels for molecular markers against the minimum distance in phenotype traits for their use in testing for essentially derived varieties. This provides a system combining phenotypic and molecular distances as a tool to improve the efficiency of distinctness evaluation (Jones et al. 2013). The BMT group of UPOV is putting efforts in harmony with the DNA working group of ISTA and the International Association of Plant Breeders (ASSISNEL) to study the implications of the use of molecular markers in testing the distinctness of varieties for granting protection.

Within UPOV's system, breeders can freely use protected varieties in breeding programmes. However, breeders of protected varieties may seek to share the ownership of essentially derived varieties once it is proven that these, except for a few distinctive DUS trait(s), conform to parental varieties in essential characteristics. DNA markers can be a good replacement for morphological traits in defining boundaries between independent and essentially derived varieties. With the advent of new breeding technologies that allow minor modification in varieties with outcomes of specific merit or utility, detecting distinctness between varieties may become increasingly challenging (Yu and Chung 2021). Extensive studies have been undertaken in maize regarding the use of SSR and SNP markers for EDV identification (UPOV BMT 2007; Yang et al. 2021), and technical guidelines to help determine EDV status using SSRs and SNPs in maize have been suggested (Rousselle et al. 2015).

5   Transgenic/Genetically Modified Seeds

Biotechnological advances have offered tremendous scope for creating novel transgenic plants to combat biotic and abiotic stresses more efficiently and to improve the yield and quality. Plants derived by transfer of genes for specific traits from diverse organisms such as bacteria, viruses, animals, etc. using molecular or recombinant DNA/genetic engineering techniques are called GMOs (genetically modified organisms)/genetically engineered/transgenic plants (Phillips 2008). The potential of transgenics to effectively address many specific problems such as resistance to pests and diseases and tolerance to drought and herbicides made it gain wider acceptance all over the world. Currently, the global area under GM crops is 190.4 million hectares cultivated in 26 countries which include 21 developing and 5 industrial countries (ISAAA 2019). Major traits which have been utilized for developing transgenics include insect resistance, herbicide tolerance, and improved nutritional quality. GM crops that have been commercialized in the recent past include tomato, corn, soybean, cotton, canola, rice, potato, squash, melon, and papaya, of which herbicide-tolerant soybeans; insect- and herbicide-tolerant corn; bollworm-resistant, herbicide-tolerant cotton; and herbicide-tolerant canola are the major ones.

6   Methods of GM Seed Testing

There are four different levels of GM assessment in seed samples as mentioned below:

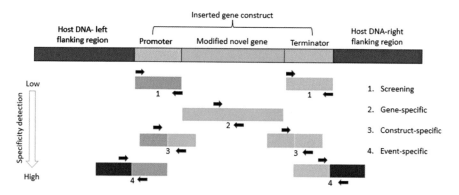

Fig. 2 Different levels of GM seed testing (Shrestha et al. 2010)

- GM detection: Screening a seed sample for detecting the presence of transgenic seeds. Primers are used to detect a general genetic element such as the constitutive promoter or a selection marker, which is frequently found in all GMOs, thereby detecting their presence within a seed lot.
- Gene-specific detection: Identification of a specific GMO by testing for a specific transgene, using primers specific for the gene sequences. The presence of amplified fragments indicates that the seeds carry unique transgene being tested for.
- Gene construct-specific detection: Identification of specific gene construct in a seed lot. This is done by using primers specific to sequences of promotor/terminator and part of the gene.
- Event-specific detection: Is the detection of a gene event for which one should know about the flanking sequences of a targeted gene construct, i.e. host genome sequence which is close to the construct. The event is identified by using primers designed to detect the unique integration site of a specific GMO (Fig. 2).

The two major purposes of GM seed testing include:

- GM trait purity: Tests the extent of non-GM seeds in a GM seed lot.
- Adventitious presence (AP) of GMOs: Detects the presence of GM seed in the non-GM seed lot. Mostly required for labelling in the global seed trade.

Hence, both qualitative and quantitative analytical methods are used for GM seed testing, which could be based on PCR, based on protein assays, or based on trait expression level (bioassays).

PCR-based methods	Protein-based methods	Bioassays
Conventional PCR using DNA (endpoint qualitative PCR for the presence of gene)	Dipstick assays (lateral flow strip test): Tests for the presence of gene based on antigen-antibody reaction	Scoring based on trait expression in a specific condition provided artificially

(continued)

PCR-based methods	Protein-based methods	Bioassays
Real-time PCR using RNA/DNA: Follows progressive detection and is used to determine the quantity of DNA or gene copy number	ELISA test: Tests for the transgene presence and level of expression based on antigen-antibody reaction	Seed soak bioassays

6.1   Lateral Flow Strip Test

The lateral flow strip test is based on immunoassay which uses a capillary paper immobilized with antibodies. The paper is inserted into a crude extract which allows the antigen to flow along the paper strip. The antigen in the extract binds with antibodies labelled with colloidal gold on the dipstick so as to form a distinct coloured band whenever antigen-antibody binds to each other (ISTA Rules 2021).

It is a qualitative rapid visual test and has been widely used for regulation of Bt cotton seed marketing in India (Kranthi 2013) (Fig. 3).

6.2   ELISA (Enzyme-Linked Immunosorbent Assay) Test

ELISA is a sensitive immunoassay which uses solid-phase enzyme. In this immuno-assay, the amount of unknown analyte in the sample is measured by adding the labelled antibodies. A capture antibody is coated on multi-well plate, each well loaded with crude seed extract. The labelled antibodies get attached with the analyte in the sample (antigen). The extra unbound labelled antibodies are washed away, and only antigen-bound labelled antibodies are present in the plate. When the plates are washed with a chromogenic substrate, there will be a reaction with the enzyme labelled antibodies wherever the antigen-antibody binding occurs and gives a fluorescence that is read by an ELISA reader. The intensity of the fluorescence is proportional to the quantity of the target protein (ISTA Rules 2021). In sandwich ELISA, the antigen (sample to be analysed) is sandwiched between two antibodies

Fig. 3 Dipstick method of testing Bt in seeds (Kranthi 2013)

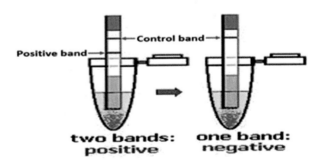

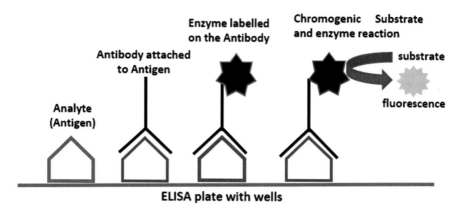

Fig. 4 Schematic representation of the direct ELISA method (Kausar et al. 2017)

for more specific reactions (Kausar et al. 2017). It is very important to note that washing step after every reaction is very crucial after every reaction so as to wash away any extra materials. ICAR-CICR, Nagpur, Maharashtra, India, has developed simple ELISA test kits to enable farmers, seed testing officers, researchers, and seed companies to detect Bt seed quality (Kranthi 2013) (Fig. 4).

6.3   Bioassays

Bioassays are tests based on visual assessment of expression of a trait under specific growing conditions. For seed testing purpose, bioassays are used particularly for testing the herbicide tolerance trait in crops. Seeds are first exposed to the herbicide and then tested for germination and growth. If the seeds germinate and grow normally, they are scored as positive. The concentration of herbicide needs to be appropriate to obtain the intended difference in expression. Seed soak bioassays have been reported by researchers for identifying/detecting herbicide-tolerant seeds in soybean (Torres et al. 2003). Seed soak bioassays will only determine the GM trait and not the event.

6.4   Conventional PCR

Conventional PCR is employed for the general screening of GM seeds using primers that recognize common DNA which most GMOs harbour, e.g. Ca MV 35 S promoter and *nos* terminator. It can also be used for determining specific transgene/event purity. The procedure involves use of primers specific to the gene or event depending on the requirement. The test is positive if a band of appropriate size is observed on the gel. Multiplex PCR, a variant of conventional PCR, involving simultaneous amplification of multiple target sequences in a test sample is also employed for faster GM seed detection. A multiplex PCR assay was developed for the simultaneous amplification of *transgene, (CaMV)* 35S promoter, selectable marker gene, and *nopaline synthase* (*nos*) terminator along with *β-fructosidase* gene, an endogenous gene of Solanaceae family, for routine testing to detect GM tomato seeds in its germplasm collection (Randhawa et al. 2011). Multiplex PCR assays are also available to detect various GM events of maize (Degrieck et al. 2005) and cotton (Nadal et al. 2009).

6.5   Real-Time Polymerase Chain Reaction (RT-PCR/qPCR)

GMO quantification based on event-specific primers can be achieved through real-time PCR. It consists of amplification, simultaneous detection, and quantification of targeted DNA. In real-time PCR, DNA amplification activates fluorochromes attached to the primers, thus increasing the fluorescent signal with amplicons produced. This activation can be measured in real time giving an estimate of the DNA molecules being amplified in each cycle. In qualitative RT-PCR, positive score is given when the fluorescence detected is above the baseline. In quantitative RT-PCR, the target quantity is measured against a standard curve produced from certified reference material. There are several advantages real-time PCR has such as rapid cycling to reduce DNA amplification time, completion of PCR in a closed system to reduce the risk of cross-contamination, post-PCR electrophoresis not needed, and possibility of multiplexing by using probes containing different reporter dyes with distinct spectra (Fig. 5).

7   ISTA and GM Seed Testing

The coexistence of GM and non-GM crops has raised a concern in many European countries, and the law requires that all GM food be traceable to its origin. Any product, food, or seed with GM content greater than 0.9% needs to be labelled. The GM quantification studies are done between 0% and 0.9% labelling thresholds, since the threshold levels vary within this range in different countries and regions around the world. When seeds are traded across borders between countries with different thresholds (*http://norden.diva-portal.org*), a reliable testing for determining these threshold levels has to be performed (Degrieck et al. 2005). As part of its role in harmonization of GM seed testing procedures, ISTA has been conducting proficiency test programmes since 2002 and workshops on GMO testing since 2001 and included a chapter on GM testing in the ISTA Seed Rules since 2014. ISTA developed methods which can be used for both testing adventitious presence (AP) of GM seeds and GM trait purity testing of seed lots (ISTA Rules 2021). ISTA international certificates, which can only be issued by ISTA-accredited laboratories, guarantee the identity of the seed lot with a single reference, the traceability of the analysis, the competence of the laboratory that did the analysis, the use of validated methods and standard units, and the use of standard reporting

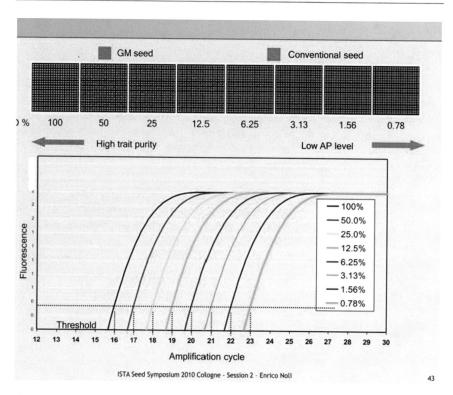

Fig. 5 RT-PCR quantification of DNA using different spectral dyes with distinct spectra (Noli 2010)

languages. Today, the ISTA Orange International Seed Lot Certificate (OIC) is widely used for international trade. Certified reference materials are used for the calibration or quality control of GMO quantification measurements typically carried out by quantitative real-time polymerase chain reaction (qPCR). Numerous sets of reference materials for different GM events in maize, soybean, potato, sugar beet, and cotton are offered for testing in the laboratories worldwide (www.ec.europa.eu). GM seed testing by ISTA is undertaken by (1) qualitative assay to assess trait purity in a GM seed sample by testing individual seeds in a sample, (2) semi-quantitative assay to assess adventitious presence of GM seeds in conventional seed lots by testing various seed bulks (sub-samples) within a seed lot, or (3) quantitative assay for quantitative assessment of adventitious presence (quantity of GM seed) in a non-GM seed lot done by analysing one single seed bulk from a seed lot (Enrico 2010). However, both AP GM seed testing and GM purity testing can be done on individual seeds or seed bulks.

The limit of detection (LOD) for quantitative tests and the limit of quantification (LOQ) for qualitative tests are critical criteria in deciding the sample size to be drawn (Broeders et al. 2014). LOD is defined as the smallest number of target seeds that has been demonstrated to be detected with a given level of confidence. The limit of quantification is the smallest amount of target analyte that has been demonstrated to be reliably measured with acceptable levels of accuracy and precision (ISTA Rules 2021). Obtaining a representative seed sample for testing GM seeds is important as in case of other quality parameters. Working sample size depends on the threshold requirements, method capability, and statistical confidence level and can be determined using appropriate statistical tools such as SeedCalc 8.0 (Remund et al. 2001). Testing methods developed using this tool can be used to estimate genetic purity of a seed lot, as well as a criterion to accept/reject a lot. The sample submitted to the laboratory must be at least the size of working sample and preferably larger than the working sample for accurate estimation.

An alternate PCR-based approach to test trait purity in bulk samples has been developed in which the absence of transgenic DNA is detected. For this, the insertion site of a transgene is characterized, and the corresponding sequence of the wild-type (wt) allele is used as primer binding site for amplification. This method could quantify the non-GM contamination as well as GM trait purity in RR soybean (Battistini and Noli 2009). Whatever the method is, ISTA suggests uniformity of result, and hence, a performance-based approach is followed for giving accreditation to any laboratory in GM seed testing. For detailed understanding of getting laboratory accreditation of ISTA in GM seed testing, readers may refer to *ISTA Principles and Conditions for Laboratory Accreditation Under the Performance-Based Approach.*

8   Molecular Markers in Germplasm Conservation and Maintenance

With the widespread cultivation of GM crops, conservation and exchange of germplasm have become a challenge in terms of determining the adventitious presence of GM seeds in conventional seeds. Conventional singleplex and multiplex PCR assays and real-time PCR targeting common screening elements or specific GM targets can be efficiently employed to check the unintended presence of GM seeds in any lot. DNA-based diagnostics including PCR/real-time PCR assays have been successfully employed to detect the adventitious presence of transgenic seeds in ex situ brinjal, okra, and cotton accessions conserved in National GeneBank (NGB) at the Indian Council of Agricultural Research-National Bureau of Plant Genetic Resources (ICAR-NBPGR), New Delhi (Kuwardadra et al. 2020; Randhawa et al. 2011). With technology advancement, the organization developed many cost- and time-efficient DNA-based GM detection technologies for the simultaneous detection of GM events in various crops (Randhawa et al. 2016). These include multiplex PCR, real-time PCR, visual LAMP, real-time LAMP, and multitarget TaqMan real-time PCR plate methods.

Minimizing the inclusion of duplicate accessions is important in germplasm conservation. It is estimated that there is an average of 50% duplication in different collections (Singh and Singh 2019). Molecular markers can be used for the unambiguous identification of duplicate accessions which can be safely removed from the holding for better maintenance. Genetic distance between the accessions can be determined from Nei's dissimilarity matrix, and the ones which are having mean distance less than the threshold value can be considered as potential duplicates (Das et al. 2020).

9   Molecular Techniques for Seed Health Testing

Modern molecular methods hold great potential for improving pathogen detection in seeds, as it has many advantages over conventional assays in terms of specificity, sensitivity, rapidity, ease of implementation, and applicability.

9.1   Polymerase Chain Reaction (PCR)

The PCR technique had been successfully exploited for the detection of some of the seed-borne pathogens like *Ascochyta lentis* from lentil seeds (Hussain et al. 2000), *Magnaporthe grisea* (rice blast disease) (Chadha and Gopalakrishna 2006), *Tilletia indica* (karnal bunt disease) (Thirumalaisamy et al. 2011), and soybean yellow mottle mosaic virus (SYMMV) in French bean and mung bean seeds (Nagamani et al. 2020) (Fig. 6).

9.2   Bio-Polymerase Chain Reaction (bio-PCR)

Bio-PCR improves the efficiency and sensitivity of PCR by allowing target pathogen populations to increase in a pre-enrichment phase, before DNA extraction and PCR. This results in higher quantities of target DNA, which ultimately results in higher sensitivity. During the incubation and enrichment phase on artificial media, inhibitory compounds are adsorbed or diluted and do not interfere with DNA amplification. The seed samples are washed initially and crushed in suitable buffer to extract seed-borne bacteria. Aliquots are spread onto semi-selective media and incubated for 2–3 days. Colonies are then harvested for DNA extraction, and PCR is conducted with specific primers. In the case of seed-borne fungi, seeds are incubated under conditions of high relative humidity to increase target fungal mycelium mass before DNA extraction and PCR (Walcott 2003).

The advantages of Bio-PCR are that only viable colonies are detected, as the target organisms must grow on selective medium before it can be detected by PCR, and there is no need to identify the pathogen based on colony morphology as specific primers are used for amplification. Bio-PCR has been developed for the detection of bacterial fruit blotch of watermelon (*Acidovorax avenae* subsp. *citrulli*), halo blight (*Pseudomonas syringae* subsp. *phaseolicola*) of beans (*Phaseolus vulgaris*), bacterial ring rot (*Clavibacter michiganensis* subsp. *sepidonicum*) of potato (*Solanum tuberosum*), and black rot (*Alternaria radicina*) of carrot (*Daucus carota*) (Pryor and Gilbertson 2001; Schaad et al. 1999). The disadvantage of Bio-PCR includes the

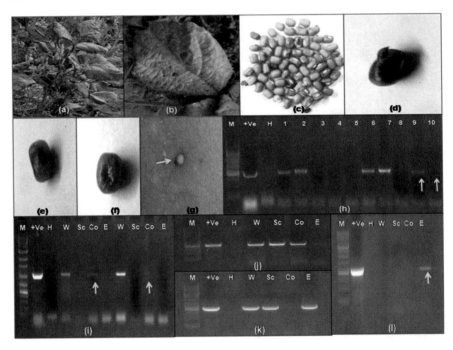

Fig. 6 Confirmation of seed transmission of SYMMV in mung bean cv. Pusa Vishal. (**a**) Symptomatic mung bean plants showing less pod formation and stunted growth of the plants. (**b**) Infected leaf with mottling and puckering symptoms. (**c–f**) Seeds from infected plants showing brownish discolouration. (**g**) ISEM confirmation of SYMMV from infected mung bean seed. (**h**) PCR amplicons (1065 bp) obtained with coat protein (CP)-specific primers NS1 and NS2 (lanes 1–10 indicate RT-PCR from single whole infected seeds). (**i**) Detection of SYMMV through RT-PCR with CP primers in group of two (initial four lanes) and five seeds (last four lanes) where amplification was observed in whole seed and cotyledons. (**j–l**) Detection of SYMMV with CP primers in group of five seeds from various seed tissue parts (W, whole seed; Sc, seed coat; Co, cotyledons; E, embryo; H-RT, PCR from healthy seed; +ve, leaf tissue infected with SYMMV; M, GeneRuler 1 kb and 1 kb plus DNA ladders)

requirement of semi-selective medium for each pathogen for which specific knowledge about the nutritional requirements and chemical tolerances of the target organism is required. Also, the method requires 2–3 days for bacteria and 5–7 days for fungi to grow, thereby significantly increasing the time required for assay completion. Another drawback of Bio-PCR is that it is limited primarily to readily culturable bacteria and fungi and can't be used for obligate parasites (e.g. viruses).

9.3   Multiplex Polymerase Chain Reaction (mPCR)

Multiplex PCR is a more reliable, fast, and cost-effective method for routine detection of seed-borne pathogens. Multiplex PCR (DNA targets) and multiplex RT-PCR (RNA targets) are useful for the simultaneous detection of multiple pathogens (broad-spectrum detection) in a single reaction containing more than one set of primers. Multiplex PCR was utilized for the simultaneous detection of three bacterial seed-borne diseases, viz. bacterial grain rot (*Burkholderia glumae*), bacterial leaf blight (*Xanthomonas oryzae* pv. *oryzae*), and bacterial brown stripe (*Acidovorax avenae* subsp. *avenae*), based on 16S and 23S rDNA and transposase A gene sequence in rice (Kang et al. 2016). Multiplex TaqMan real-time PCR assay was used for the detection of spinach seed-borne pathogens, viz. *Peronospora farinosa* f.sp. *spinaciae*, *Stemphylium botryosum*, and *Verticillium dahliae*, by using the primers based on internal transcribed spacer, intergenic spacer, and the elongation factor 1 alpha. Sensitivity of multiplex PCR is influenced by the number of target pathogens to be detected where a number of different primers are more important than the total amount of primer in the reaction mixture. Some of these limitations can be overcome with more precise specificity and sensitivity by improving the quality of nucleic acid extraction procedure and modification of PCR technology with the use of magnetic nanobeads and dual priming oligonucleotide primers or a nested reaction (Kwon et al. 2014).

9.4 Nested PCR

Nested PCR involves two pairs of amplification primers and two successive rounds of PCR. Initially, one primer pair is used in the first round of PCR with 15–30 cycles. The products of the first round of amplification are then subjected to a second round of amplification using the second set of primers which anneal to a sequence internal to the sequence amplified by the first primer set.

Nested PCR was successfully employed for the detection of some of the seed-transmitted pathogens, viz. tomato black ring virus (TBRV), Arabis mosaic virus (ArMV), cherry leafroll virus (CLRV), and grapevine fanleaf virus (GFLV), tobacco ring spot virus, and *Ustilaginoidea virens* (false smut disease of rice) (Danesh et al. 2014). The advantage of nested PCR is its increased sensitivity and specificity due to higher number of total cycles and annealing of the second primer set to sequences found only in the first round products. However, nested PCR assays are time-consuming and have an increased risk of cross-contamination which can create false-positive results.

9.5 Real-Time PCR

Inspite of many advantages real-time PCR has not been used much for the detection of seed-borne pathogens due to the requirement of expensive thermal cyclers that are equipped to detect fluorescence. Detection of seed-borne pathogens through real-time PCR was reported for *Didymella bryoniae* causing gummy stem blight of cucurbits (Ling et al. 2010), *Clavibacter michiganensis* subsp. *michiganensis* causing bacterial canker of tomato (Han et al. 2018), and soybean yellow mottle mosaic virus (Nagamani et al. 2020).

9.6   Loop-Mediated Isothermal Amplification (LAMP)

Loop-mediated isothermal amplification (LAMP), developed by Notomi et al. (2000), is a simple, cost-effective, and rapid method for the specific detection of genomic DNA that enables the synthesis of large amounts of DNA in a short period of time without the use of thermal cycler. LAMP technology uses a pair of four or six oligonucleotide primers and a thermophilic DNA polymerase for DNA amplification. LAMP products can be visualized by gel electrophoresis using magnesium pyrophosphatase which enhances the precipitation of amplified DNA or with a real-time turbidity reader or with the addition of an intercalating dye, such as SYBR Green I, which produces a colour change in the presence of target phytopathogen.

LAMP was used successfully for the detection of *Fusarium graminearum* (head blight disease) using the primers based on galactose oxidase gene (*gaoA*) (Niessen and Vogel 2010). Reverse transcription LAMP (RT-LAMP) assay was used for the detection of tomato brown rugose fruit virus in tomato and pepper seeds (Rizzo et al. 2021).

9.7   Microarray Technology

DNA chip or microarray technology has been applied to detect seed-borne pathogens. Microarray technology depends on the unique ability of nucleic acid molecules to hybridize specifically with molecules of complimentary sequences. In microarray technology, hundreds to thousands of oligonucleotides will be attached to specific locations on each chip. These oligonucleotides can be complementary to DNA sequences unique to certain pathogens and hence can be used to detect pathogens present in the seed sample. The DNA or RNA is extracted from the seed sample to be tested, amplified, and digested into smaller fragments that are labelled with fluorescent markers and hybridized with oligonucleotides fixed to the DNA chip. After hybridization, the chip is washed thoroughly, and fluorescence, which is directly proportional to the amount of nucleotide retained, is measured. If the pathogen of interest is present in the seed sample, then the oligonucleotide probe at the position on the chip that corresponds to that pathogen will display fluorescence.

DNA chip technology, which helps in the simultaneous detection of a wide range of pathogens within a short period of time (6 h), has great scope of application, and several such DNA chip assays for seed pathogen detection are already available.

10   Challenges in Using Molecular Techniques for Seed Quality Testing

Variety identification, seed genetic purity, and seed health testing are important parameters determining the seed quality. Being efficient, time-saving, less labour-intensive, reproducible, and amenable to high-throughput systems, the molecular marker systems would play an important role in cultivar identification and seed genetic purity testing.

Another important factor in genetic purity determination using molecular markers (e.g. SSR) is the number of core primers to be employed, which depends on the purpose of seed quality test. While one robust marker shall be enough for testing hybrid purity, multiple markers with high discrimination and repeatability are needed for varietal purity testing (ISTA 2021). Jamali et al. (2019) reviewed the potential deployment of DNA markers in plant variety protection and registration and summarized their efficacy, particularly in case of establishing the status of essentially derived varieties (EDVs). However, the number of markers required for establishing the EDV status may be even more. There is a need to standardize and identify minimum number of SNP/SSR markers, which have high reproducibility and can differentiate a good number of cultivars/genotypes precisely. These selected SSR/SNP markers can be analysed using high-throughput platforms, and these platforms also need to be flexible to accommodate variable number of samples.

Thorough mapping of crop genome by molecular markers that can adequately discriminate among elite adapted germplasm is required before these can be used in variety discrimination. Selection of markers has to be done using a large set of genotypes because a set of markers highly discriminating one set of genotypes with high PIC values need not necessarily be effective in discriminating another set of inbred lines. ISTA suggests that the identified markers are ought to produce sharp bands without null alleles and give similar allele patterns across repeats. Only common marker sets prescribed by ISTA have to be used in routine seed testing of a specific crop.

Tight linkages between the molecular marker loci and the loci expressing a morphological trait will facilitate its efficient use in purity testing/variety discrimination/protection. For example, molecular markers tightly linked to male sterile genes would facilitate an efficient and rapid transfer of *ms* genes into different genetic backgrounds through marker-assisted backcrossing, hybrid seed production, and genetic purity testing of hybrid seeds (Naresh et al. 2018).

Information about the homozygosity of polymorphic molecular marker loci in the parental inbred line is of importance, since segregation of these may create false interpretations of the purity of hybrid seed lots. The residual heterozygosity of useful marker loci needs to be kept at a minimum by adopting good maintenance breeding practices (Santhy et al. 2019). Hence, it is pertinent that there should be only minimum polymorphic loci available for purity checking so that it is easier to maintain the parental lines, and F1 profiles are also not varying during purity testing. Maintaining the genetic constitution of the reference material is of paramount importance when using markers in routine seed testing programmes. It is also important to ensure that the results obtained by the laboratories quantifying GMOs are comparable and results are based on the testing of a representative seed sample with an adequate sample size.

In seed health testing, it will be critical to rigorously evaluate molecular detection assays for their precision and accuracy through high-throughput testing of naturally infested seeds before adopting these for routine seed testing. To ensure that these detection assays work, they must be validated through stringent multi-laboratory tests which would evaluate their reproducibility, repeatability, and reliability.

References

Abd El-Moghny AM, Santosh HB et al (2017) Microsatellite marker-based genetic diversity analysis among cotton (Gossypium hirsutum) accessions differing for their response to drought stress. J Plant Biochem Biotechnol 26(3):366–370

Ali MA, Seyal MT et al (2008) Hybrid authentication in upland cotton through RAPD analysis. Aust J Crop Sci 2(3):141–149

Arun Kumar MB, Dadlani M et al (2014) Identification and validation of informative SSR markers suitable for ensuring the genetic purity of brinjal (Solanum melongena L.) hybrid seeds. Sci Hortic 171:95–100

Babu BK, Mathur RK et al (2017) Development, identification and validation of CAPS marker for SHELL trait which governs dura, pisifera and tenera fruit forms in oil palm (Elaeis guineensis Jacq.). PLoS One 12(2):e0171933

Battistini E, Noli E (2009) Real-time quantification of wild-type contaminants in glyphosate tolerant soybean. BMC Biotechnol 9:16

Bora A, Choudhury PR et al (2016) Assessment of genetic purity in rice (Oryza sativa L.) hybrids using microsatellite markers. Biotech 6:50

Bradford KJ (2006) Methods to maintain genetic purity of seed stocks. Agricultural biotechnology. In: California Series Publication 8189

Broeders S, Huber I et al (2014) Guidelines for validation of qualitative real-time PCR methods. Trends Food SciTechnol 37(2):115–126

Butler JM (2012) Random amplified polymorphic DNA. In: Advanced topics in forensic DNA typing methodology. Elsevier Academic Press, San Diego, CA, pp 473–495

Chadha S, Gopalakrishna T (2006) Detection of Magnaporthe grisea in infested rice seeds using polymerase chain reaction. J Appl Microbiol 100:1147–1153

Danesh YR, Tajbakhsh M et al (2014) Using nested PCR for detection of seed borne fungi Ustilaginoidea virens in rice fields of Iran. In: Turkey 5th seed congress. pp 670–672

Daniel IO et al (2012) Application of SSR markers for genetic purity analysis of parental inbred lines and some commercial hybrid maize (Zea mays L.). Aust J Exp Agric 2(4):597–606

Das A, Rajesh KS et al (2020) Identification of duplicates in ginger germplasm collection from Odisha using morphological and molecular characterization. Proc Natl Acad Sci India Sect B Biol Sci 90(5)

Degrieck I, de Andrada Silva E et al (2005) Quantitative GMO detection in maize (Zea mays L.) seed lots by means of a three-dimensional PCR based screening strategy. Seed Sci Technol 33(1):131–143

Ferriol M, Picó B, Nuez F (2003) Genetic diversity of a germplasm collection of Cucurbita pepo using SRAP and AFLP markers. Theor Appl Genet 107:271–282

Grzebelus D et al (2008) The use of AFLP markers for the identification of carrot breeding lines and F1 hybrids. Plant Breed 120(6):526–528

Han S, Jiang N et al (2018) Detection of Clavibacter michiganensis subsp. Michiganensis in viable but nonculturable state from tomato seed using improved qPCR. PLoS One 13(5):e0196525

Hussain S, Tsukiboshi T, Uematsu T (2000) Quick detection of Ascochyta lentis from lentil seeds using polymerase chain reaction (PCR) based techniques. Pak J Bot 32:45–56

Hwu K-K (2013) Overview of DNA technologies: current uses and applications. In: 30th ISTA Seed Congress held at Antalya, Turkey

ISAAA (2019) Global status of commercialized biotech/GM crops in 2019: biotech crops drive socio-economic development and sustainable environment in the new frontier (https:/// resources/publications/pocket/16/)

ISTA (2021) International rules for seed testing (2021) Full Issue I–19-8 (300). https://doi.org/10.15258/istarules

Jamali SH, Cockram J, Lee T (2019) Insights on deployment of DNA markers in plant variety protection and registration. TheorAppl Genet 132(7):1911–1929

Jang I, Moon JH, Yoon JB et al (2004) Application of RAPD and SCAR markers for purity testing of F1 hybrid seed in chili pepper (Capsicum annuum). Mol Cells 18(3):295–299

Jiang GL (2013) Molecular markers and marker assisted breeding in plants. In: Andersen SB (ed) Plant breeding from laboratories to fields. Intech Open. https://www.intechopen.com/books/3060, London. https://doi.org/10.5772/3362

Jones H, Norris C et al (2013) Evaluation of the use of high-density SNP genotyping to implement UPOV model 2 for DUS testing in barley. Theor Appl Genet 126:901–911

Josia C, Mashingaidze K et al (2021) SNP-based assessment of genetic purity and diversity in maize hybrid breeding. PLoS One 16(8):e0249505

Kang IJ, Kang MH et al (2016) Simultaneous detection of three bacterial seed-borne diseases in rice using multiplex polymerase chain reaction. Plant Pathol J 32(6):575–579

Kausar M, Haleema S, Muhammad HB (2017) Protein based detection methods for genetically modified crops. https://doi.org/10.5772/Intechopen.75520

Khajudparn P, Prajongjai T et al (2012) Application of ISSR markers for verification of F1 hybrids in mungbean (Vigna radiata). Genet Mol Res 11(3):3329–3338

Kim M, Jung JK et al (2021) Genome-wide SNP discovery and core marker sets for DNA barcoding and variety identification in commercial tomato cultivars. Sci Hortic 276:109734

Kranthi KR (2013) Testing seed quality of Bt cotton. CAI Weekly Publication No. 29

Kumar V, Sharma S et al (2008) Genetic diversity in Indian common bean (Phaseolus vulgaris L.) using random amplified polymorphic DNA markers. Physiol Mol Biol Plants 14(4):383–387

Kuwardadra SI, Bhat KC et al (2020) Monitoring adventitious presence of transgenes in brinjal collections from the regions in India bordering Bangladesh: a case report. Genet Resour Crop Evol 67:1181–1192

Kwon JY, Hong JS et al (2014) Simultaneous multiplex PCR detection of seven cucurbit infecting viruses. J Virol Methods 206:133–139

Law JR, Donini P, RMD K et al (1998) DNA profiling and plant variety registration, III: the statistical assessment of distinctness in wheat using amplified fragment length polymorphisms. Euphytica 102:335–342

Ling KS, Wechter WP et al (2010) An improved real-time PCR system for broad-spectrum detection of Didymella bryoniae, the causal agent of gummy stem blight of cucurbits. Seed Sci Technol 38(3):692–703

Liu L et al (2007) Evaluation of genetic purity of F1 hybrid seeds in cabbage with RAPD, ISSR, SRAP, and SSR Markers. Hort Sci 42(3):724–727

Madhuchhanda P, Ngangkham U et al (2020) A multiplex PCR system for testing the genetic purity of hybrid rice (Oryza sativa L.). Indian J Genet Plant Breed 80(2):213–217

Nadal A, Esteve T, Pla M (2009) Multiplex polymerase chain reaction-capillary gel electrophoresis: a promising tool for GMO screening- assay for simultaneous detection of five genetically modified cotton events and species. J AOAC Int 92(3):765–772

Nagamani S, Tripathi A et al (2020) Seed transmission of a distinct soybean yellow mottle mosaic virus strain identified from India in natural and experimental hosts. Virus Res 280:197903

Naresh V, Yamini KN et al (2009) EST-SSR marker-based assay for the genetic purity assessment of safflower hybrids. Euphytica 170(3):347–353

Naresh P, Lin S et al (2018) Molecular markers associated to two non-allelic genic male sterility genes in peppers (Capsicum annuum L.). Front Plant Sci 9:1343

Niessen L, Vogel RF (2010) Detection of Fusarium graminearum DNA using a loop-mediated isothermal amplification (LAMP) assay. Int J Food Microbiol 140(2–3):183–191

Enrico Noli (2010) New Tools for measuring genetic quality in seed. In: 29th ISTA Seed Congress Cologne 2010. https://www.yumpu.com/en/document/view/8667848/aspects-of-purity-international-seed-testing-association

Notomi T, Okayama H, Masubuchi H et al (2000) Loop-mediated isothermal amplification of DNA. Nucleic Acids Res 28(12):e63

Pallavi HM, Gowda R, Shadakshari YG (2011) Identification of SSR markers for hybridity and seed genetic purity testing in sunflower (Helianthus Annuus L.). Helia 34(54):59–66

Park G, Choi Y et al (2021) Genetic diversity assessment and cultivar identification of cucumber (Cucumis sativus L.) using the Fluidigm single nucleotide polymorphism assay. Plants (Basel) 10(2):395

Parthiban S, Govindaraj P, Senthilkumar S (2018) Comparison of relative efficiency of genomic SSR and EST-SSR markers in estimating genetic diversity in sugarcane. 3 Biotech 8(3):144

Peng WANG, Zhejuan TIAN et al (2021) Establishment and application of a tomato KASP genotyping system based on five disease resistance genes. Yuan Yi Xue Bao 48(11):2211

Phillips T (2008) Genetically modified organisms (GMOs): transgenic crops and recombinant DNA technology. Nat Educ 1(1):213

Pryor BM, Gilbertson RL (2001) A PCR-based assay for detection of Alternaria radicina on carrot seed. Plant Dis 85:18–23

Randhawa G, Chhabra R, Singh M (2011) PCR based detection of genetically modified tomato with AVP1D gene employing seed sampling strategy. Seed Sci Technol 39(1):112–124

Randhawa G, Singh M, Sood P (2016) DNA-based methods for detection of genetically modified events in food and supply chain. Curr Sci 110(6):1000

Ravishankar KV, Dinesh MR (2015) Development and characterization of microsatellite markers in mango (Mangiferaindica) using next-generation sequencing technology and their transferability across species. Mol Breed 35:93

Ravishankar KV, Muthaiah G et al (2018) Identification of novel microsatellite markers in okra (Abelmoschus esculentus (L.) Moench) through next-generation sequencing and their utilization in analysis of genetic relatedness studies and cross-species transferability. J Genet 97(1): 39–47

Remund K, Dixon D et al (2001) Statistical considerations in seed purity testing for transgenic traits. Seed Sci Res 11(2):101–120

Rizzo D, Lio DD et al (2021) Rapid and sensitive detection of tomato brown rugose fruit virus in tomato and pepper seeds by reverse transcription loop-mediated isothermal amplification assays (real time and visual) and comparison with RT-PCR end-point and RT-qPCR methods. Front Microbiol 12:640932

Rousselle Y, Elizabeth J et al (2015) Study on essential derivation in maize: III. Selection and evaluation of a panel of single nucleotide polymorphism loci for use in European and North American germplasm. Crop Sci 55:1170–1180

Santhy V, Meshram M (2015) Widening the character base for distinctness in cotton: emerging perspective. Curr Sci 109(11)

Santhy V, Mohapatra T et al (2000) DNA markers for testing distinctness of rice varieties. Plant Var Seeds 13:141–148

Santhy V, Dadlani M et al (2003) Identification of the parental lines and hybrids of rice (Oryza sativa L.) using RAPD markers. Seed Sci Technol 31(1):187–192

Santhy V, Meshram M et al (2019) Molecular diversity analysis and DNA fingerprinting of cotton varieties of India. Ind J Genet Plant Breed 79(4):719–725

Schaad NW, Berthier-Schaad Y et al (1999) Detection of Clavibacter michiganensis subsp. sepedonicus in potato tubers by BIO-PCR and an automated real-time fluorescence detection system. Plant Dis 83:1095–1100

Selvakumar P, Ravikeshavan R et al (2010) Genetic purity analysis of cotton (Gossypium spp.) hybrids using SSR markers. Seed Sci Technol 38:358–366

Shengrui Z, Li B et al (2020) Molecular-assisted distinctness and uniformity testing using SLAF-sequencing approach in soybean. Genes (Basel) 11(2):175

Shrestha HK, Hwu K-K, Chang M-C (2010) Trends Food Sci Technol 21(9):442–454

Singh BD, Singh AK (2019) Fingerprinting and gene cloning. In: Singh BD, Singh AK (eds) Marker assisted plant breeding principles and practices. Springer, New Delhi

Smith JSC, Register JC (1998) Genetic purity and testing technologies for seed quality: a company perspective. Seed Sci Res 8(2):285–294

Tatiana SA, Guchetl Z et al (2006) Development of marker system for identification and certification of sunflower lines and hybrids on the basis of SSR-analysis. Helia 29(45):63–72

Thirumalaisamy PP, Singh DV et al (2011) Development of species-specific primers for detection of Karnal bunt pathogen of wheat. Indian Phytopathol 64:164

Torres AC, Nascimento et al (2003) Bioassay for detection of transgenic soybean seeds tolerant to glyphosate. Pessqui Agropecu Bras 38(9):1053–1057

UPOV (2007) Working group on biochemical and molecular techniques, and DNA-profiling in particular

Utami DW, Rosdianti I et al (2016) Utilization of 384 Snp genotyping technology for seed purity testing of new Indonesian rice varieties, Inpari Blas and Inpari Hdb. SABRAO J Breed Genet 48(4):416–424

Walcott R (2003) Detection of seed borne pathogens. Hort Technol 13(1):40–47

Yang CJ, Russell J, Ramsay L et al (2021) Overcoming barriers to the registration of new plant varieties under the DUS system. Commun Biol 4:302

Yu JK, Chung YS (2021) Plant variety protection: current practices and insights. Genes (Basel) 12(8):1127. https://doi.org/10.3390/genes12081127

Zhang J, Yang J et al (2020) A new SNP genotyping technology target SNP-seq and its application in genetic analysis of cucumber varieties. Sci Rep 10(1):1–11

Seed Quality Enhancement

Elmar A. Weissmann, K. Raja, Arnab Gupta, Manish Patel, and Alexander Buehler

Abstract

With the expansion of precision and intensive agriculture, seed quality, which alone could contribute up to 15–20 per cent in terms of crop productivity, has assumed greater importance. Every care is taken not only to produce high-quality seeds and maintain the same through various stages of production, from growing conditions of the seed crop to pre-harvest, harvest, processing and storage activities, but also to adopt certain technologies to further improve the performance of seeds upon sowing of the crop under a wide range of environments. These technologies, collectively known as 'enhancement', cover a variety of methods applicable to different crop species and aimed at meeting specific requirements, such as unfavourable growing conditions of hard/acidic/sodic soil, high or low temperature, excess or deficient rainfall, etc., as well as the stress imposed by the presence of pests and diseases that affect crop performance, particularly during the early vegetative stage resulting in poor seed emergence,

E. A. Weissmann (✉)
HegeSaat GmbH & CoKG, Singen, Germany
e-mail: elmar.weissmann@eaw-online.com

K. Raja
Department Seed Science & Technology, TNAU, Coimbatore, Tamil Nadu, India

A. Gupta
Wageningen University & Research, Wageningen, The Netherlands
e-mail: arnab.gupta@wur.nl

M. Patel
Incotec India Pvt Ltd, Ahmedabad, Gujarat, India
e-mail: manish@incotecasia.com

A. Buehler
Operational Excellence – Seed Applied Solutions EME, Bayer Crop Science, Langenfeld, Germany
e-mail: alexander.buehler@bayer.com

© The Author(s) 2023
M. Dadlani, D. K. Yadava (eds.), *Seed Science and Technology*,
https://doi.org/10.1007/978-981-19-5888-5_16

crop establishment and vegetative growth. Selection of ~~the~~ appropriate and cost-effective technologies is important in accruing the best results from seed enhancement.

Keywords

Seed quality · Seed performance · Seed coating · Seed priming · Seed sorting

1 Introduction

As stated by Hay (2019), 'enhancing germination and reducing the loss in germination as much as possible, are fundamental to the successful use of seeds'. Hence, quality seed is desired not only to have high genetic and physical purity but also to be capable to produce vigorous seedlings with a well-developed shoot and root system that can result in good stand establishment and growth of the crop in a range of environmental conditions, achieving its potential yield. Many of these desired attributes, such as tolerance against major biotic and abiotic stress factors, can be achieved through genetic improvement by applying conventional breeding and biotechnological manipulations. However, genetic improvement using conventional breeding techniques in major food crops is fast reaching a 'yield plateau' with limited accessibility of plant germplasm narrowing the genetic base, while the acceptance of genetic manipulations, including GM and gene editing, is resisted in most countries at present. Hence, 'seed enhancement' technologies offer a means to compliment the genotype in such a manner that the planting value of the seed is enhanced over a wider range of growing conditions. The primary objective of seed enhancements is not only to optimize the inner physiological ability of the seed that normally does not express under normal sowing-cropping practices (Black and Peter 2006; Patel and Gupta 2012) but also to provide ease of handling seeds and better protection against biotic and abiotic stresses, especially in the early vegetative growth stages.

1.1 Scope of Seed Quality Enhancement

Seed quality enhancement is defined by Taylor et al. (1998) as 'post-harvest treatments that improve germination or seedling growth, or facilitate the delivery of seeds and other materials required at the time of sowing'. It includes a range of treatments applied to seeds between harvest and next sowing particularly to improve their performance. Considering the vast developments in seed enhancement technologies in the last 25 years, the definition can be expanded as 'such post-harvest treatments that improve germination and seedling growth, or facilitate the delivery of seeds and other essential substances required at the time of sowing for better performance'. This definition includes three broad groups of enhancement technologies. These are a) seed conditioning (by physical and chemical means), b) pre-sowing hydration treatments (priming) and c) coating, pelleting and encrusting technologies. Of these, pre-sowing hydration treatments include non-controlled systems (methods in which water is freely available and not restricted by the environment) as well as controlled (methods that regulate seed moisture content preventing the completion of germination) water uptake systems (Taylor et al. 1998).

1.1.1 Advantages Expected from Enhancement Techniques

One or more of the following advantages are expected from seed enhancement treatments that are performed singly or in combination:

- Ease in precision planting and handling.
- Early and uniform emergence due to improved germination and rapid seedling growth.
- Reduced seed rate.
- Better nursery management.
- Delivery of supplementary nutrients and growth stimulants needed for better performance after sowing.
- Protection against pests (including weeds) leading to better stand establishment.
- Removal of weak- or poor-performing seeds using non-traditional upgrading techniques.
- Tagging of seeds with visible pigments or other marker substances for traceability and identity preservation.

2 Types of Seed Quality Enhancement

Seed quality may be enhanced to various levels through different methods. These can be classified based on the mode of application.

2.1 Physical Methods

While the purpose of seed processing/conditioning itself is to upgrade the overall seed quality, some of the physical processes that could have a direct impact on the planting value or are performed specifically to enhance the quality further are:

- Size grading
- Brushing and scarifying
- Gravity and liquid density separation
- Colour sorting
- X-ray separation
- Imaging techniques, viz. near-infrared spectroscopy imaging, thermal imaging, magnetic resonance imaging, chlorophyll fluorescence sorting, multispectral analysis and scanning electron microscopic imaging.

2.2   Film Coating, Pelleting and Encrusting

Several technologies are based on the application of one or more thin coats of an inert polymer singly or in combination with active substances to cover the seed fully or partially for various purposes, which may or may not significantly increase the seed size or weight. These include:

- Coating for better flowability.
- Coating for binding pesticides or micronutrients.
- Coating for slow release of active chemicals.
- Coating for branding.
- Embedding biologicals on the seed surface.
- Encrusting.
- Pelleting with water-soluble inert substances.
- Optimizing seed for aerial seeding such as Seed Bomb™.

2.3   Seed Treatment for Protection Against Pests and Diseases

Seed dressings with pesticides offer the earliest and simplest technology for enhancing seed performance upon sowing. With the advancements in this field, in addition to the basic technologies, several physical and chemical methods are employed which are environment-friendly and offer higher precision. These include:

- Chemical seed dressing.
- Steam treatment/ThermoSeed™.
- Virus removal treatments.
- Hot water treatment.
- Dry heat treatment.
- Magnetic treatment.
- Electromagnetic treatment.
- Radiation treatment.
- Plasma treatment.

2.4   Physiological Enhancements

Physiological treatments for seed quality enhancement are based on metabolic advancement of seed that enables it to a better start upon sowing. These are collectively termed as 'seed priming', the word first coined by Malnassy (1971) and popularized by Heydecker (Heydecker 1974; Heydecker and Coolbear 1977). Seed priming techniques are essentially based on restricted hydration of the seed, which allows early events of germination but stops shortly before radicle protrusion not advancing to a dehydration-sensitive stage (Basu 1994; McDonald 2000). A variety of priming technology is in vogue for commercial applications and suited to various purposes.

2.4.1 Pre-sowing Priming
- Hydropriming.
- Osmopriming.
- Halopriming.
- Biopriming.
- Solid matrix priming.
- Pre-germination.
- Endophytic treatment.
- Nanopriming.

2.4.2 Mid-storage Correction
- Hydration-dehydration with or without chemicals.
- Halogenation.

2.5 Other Technologies

2.5.1 IP Protection and Microbranding
- Fluorescent marker coating.
- Microprinting.
- DataDots™.

3 Methods of Seed Enhancements

3.1 Physical Enhancements

3.1.1 Seed Cleaning, Processing and Quality Upgradation
The fundamental processing operations such as pre-cleaning, cleaning and grading enhance seed quality by way of improving its purity, appearance and vigour. The principles of grading and upgrading operations vary with the machine and mostly depend on the physical traits of the seeds such as size, shape, length, density, colour, texture, etc. Among the different machines, the grader works based on the seed size and shape which is known as basic cleaning of the seed. The indented cylinder separates the seed based on its length, the spiral separator based on seed shape, and the gravity separator based on density (Patil and Bansod 2014). Special operations like needle separator pick out the seeds with cracked seed coat and cut parts. Apart from this, many sophisticated advanced technologies or machines are there such as liquid density separator (selects well-filled and high-density seeds) (Koning et al. 2011); X-ray separation (selects well-filled seeds with mature embryo; https://www. incotec.com/en-gb/seed-technologies/seed-upgrading); Q_2 scanner, which selects vigorous seed based on respiratory efficiency (Bradford et al. 2013 and Centor Europe BV n.d.); magnetic resonance imager; chlorophyll fluorescence sorter (selects fully mature seeds based on chlorophyll breakdown); and multispectral analyser (https://neutecgroup.com/news-events/what-s-new/136-seed-and-grain-

analysis-by-multispectral-imaging-video), involving sensing the seed components and constituents (Jalink et al. 1998, 1999; Mortensen et al. 2021; El Masry et al. 2019) are being tested and used to enhance the quality of seed. However, many of these are yet to get scientific validation and hence in limited use.

3.2   Film Coating

Coatings are often applied to commercial seed lots to incorporate active ingredients (ai) for seed protection through targeted delivery and to enhance germination through the application of stimulators, besides improving the ease of seed handling (Pedrini et al. 2018). This is particularly useful for the targeted delivery of pesticides to reduce the use of chemicals in agriculture (Hay 2019). In the simplest terms, film coating is the application of a thin coat of (a) polymeric substance(s) to the seed in such a way that it does not significantly alter its size, weight and shape and does not affect its performance per se. The coating material, thus applied to the seed as a thin layer (film), not only improves the ballistic properties of the seed but also acts as a carrier to apply seed protectants and plant growth stimulators (fungicide, insecticide, hormones, micronutrients, etc.) directly on the seed surface, so that it is most effective to the germinating embryos, as well as makes the active ingredients available to growing seedlings for a longer period by way of the slow release from the applied coating. The colouring agent, usually used in a film coating, also improves the market value and provides a distinct identity to a seed brand. Many branded polymers such as Seedworx™, Ezi-cote™, etc. are available in the market for seed coating and used for coating vegetable seeds that have been tested for appearance, dust reduction, flowability, plantability and many other functions of value (apsaseed.org) (Fig. 1).

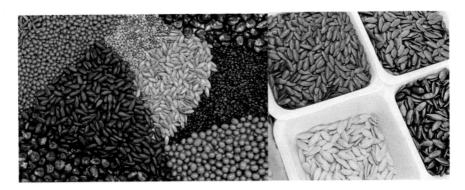

Fig. 1 Seed film coating with different colour polymers. https://www.openpr.com/news/1883041/ new-excellent-growth-of-seed-coating-materials-market-by-2019, accessed April 2022

3.2.1 Advantages of Film Coating

- Better appearance and flowability of seed.
- Accurate and controlled dosage of pesticides with a high level of precision on individual seeds is possible.
- Low dust formation enables safe and easy handling of seeds during packing, storage and transport.
- Slow release of applied molecules enhances the protection level of the germinating seedlings.
- Provides a unique identity to a company for their seeds in the market coated in a special colour. Coating in specific colour may also help to avoid mixture of varieties while handling, packing and storing.
- Application of plant protectants, microbial inoculants, micronutrients and other stimulants through seed coating enhances the overall performance of the seed.
- Low or no damage due to storage pests especially in cereals and leguminous seeds.

3.2.2 Trends in Seed Film Coating

The global market is aiming at 'More from Less', which triggers the use of precision farming wherein one can deliver required molecules to plants when it is needed most. By using pesticides through film coating, the number of pesticide sprays can be reduced significantly in open fields. Crop-specific nutritions can also be loaded on seeds to avoid large-scale application of such chemical nutrients to the soil. There is a trend in the global market to apply pesticides and nutrients as much as possible to the seed itself so that the post-emergence operations related to plant protection and nutrition can be minimized, besides minimizing the environmental pollution. By using fully automated seed treaters with the latest technology of rotary coating along with online seed dryers, it is possible to handle a few hundred metric tons per day in seed film coating. Also, there is a trend to use a specialized coating on seeds to control counterfeit and create a brand image among farmers.

Many countries in the global market have strengthened their seed treatment policy through new seed quality regulations in the treated seed. The European Seed Association has its agency ESTA (European Seed Treatment Assurance Industry Scheme) which monitors the quality of seeds that are delivered in the market and also keeps an eye on the environmental/public health repercussions for using treated seeds (ESTA 2020). There are some reports (Copeland and McDonald 2001) on using water-impermeable plastic film coating to delay germination up to a certain time to coincide with favourable conditions or for achieving synchronous germination in the parental lines of hybrid seed production (Johnson et al. 1999). This technology already has a commercial application called the Pollinator Plus (R) which is an Incotec-Croda proprietary, derived from research made at Purdue University. This involves coating the seeds of male parental lines of maize hybrids in such a way that they germinate and subsequently flower for a wider duration so that the female lines get enough time for fertilization. This technology was also called IntelliCOAT™ and was previously owned by the Landec Inc. (Landec 2011). Organic coating and the use of biologicals are new demands to support upcoming organic farming and sustainable agricultural practices. For the same purpose, research on the development of microplastic-free polymers is also being tested for seed coating (Pedrini et al. 2017).

3.3   Seed Pelleting

Seed pelleting is a process of enclosing the seed with an inert material to produce a globular unit of standard size to facilitate precision sowing. It is mainly useful for mechanized sowing of seeds which results in an even distribution of small seeds in the field and reduces the seed rate as compared to traditional planting methods. Pelleting technology is most commonly applied in small, minute, lightweight and expensive seeds of vegetables and flowers or those with appendages that hinder seed handling. Some examples of plant species where seed pelleting is commonly employed are leek, celery, onion, lettuce, carrot, monogerm sugar beet, chicory, forage grasses, flowers, etc.

The basic components of pelleting are fillers, binders, active ingredients (nutrients, plant protectants, etc.) and water. Two types of materials are used for commercial seed pelleting, that is, split type or melting type. In both cases, it is preferred to use inert substances as filler material that have good water-absorbing capacity, remain firm on drying and disintegrate upon soaking.

Besides improving the seed size and shape, similar to film coating, with pelleting, it is possible to add various nutrients, bio-regulators and plant-protective chemicals along with the inert material as per the need of specific species. Different kinds of pelleting materials, viz. LightKote™, HeavyKote™, Split Pill™, SplitKote™, etc., are commercially available for meeting the specific needs, including the biodegradability and the final size of the pellets. The size of the pellets normally varies between 2 and 5 mm, while the size ratio varies from 2 to 15 times, and the weight increases from 4 times to 10 times of the normal raw seed. For example, for 1 kg of tomato seeds with an average seed size of about 2 mm, the weight of the pelleted seed will increase ten times to 10 kg after pelleting of about 4 mm size (Fig. 2).

During pelleting, nutrients, plant growth stimulants, plant-protective chemicals and bioinoculants can also be added along with the inert material integrating different technologies during processing. This highly sophisticated seed can be named as 'designer seed' (Fig. 3).

To obtain seed pellets with uniform size and shape, the measured quantity of pelleting and binder materials and other additives, if any, are added to the rotating drum of an automatic (Fig. 4) or manually operated machine. The multilayer coatings are made by adding the materials sequentially as per their compatibility and mode of action, with the required number of such actions. After processing, the pellets need to be calibrated based on size and weight to make available uniform seed pellets for commercial use (seedquest.com).

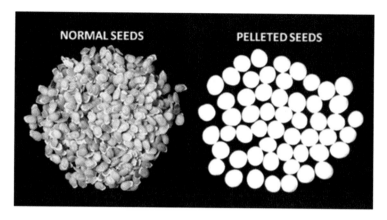

Fig. 2 Size increase in pelleted seeds. https://www.bighaat.com/blogs/kb/pelleted-seeds-v-s-normal-seeds, accessed April 2022

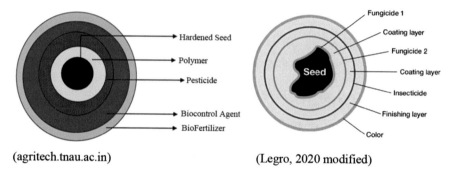

(agritech.tnau.ac.in)   (Legro, 2020 modified)

Fig. 3 Designer seed technology

3.3.1   Selection of the Pelleting Material

It is important to select the appropriate inert material, hormones, nutrients and plant-protective chemicals to avoid problems of losing viability and vigour of pelleted seed. Therefore, the following criteria should be considered to identify the right type of pelleting material.

- It should be porous and should allow the water to enter it while the pellets are sown in the field. It should also be easily dissolvable with water.
- The choice of the pelleting material is also determined by the type of seed and conditions at sowing, where the seed is likely to be sown. Irrigated or rainfed situations, for instance, are different, and therefore, the pelleting materials are to be chosen accordingly.
- Materials to be added to the seed should not release harmful dust during handling and sowing.

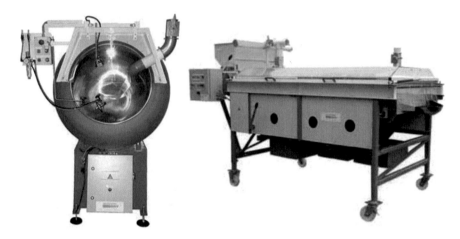

Fig. 4 Seed pelletizer and pellet sorter. (https://www.seedprocessing.com/hemp-pelleting/ accessed on April 2022)

- The choice of adhesive is also important to produce dust-free pellets, without having any adverse effect, per se, on the performance.
- The nutrients, growth regulators and stimulants, plant-protective chemicals and other additives must be compatible with the filler as well as the binder materials. Appropriate dosages need to be standardized to avoid excess and overdosing.

3.3.2   Advantages of Seed Pelleting

- Singulation of seed is achieved during pelleting which leads to easy sowing and reduction in seed rate.
- Small seeds can be made larger to facilitate mechanization.
- Irregularly shaped seeds can be coated to make them smooth and thus suitable for mechanized sowing.
- Accurate application of chemicals is possible, reducing wastage of chemicals.
- Phenomenon of dust-off of additives, especially pesticides, is reduced by seed pelleting.
- Pelleting is the best carrier for all kind of additives to seed in the safest way. The materials can be applied as several layers without affecting their function.
- Slow release of loaded chemicals results in longer periods of their beneficial effects to the growing seedlings creating visible difference between the performance of pelleted and raw seed.
- Post-sowing applications can be minimized based on loading dosages of plant-protective materials in pellets.
- Pelleting in combination with priming technology gives a unique advantage to seed, facilitating faster and uniform germination and healthy seedling growth.

3.4   Seed Encrustation

Encrustation is useful with seeds of irregular surface, where grooves and cavities on the seed surface are filled with a special type of filler material in order to make the seed smoother and more uniform in shape as well as to increase the reasonable level of weight (Pedrini et al. 2018). Encrustation of seed is more useful for small and lightweight seeds which are not very expensive and bulky in nature, such as pasture and forage seeds; agronomic crops like sunflower, maize, etc.; and open pollinated vegetables like carrot, onion, endives, fennel, etc. There is no drastic effect on the size of the seed while encrustation, but weight increases from two to four times. This means that after encrustation, 1 kg of carrot seed will become 2 to 4 kg in weight. There is possibility to load enough amounts of various plant-protective chemicals and nutrients in encrustation process. Compared to pelleting, encrustation is much faster and, therefore, more economical. In the USA and Australia, many pasture companies are selling their grass seeds in encrusted forms since most of the sowing operations are done using aircraft which need enhanced seed weight, i.e. encrusted seed. Mechanized sowing with high level of accuracy is possible using encrusted seed which is required in certain crops like onion, carrot, sugar beet or chicory, which helps to achieve uniform and balanced tuber growth. Encrusting technology, wherever applicable, is almost as good as pelleting, even at a lower cost. In countries like India, where seed sale price is relatively low, such technology has good potential. Currently, encrusting technology has been adopted in onion seed at a large scale, and direct sowing of seed is becoming popular in many parts of India. It reduces production cost by saving on labour and seed rate per acre and also reduces the overall crop duration. Trials in sugar beet and chicory are in progress, and it is expected that soon the growers will prefer to use encrusted seeds for direct sowing in few crops where labour is getting problematic and expensive.

In many industrialized countries, small-sized seeds having all good performance parameters are encrusted to enable the use of pneumatic single seed planters. Loading of various chemicals and biologicals is feasible in encrusted seeds to support better growth after germination. For example, Seed Innovations™ encrustation technology is used to achieve precision planting and to enable the application of plant protection products and additives to get maximum plant performance (seedinnovations.co.nz, Fig. 5).

3.5   Hot Water Treatment

Hot water treatment is an age-old practice to destroy surface-borne and seed-borne pathogens at temperatures high enough to kill pathogens and cold enough to safeguard the viability of the seed as described by Suryapal et al. (2020). In many investigations, it was noted that even with longer treatment times, hot water treatment with a temperature of 40 °C had no significant effect on the seed-borne pathogens. However, on most crops, the hot water treatments at temperatures above 50 °C or 53 °C for 10–30 min have a good phytosanitary effect. In the majority of cases, these treatment conditions did not affect seed germination. To

Fig. 5 Encrustation of onion seeds (https://www.seedinnovations.co.nz/resources/news/encrusting-carrot-and-onion-seed.html, accessed April 2022)

Table 1 Pathogens in wheat and barley susceptible to water and steam treatment (Modified after Bänziger et al. 2022 and Forsberg et al. 2005)

Botanic name	Common name	Wheat	Barley
Fusarium spp./*Microdochium nivale*	Snow mould/leaf blotch	x	
Septoria nodorum	Nodorum blotch	x	
Tilletia caries	Common bunt	x	
Tilletia controversa	Stink smut	x	
Ustilago tritici	Loose smut	x	
Drechslera graminea	Leaf stripe		x
Ustilago hordei	Covered smut		x
Ustilago nuda	Loose smut		x

reduce the adverse effect of higher temperatures (~53 °C) on germination, comparatively shorter treatment time must be used, especially on sensitive crops like cabbage, etc. (Nega et al. 2003). This eco-friendly technology works best for small seeds, particularly for vegetable crops. It is not applicable to old or very large seeds. Similarly, it is not effective for primed, coated, pelleted and pesticide-treated seed.

3.6 Steam Treatment

Steam treatment has been developed to disinfect seeds from pathogens (Forsberg et al. 2015). In thermal seed treatment, seeds are exposed to hot humid air for a fraction of seconds to disinfect/disinfest them from seed-borne microorganisms. This is a patented technology being adopted at commercial level in many industrialized countries for the effective control of microorganisms and a safer environment. The base unit disinfects seeds with active steam, enabling effective, gentle and targeted heat transfer eliminating broad spectrum of fungi and bacteria (Table 1).

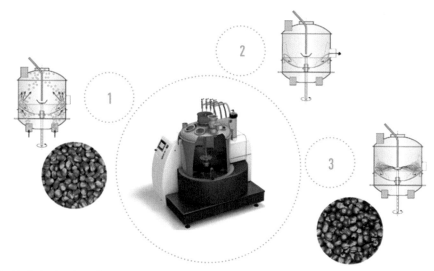

1 - Hygienisation through active steam treatment
2 - Re-Drying
3 - Additional application of protective components

Fig. 6 Steam treatment, working principle of PETKUS HySeed bio ®. (http://www.petkus.com/products/-/info/coating/hygienisierung/multicoater-cm-hyseed-bio, accessed April 2022)

The main unit of a steam treatment equipment includes a control cabinet, the core unit, steam generator and the dry generator unit (Fig. 6). Within the core unit, a sanitization/hygienization through active steam treatment is followed by re-drying and optional additional application of protective components. Compared to hot water disinfection, this method is more energy-efficient, as steaming enables a faster drying process (Ascard et al. 2007).

3.7 Dry Heat Treatment

Dry heat treatment is a common physical seed treatment used to eradicate the fungal and bacterial seed-borne pathogens. Also, it is an effective and eco-friendly method for inactivating seed-borne viruses. This treatment is mostly followed for vegetable crop seeds. In cucumber, 70 °C dry heat treatment for 90 min eradicates *Cladosporium cucumerinum*, *Didymella bryoniae* and *C. orbiculare* (Shi et al. 2016). Similarly, exposure of Andean lupin seeds to dry heat for 8 or 12 h reduced pathogen transmission up to 85% from seed to the seedlings (Falconi and Viviana 2016). However, care is needed to perform this treatment to maintain the germinability and vigour of seeds.

3.8   Other Potential Treatments

Many potential treatments, both physical and chemical, have been reported to improve seed performance. However, these are yet to be standardized for commercial application (e.g. Dubinov et al. 2000; Kopacki et al. 2017; Legro 2020; Rochalska et al. 2011; Sivachandiran and Khacef 2017).

3.8.1   Magnetic and Electromagnetic Seed Enhancement Treatment

Exposure of seeds to different flux density of magnetic field is reported to promote rapid germination, uniform crop stand, enhanced yield and resistance to disease incidence (Pietruszewski and Martinez 2015). The magnetic seed treatment is a non-invasive physical stimulant for a specific duration of time to induce physiological changes in seed (Pietruszewski and Martinez, 2015). It is reported to control free radicals, increase activity of many enzymes and seed vigour (Vashisth and Nagarajan 2008), increase plant hormones IAA and GA in germinating seeds (Podlesna et al. 2019) and reduce pathogenic diseases (Afzal et al. 2015). It is often effectively used as a pre-sowing treatment for mitigating both biotic and abiotic stresses such as drought, salinity, diseases and pests during germination and early crop growth (Kataria 2017). Exposure of seeds to pulsed electromagnetic field (EMF) at low-frequency levels is also found effective. The positive impacts of EMF treatment include better seed germination (Gorski et al. 2019), seedling vigour (Isaac et al. 2011), tolerance to unfavourable environmental conditions (Pietruszewski and Kania 2010; Balakhnina et al. 2015) and plant growth and yield (Efthimiadou et al. 2014).

3.8.2   Radiation Treatments

Low doses of gamma ray, high-energy electrons, ultrasonic radiation and microwave and UV radiation are also used as an alternative seed treatment for the management of microbial infection (Brown et al. 2001). Studies reveal that recommended dose of gamma radiation (up to 20 Gy) stimulated seed germination without causing collateral DNA damage (Nesh et al. 2019).

3.8.3   Electron Treatment

'The electron treatment of seeds is a promising technology that is based on the biocidal ionizing effect of low-energy electrons (wavelength below 100 nm). The accelerated electrons are generated following the principle of the Braun tubes. When high electrical voltages are applied between a cathode and anode, electrons are emitted from the cathode and are accelerated in the direction of the electron exit window by the difference in the electrical charge. While treating seeds, the applied dose and electron energy are critically monitored. The dose can be determined by regulating the current strength, and the electron energy can be adjusted with the acceleration voltage. During the electron treatment of seeds, the lethal dose is crucial to combat the existing pathogens. The electron energy is a measure of the kinetic motivity of electrons. When electrons penetrate matter, they lose their energy. Once the energy is spent, they do not penetrate further into the material. This fact is used to

Technology Description

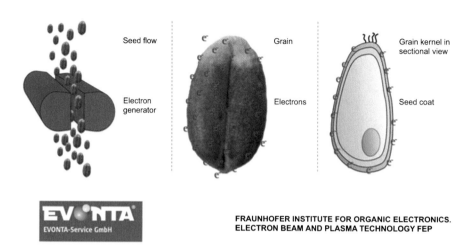

Fig. 7 Principle of electron beam treatment (https://www.evonta.de/Application/46511/, accessed April 2022)

precisely control the sphere of action during electron treatment and to ensure that it does not exert harmful effects on the embryo and the endosperm. Harmful organisms hit by accelerated electrons in the effective range are killed' (cited from Rögner 2018; see Fig. 7).

A splitting of molecular chains in microorganisms results in killing the pathogens, irrespective of their nature (Vishwanath et al. 2016). The method is reported to be effective against various fungal spores, bacteria and viruses (Jahn et al. 2005). This method does not use radiation and is acknowledged as a biological control method according to EU regulation 2092/91 (https://deutsche-saatguterzeuger.de/2017/0 6/10/b1-2/, accessed April 2022).

4 Physiological Enhancements

4.1 Pre-sowing Seed Priming

Seed priming is the process of controlled hydration of seeds to a level that permits pre-germinative metabolic activities, but prevents the emergence of the radical (Heydecker 1973). The hydrated primed seeds can be reverted back to a safe moisture content before its distribution and planting. Seed priming, if performed carefully, is a simple and effective technique to get speedy and uniform emergence and better performance.

Materials such as plain water or dilute solutions of salts, growth regulators, vitamins, bioinoculants, plant products, leaf extracts, etc. can also be used as priming agents. However, optimization of the correct method of seed priming for a seed species is critical and needs thorough standardization concerning priming media, temperature, duration, additives and post-hydration drying technology of the treatment (e.g. Waqas et al. 2019; Malek et al. 2019).

4.2   Types of Seed Priming

4.2.1   Hydropriming

It is a controlled hydration process in which the seed is soaked in water for a predetermined period and then re-dried to its initial moisture content (Pill and Necker 2001). Besides enhancing the germination and vigour of seed under normal conditions, hydropriming is reported to improve seed performance under salinity stress and drought-prone environments (Pill and Necker 2001). The process of hydration is time-regulated so that the seed goes through phases I and II of germination allowing metabolic activation and some degree of cellular repair enhancing seed performance, but must be stopped and dried back to prevent the onset of phase III of germination. To avoid rapid absorption of water, which could be harmful in some seeds especially legumes, spraying water on seeds is practised. The patented technology 'drum priming' is used by seed companies for hydropriming commercial seed lots, wherein a batch of seeds is placed in a large rotary drum while water is sprayed through a jet moistening the seed uniformly. In this way, seeds are constantly upturned and hydrated with only the desired amount of water, keeping them moist for a certain period (depending on the species), and the sudden inrush of water, as in soaking, is avoided. After the desired period, seeds are dried in the same rotary drum.

4.2.2   Osmopriming

Soaking seeds in aerated osmotica (mannitol or polyethylene glycol) to keep the seed hydrated only up to a certain level for a certain period of time, followed by washing and drying, is known as osmopriming. In this method, the desired water potential is maintained with the addition of osmotic substances in the required concentration, thus restricting the water uptake to a certain point and not allowing the seed to advance beyond phase II. Polyethylene glycol (PEG) is the most preferred osmoticum for its chemically inert nature, though it has a low oxygen solubility. It has been successfully applied in many vegetable species (Varier et al. 2010). Patented technology of membrane osmopriming is reported to be effective in high-value, small volumes of vegetable and flower seeds, especially those which are mucilaginous (Rowse et al. 2001).

4.2.3   Solid Matrix Priming

Another way of controlled hydration is by raising the seed moisture by putting it in solid, porous matrices maintained at the desired level of water potential. This consists of mixing seeds with an organic or inorganic carrier maintained at 0.4 to 1.5 MPa at 15 °C for achieving a moisture content sufficient for metabolic processes to continue, but not allowing radicle protrusion. Seed water potential is regulated by the matric potential of the seed during priming; the water is largely held by the carrier resulting in slow imbibition. Matric carriers commonly used are calcinated clay, vermiculite, peat moss, sand, microgel, diatomaceous earth, ligneous shale, etc. The amount of water to be added is determined by the weight of seed and final seed moisture content targeted (Varier et al. 2010).

4.2.4 Halopriming

It involves soaking seeds in low-concentration salt solutions of chlorides, sulphates, nitrates, etc. (Gour et al. 2022). This results in hardening the seed improving their performance under salt-stressed conditions. The seed size, structure, biochemical constitution, position of seed-protecting layers, type of salt and soaking time are the factors influencing ion penetration into the embryo.

4.2.5 Biopriming

Soaking seeds in an aqueous solution/suspension containing beneficial microorganisms such as various strains of fungal and bacterial species of *Azospirillum*, *Rhizobium*, *Bacillus*, *Pseudomonas*, *Trichoderma*, etc. is effective in facilitating the advancement of pre-germinative metabolism, better seedling growth (Bennett and Whipps 2008) and protection against harmful microbes in the soil. This involves imbibition of seed with bacterial inoculation (Callan et al. 1990). This treatment not only increases the uniformity and speed of germination but also guards the germinating seed against many soil- and seed-borne pathogens. However, infected seed imbibition during priming may result in a stronger microbial growth and consequently impairment of plant health. Applying antagonistic microorganisms during priming is an ecological approach to overcome this problem (Reddy 2013). In many cases, especially in vegetables, biopriming is an effective approach to disease management (Müller and Berg 2008).

Endophytic Association

Many of the applied bacterial, fungal and mycorrhizal species develop an endophytic association with seed and help considerably in overcoming the negative consequences imposed by various stress factors. Use of endophytic microorganisms could serve as a viable option to circumvent the limitations associated with seed. Endophytes may enhance host growth and nutrient acquisition and improve the plant's ability to tolerate biotic and abiotic stresses. In case of wheat, fungal entomopathogens such as *Beauveria bassiana* and *Metarhizium brunneum* can promote plant growth through seed treatment and suppress disease pathogens following plant colonization as well (Lara 2018). Similarly, faba bean seeds treated with endophytes *Trichoderma asperellum*, *Gibberella moniliformis* and *Beauveria bassiana* had a significantly lower number of aphids when compared to untreated controls. The endophytic fungus Piriformospora indica induced alteration in plant metabolites under drought stress is reported (Ghaffari et al. 2019). Endophytic bacterial species equipped with plant growth-promoting traits may induce tolerance to salt stress by modulating the morphological, physiological and biochemical characteristics of plants.

4.3   Pre-germination

Pre-germination describes the process of incubating seed under favourable germination conditions up to the point of radical protrusion, followed by sowing in the field. During direct sowing in the field, distinguishing germinable seeds from non-germinable ones is not possible, especially in case of farm-saved seeds. This often leads to uneven germination and poor crop stand (Bidhan 2013). However, use of pre-germinated seeds, based on the length of the emerged radicles, automatically eliminates the poor-quality seeds and results in rapid and uniform field emergence and good plant stand. The pre-germination technique also reduces overall seed rate. Seedlings from pre-germinated seeds were found heavier and more vigorous than the seedlings from dry seeds (Ghate and Phatak 1982). Ridge gourd seeds soaked in distilled water for 20 h followed by incubation for 1 day produced vigorous seedlings which could overcome germination-related problems (Abinaya et al. 2020). However, pre-germinated seed has a very short shelf life due to its desiccation sensitivity and needs to be sown immediately, restricting its application.

4.4   Mid-storage Correction by Hydration-Dehydration

The mid-storage correction treatments improving seed vigour and subsequent storability were standardized for several field crops and vegetables, primarily as an on-farm treatment by Prof. R.N. Basu and his group in Calcutta University, India (Basu et al. 1975; Basu and Dasgupta 1978; Basu 1994). It is the process of hydrating (partially or fully) the low and medium vigour seeds in water for short durations with or without added chemicals, so as to raise the seed moisture content to 25–30%, maintaining for few hours and drying back the seeds to a safe moisture content for dry storage. Post-hydration drying is a crucial step in this process, which determines the shelf life of treated seed. The quality enhancement in such treatments results from:

- Counteraction of free radicals and lipid peroxidation reactions.
- Cellular repair in hydrated state.
- Germination advancement and embryo enlargement during hydration.

Removal of toxic metabolites accumulated in seed could also play a role in treatments of soaking for a longer duration.

The dry seed treatments with diverse substances and methods are also reported to be effective in maintaining high vigour and viability during storage. These include halogenation in closed environment or treatment of seed with halogenated plant-based powders such as *Albizia amara*, tamarind, neem, *Colocasia*, etc. abstracting free radicals in the cellular environment (Basu and Punjabi 2022; Dharmalingam and Basu 1978).

4.5 Emerging Seed Enhancement Treatments

Application of nanotechnology for seed quality enhancement is an emerging trend in the seed sector. It is often referred as the 'third-generation' seed treatment for quality enhancement. Seed coating or priming with nanoparticles, detection of seed quality by e-nose technology (sensing volatiles released by the seed), nanobarcoding, etc. are some of the latest applications of nanotechnology in the seed sector.

Seed priming with nanoparticles or nanopriming involves soaking seeds in solutions of nanoparticle/compounds, followed by drying (Panda and Mondal 2020). It is reported to augment the performance of seeds in many ways such as enhancing α-amylase activity, increasing soluble sugar content to support early seedling growth, upregulating the expression of aquaporin gene in germinating seeds, increasing stress tolerance controlling reactive oxygen species (ROS) production and creation of nanopores for enhanced water uptake, to mention a few.

Use of several metal-based nanoparticles (NPs) such as silver nanoparticles (AgNPs), gold nanoparticles (AuNPs), iron nanoparticles (FeNPs and FeS2NPs), copper nanoparticles (CuNPs), titanium nanoparticles (TiO2NPs), zinc nanoparticles (ZnNPs and ZnONPs), cerium oxide nanoparticles (CeNPs) and carbon-based NPs (such as fullerene and carbon nanotubes) in pre-sowing seed treatments is reported to promote germination, early seedling growth and environmental stress tolerance (Mahakham et al. 2017; Guha et al. 2018). Priming with iron nanoparticles has significantly improved germination, rate of emergence and subsequent growth in watermelon seed by triggering metabolic processes during the early phase of seed germination (Kasote et al. 2019).

Synthesis of plant-based nanoparticles is a further refinement of nanotechnology that uses sustainable manufacturing processes to produce safe and innocuous nano-scale biomaterials for agricultural applications. Vurro et al. (2019) attempted to develop silver nanoparticles (AgNPs) by using onion peel extract, as silver has anti-bactericidal and anti-fungicidal properties. Similarly, turmeric oil nanoemulsions and silver nanoparticles synthesized from agro-industrial by-products used as nanopriming agents showed enhanced seed germination, growth and yield in watermelon while maintaining fruit quality (Acharya et al. 2020).

5 IP Protection and Microbranding

Besides enhancing the performance of seed, coating with additional markers using fluorescent substances can be helpful in avoiding seed adulterations. Seeds with a coating of rhodamine B show red fluorescence (Guan et al. 2013). Such coatings provide a means for a unique hidden identity that establishes the ownership and can be effective in preventing counterfeits (Thiphinkong and Sikhao 2021). Similarly, a non-ionic fluorescent tracer, coumarin 120, was used for differentiating the varieties in soybean (Taylor and Salanenka 2012). Uptake of coumarin 120 and rhodamine B was measured more in the hypocotyl and root, with lesser amounts in the epicotyl and true leaves (Wang et al. 2020). Specialized application of such technologies is being made for microbranding. In this innovative application, very small pieces of microprinted plastic are mixed with the film coat liquid which remain on the surface of the seed. When needed, the seed surface can be observed under a magnifier, to check the brand name/code. This technology is marketed as 'DataDots' and currently in use in the seed industry in the South and South-East Asia (DataDot n.d.; https://www.datadotdna.com/au/solutions/authentication2/datadot/).

6   Conclusion

Manifold improvements in seed performance can be achieved by applying enhancement technologies singly or in combination, using physical, chemical or physiological methods. Nevertheless, an effective technology can be successful only if it is affordable. As the seed is considered to be an effective delivery system and carrier of new technology, it is appropriate to have the best suitable quality enhancement treatments disseminated through the seed, which are both environment-friendly and affordable.

References

Abinaya K, Sundareswaran S, Raja K (2020) Effect of pre-germination technique on enhancing seed and seedling vigour in ridge gourd (*Luffa acutangula* Roxb.). Madras Agric J 107(4–6): 119–124

Acharya P, Jayaprakasha GK, Crosby KM, Jifon JL, Patil BS (2020) Nanoparticle-mediated seed priming improves germination, growth, yield and quality of watermelons (*Citrullus lanatus*) at multi-locations in Texas. Sci Rep 10:5037. https://doi.org/10.1038/s41598-020-61696-7

Afzal I, Noor MA, Bakhtavar M, Ahmad A, Haq Z (2015) Improvement of spring maize performance through physical and physiological seed enhancements. Seed Sci Technol 43:238–249

Ascard J, Hatcher PE, Melander B, Upadhyaya MK (2007) Thermal weed control. In: Upadhyaya MK, Blackshaw RE (eds) Non-chemical weed management: principles, concepts and technology. CABI Publishing, Wallingford, pp 155–175

Balakhnina T, Bulak P, Nosalewicz M, Pietruszewski S, Włodarczyk T (2015) The influence of wheat *Triticum aestivum* L. seed pre-sowing treatment with magnetic fields on germination, seedling growth and antioxidant potential under optimal soil watering and flooding. Acta Physiol Plant 37:59. https://doi.org/10.1007/s11738-015-1802-2

Bänziger I, Kägi A, Vogelsang S, Klaus S, Hebeisen T, Büttner-Mainik A, Sullam K (2022) Comparison of thermal seed treatments to control snow Mold in wheat and loose smut of barley. Front Agron 3. https://doi.org/10.3389/fagro.2021.775243

Basu RN (1994) An appraisal of research on wet and dry physiological seed treatments and their applicability with special reference to tropical and sub-tropical countries. Seed Sci Technol 22: 107–126

Basu RN, Dasgupta M (1978) Control of seed deterioration by free radical controlling agents. Indian J Exp Biol 16:1070–1073

Basu RN, Punjabi B (2022) Seed invigoration- treatments and quantification of vigour. Astral. Intl. Pvt. Ltd, New Delhi, p 382

Basu RN, Bose TK, Chattopadhyay K, Dasgupta M, Dhar N, Kundu C, Mitra R, Pal P, Pathak G (1975) Seed treatment for the maintenance of vigour and viability. Ind Agric 19:91–96

Bennett A, Whipps JM (2008) Dual application of beneficial microorganisms to seed during drum priming. Appl Soil Ecol 38(1):83–89

Bidhan R (2013) Innovative technique for enhancement of seed germination and seedling establishment of cucurbits (Cucumis sativus L.). Int J Agric Sci 9(2):663–666

Black HM, Peter H (2006) The encyclopaedia of seeds: science, technology and uses. CABI, Wallingford, p 224

Bradford K, Bello P, Fu JC, Barros M (2013) Single-seed respiration: a new method to assess seed quality. Seed Sci Technol 41:420–438. https://doi.org/10.15258/sst.2013.41.3.09

Brown JE, Lu TY, Stevens C, Khan VA, Lu JY, Wilson CL, Igwegbe ECK, Chalutz E, Droby S (2001) The effect of low dose ultraviolet light-C seed treatment on induced resistance in cabbage to black rot (Xanthomonas campestris pv. campestris). Crop Prot 20:873–883. https://doi.org/ 10.1016/S0261-2194(01)00037-0

Callan NW, Marthre DE, Miller JB (1990) Bio-priming seed treatment for biological control of Pythium ultimum preemergence damping-off in sh-2 sweet corn. Plant Dis 74:368–372

Centor Europe BV (n.d.). https://agricsct.com/products/analytical-equipment/p/q2-scanner. Accessed Mar 2022

Copeland LO, McDonald MB (2001) Seed enhancements. In: Copeland LO, McDonald MB (eds) Principles of seed science and technology, 4th edn. Kluwer Academic, Boston, MA

DataDot (n.d.). https://www.datadotdna.com/au/solutions/Website. Owned by DataDot Technology Ltd. P.O Box 6245, Frenchs Forest DC, New South Wales 2086. Accessed Mar 2022

Dharmalingam C, Basu RN (1978) Control of seed deterioration in cotton (Gossypium hirsutum L.). Curr Sci 47:484–487

Dubinov AE, Lazarenko EM, Selemir VD (2000) Effect of glow discharge air plasma on grain crops seed. IEEE Trans Plasma Sci 28(1):180–183

Efthimiadou A, Katsenios N, Karkanis A, Papastylianou P, Triantafyllidis V, Travlos I, Bilalis DJ (2014) Effects of presowing pulsed electromagnetic treatment of tomato seed on growth, yield and lycopene content. ScientificWorldJournal 2014:369745. https://doi.org/10.1155/2014/ 369745

ElMasry G, Mandour N, Al-Rejaie S, Belin E, Rousseau D (2019) Recent applications of multispectral imaging in seed phenotyping and quality monitoring—an overview. Sensors 19(5): 1090. https://doi.org/10.3390/s19051090

ESTA (2020) The UK ESTA Standard, European Seed Treatment Quality Assurance Scheme for seed Treatment and Treated Seed (ESTA). https://euroseeds.eu/esta-the-european-seed-treatment-assurance-industry-scheme/. Accessed 26 Apr 2022

Falconi CE, Viviana Y (2016) Dry heat treatment of Andean lupin seed to reduce Anthracnose infection. Crop Prot 89:178–183

Forsberg G, Johnsson L, Lagerholm J (2005) Effects of aerated steam seed treatment on cereal seed-borne diseases and crop yield/Einfluss einer Saatgutbehandlung mit einem Heißwasserdampf-Luft-Gemisch auf Saatgut-übertragbare Getreidekrankheiten und Kornertrag. J Plant Dis Prot 112(3):247–256

Ghaffari MR, Mirzaei M, Ghabooli M, Khatabi B, Wu Y, Zabet-Moghaddam M, Mohammadi-Nejad G, Haynes PA, Hajirezaei MR, Sepehri M, Salekdeh GH (2019) Root endophytic fungus Piriformospora indica improves drought stress adaptation in barley by metabolic and proteomic reprogramming. Environ Exp Bot 157:197–210

Ghate SR, Phatak SC (1982) Preference of tomato and pepper seed germinated before planting. J Am Soc Hortic Sci 107:908–911

Gorski R, Dorna H, Rosinska A, Szopinska D, Wosinski S (2019) Effects of electromagnetic fields and their shielding on the quality of carrot (Daucus carota L.) seeds. Ecol Chem Eng S 26(4): 785–795

Gour T, Lal R, Heikrujam M, Singh V, Vashistha A, Agarwal LK, Kumar R, Chetri SPK, Sharma K (2022) Halopriming: sustainable approach for abiotic stress management in crops. In: Roy S, Mathur P, Chakraborty AP, Saha SP (eds) Plant stress: challenges and management in the new decade. Advances in science, technology & innovation. Springer, Cham. https://doi.org/10.1007/978-3-030-95365-2_9

Guan Y, Wang J, Tian Y, Hu W, Zhu L, Zhu S, Hu J (2013) The novel approach to enhance seed security: dual anti-counterfeiting methods applied on tobacco pelleted seeds. PLoS One 8(2): e57274. https://doi.org/10.1371/journal.pone.0057274

Guha T, Ravikumar KVG, Mukherjee A, Anita M, Kundu R (2018) Nano-priming with zero valent iron (nZVI) enhances germination and growth in aromatic rice cultivar (*Oryza sativa* cv. *Gobindabhog L.*). Plant Physiol Biochem 127:403–413

Hay FR (2019) Editorial. Seed Sci Technol 47(1):113–119

Heydecker W (1974) Germination of an idea: the priming of seeds. School of Agriculture Research, University of Nottingham, Nottingham, pp 50–67

Heydecker W, Coolbear P (1977) Seed treatments for improved performance – survey and attempted progress. Seed Sci Technol 5:353–425

Isaac AE, Hernández AC, Domínguez A, Cruz OA (2011) Effect of pre-sowing electromagnetic treatment on seed germination and seedling growth in maize (Zea mays L.). Agronomía Colombiana 29(2):405–411. http://www.scielo.org.co/scielo.php?script=sci_arttext&pid=S0120-99652011000200007&lng=en&tlng=en. Accessed 26 Apr 2022

Jalink H, Van der Schoor R, Frandas A, Van Pijlen JG, Bino RJ (1998) Chlorophyll fluorescence of Brassica oleracea seeds as a nondestructive marker for seeds maturity and seed performance. Seed Sci Res 8:437–443

Jalink H, Van der Schoor R, Birnbaum YE, Bino RJ (1999) Seed chlorophyll content as an indicator for seed maturity and seed quality. Acta Hortic 504:219–227

Johnson GA, Hicks DH, Stewart RF, Duan X (1999) Use of temperature-responsive polymer seed coating to control seed germination. Acta Hortic 504:229–236

Kasote DM, Lee J, Jayaprakasha GK, Patil BS (2019) Seed priming with iron oxide nanoparticles modulate antioxidant potential and defense linked hormones in watermelon seedlings. ACS Sustain Chem Eng 7(5):5142–5151. https://doi.org/10.1021/acssuschemeng.8b06013

Kataria S (2017) Role of reactive oxygen species in magnetoprimed induced acceleration of germination and early growth characteristics of seeds. In: Singh VP, Singh S, Tripathi DK, Prasad SM, Chauhan DK (eds) Reactive oxygen species in plants. https://doi.org/10.1002/9781119324928.ch4

Koning JRA, Bakker E, Rem P (2011) Sorting of vegetable seeds by magnetic density separation in comparison with liquid density separation. Seed Sci Technol 39:10.15258/sst.2011.39.3.06

Kopacki M, Pawlat J, Terebun P, Kwiatkowski M, Starek A, Kiczorowski P (2017) Efficacy of non-thermal plasma fumigation to control fungi occurring on onion seeds. In: Electromagnetic devices and processes in environment protection with seminar applications of superconductors (ELMECO and AoS), 2017, International conference on IEEE, Nalezcow 1-4

Landec Corporation (2011) Annual report. https://www.annualreports.com/HostedData/AnnualReportArchive/l/NASDAQ_LNDC_2011.pdf. Accessed Apr 2022

Lara RJ (2018) Seed inoculation with endophytic fungal entomopathogens promotes plant growth and reduces crown and root rot (CRR) caused by *Fusarium culmorum* in wheat. Planta 248:1525–1535

Legro B (2020) Loading the seed for performance. https://www.agrithority.com/wp-content/uploads/2020/02/LoadingSeed-Legro1.pdf. Accessed Apr 2022

Mahakham W, Sarmah AK, Maensiri S, Theerakulpisut P (2017) Nano-priming technology for enhancing germination and starch metabolism of aged rice seeds using photo-synthesized silver nanoparticles. Sci Rep 7:8263

Malek M, Ghaderi-Far F, Torabi B, Sadeghipour H, Hay F (2019) The influence of seed priming on storability of rapeseed (Brassica napus) seeds. Seed Sci Technol 47:87–92

Malnassy TG (1971) Physiological and biochemical studies on a treatment hastening the germination of seeds at low temperature. PhD thesis, Rutgers University, New Jersey

McDonald MB (2000) Seed Priming. In: Black M, Bewley JD (eds) Seed technology and its biological basis. CRC Press, Boca Raton, FL, pp 287–325

Mortensen AK, Gislum R, Jørgensen JR, Boelt B (2021) The use of multispectral imaging and single seed and bulk near-infrared spectroscopy to characterize seed covering structures: methods and applications in seed testing and research. Agriculture 11(4):301

Müller H, Berg G (2008) Impact of formulation procedures on the effect of the biocontrol agent Serratia plymuthica HRO-C48 on Verticillium wilt in oilseed rape. BioControl 53:305–316. https://doi.org/10.1007/s10526-007-9111-3

Nega E, Ulrich R, Werner S, Jahn M (2003) Hot water treatment of vegetable seed—an alternative seed treatment method to control seed-borne pathogens in organic farming. J Plant Dis Prot 2003(110):220–234

Nesh P, Doddagoudar S, Vasudevan SN, Shakunthala NM, Aswathanarayana DS (2019) Influence of gamma irradiation and seed treatment chemicals on seed longevity of Bengal gram (Cicer arietinum L.) and Black gram (Vigna mungo L.). Int J Curr Microbiol App Sci 8(1):2866–2881

Panda D, Mondal S (2020) Seed enhancement for sustainable agriculture: an overview of recent trends. Plant Arch 20(1):2320–2332

Patel M, Gupta A (2012) Seed enhancements: the next revolution. In: The Seed Times. National Seed Association of India 5(2):7–14

Patil SK, Bansod SV (2014) Study the effects of some parameters on seed grading in gravity separator. Int J Eng Res Technol 03(01)

Pedrini S, Merritt DJ, Stevens J, Dixon K (2017) Seed coating: science or marketing spin? Trends Plant Sci 22(2):106–116. https://doi.org/10.1016/j.tplants.2016.11.002

Pedrini S, Bhalsing K, Cross AT, Dixon KW (2018) Protocol development tool (PDT) for seed encrusting and pelleting. Seed Sci Technol 46:393–405. https://doi.org/10.15258/sst.2018.46.2.21

Pietruszewski S, Kania K (2010) Effect of magnetic field on germination and yield of wheat. Int Agrophys 24(3):297–302

Pietruszewski S, Martinez E (2015) Magnetic field as a method of improving the quality of sowing material: a review. Int Agrophys 29(3):377–389

Pill WG, Necker AD (2001) The effects of seed treatments on germination and establishment of Kentucky bluegrass (Poa pratensis L.). Seed Sci Technol 29:65–72

Podlesna A, Bojarszczuk J, Podlesny J (2019) Effect of pre-sowing magnetic field treatment on some biochemical and physiological processes in faba bean (Vicia faba L. spp. Minor). J Plant Growth Regul 38:1153–1160

Reddy PP (2013) Bio-priming of seeds. In: Reddy PP (ed) Recent advances in crop protection. Springer, pp 83–90. https://doi.org/10.1007/978-81-322-0723-8_6

Rochalska M, Grabowska-Topczewska K, Mackiewicz A (2011) Influence of low magnetic field on improvement of seed quality. Int Agrophys 25(3):265–269

Rögner FH (2018) Electron treatment of seeds. https://www.fep.fraunhofer.de/content/dam/fep/en/documents/Produktflyer/E04_Electrontreatmentofseeds_EN_net.pdf. Accessed 28 Dec 2021

Rowse HR, McKee JMT, Finch-Savage WE (2001) Membrane priming – a method for small samples of high value seeds. Seed Sci Technol 29:587–597

Shi Y, Meng S, Xie X, Chai A, Li B (2016) Dry heat treatment reduces the occurrence of Cladosporium cucumerinum, Ascochyta citrullina and Colletotrichum orbiculare on the surface and interior of cucumber seeds. Hortic Plant J 2(1):35–40

Sivachandiran L, Khacef A (2017) Enhanced seed germination and plant growth by atmospheric pressure cold air plasma: combined effect of seed and water treatment. RSC Adv 7:1822–1832

Suryapal S, Harshita S, Bharat NK (2020) Hot water seed treatment: a review. https://doi.org/10.5772/intechopen.91314

Taylor A, Salanenka Y (2012) Seed treatments: phytotoxicity amelioration and tracer uptake. Seed Sci Res 22:S86–S90

Taylor A, Allen PS, Bennett MA, Bradford JK, Burris JS, Mishra MK (1998) Seed enhancements. Seed Sci Res 8:245–256

Thiphinkong D, Sikhao P (2021) Effect of seed coating with fluorescent compound on quality and fluorescence of cucumber seeds. Int J Agric Technol 17(6):2429–2438

Varier A, Vari AK, Dadlani M (2010) The subcellular basis of seed priming. Curr Sci 99(4)

Vashisth A, Nagarajan S (2008) Exposure of seeds to static magnetic field enhances germination and early growth characteristics in chickpea (Cicer arietinum L.). Bioelectromagnetics 29(7): 571–578. https://doi.org/10.1002/bem.20426

Vishwanath K, Weidauer A, Pallavi HM, Rögner FH, RAMEGOWDA. (2016) Low energy electron treatment an eco-friendly tool against seed borne diseases. Seed Res 44(1):23–31

Vurro M, Miguel-Rojas C, Perez-de-Luque A (2019) Safe nanotechnologies for increasing effectiveness of environmentally friendly natural agrochemicals. Pest Manag Sci 75(9):2403–2412

Wang Z, Amirkhani M, Avelar SA, Yang D, Taylor A (2020) Systemic uptake of fluorescent tracers by soybean (*Glycine max* (L.) Merr.) seed and seedlings. Agriculture 10(6):248

Waqas M, Korres NE, Khan MD, Nizami A, Deeba F, Ali I, Hussainet H (2019) Advances in the concept and methods of seed priming. In: Hasanuzzaman M, Fotopoulos V (eds) Priming and pretreatment of seeds and seedlings. Springer, Singapore. https://doi.org/10.1007/978-981-13-8625-1_2

Emerging Trends and Promising Technologies

Malavika Dadlani

Abstract

The growing demands for improved seeds of food, vegetables, flowers, other horticultural species, feed, fibre, forage, and fuel crops are driving the global seed industry at a CAGR of 6.6%, which is expected to grow from a market size of USD 63 billion in 2021 to USD 86.8 billion by 2026. This will primarily depend on two key features, genetic enhancement of the crop variety and seed quality. The focus of varietal improvement will be not only on yield increase but also on tolerance against biotic and abiotic stresses to meet the challenges of climate change, better input use efficiency, and improvement of the nutritional value. Advanced molecular tools and techniques including gene editing are likely to be used for precision breeding. The scope of seed quality, on the other hand, will grow beyond the basic parameters of purity, germination, and health, to the ability to perform better under adverse growing conditions, supplement the nutrient deficiency in the soil, and withstand pests during seed germination and early growth stages. Seed enhancement technologies will complement the genetic enhancement, as the advancements in seed technology will become an integral part of future crop improvement programmes. This will call for more efficient and precise technologies for determining quality parameters and predicting seed longevity, which will need to be developed using advanced marker technologies, 3D and multispectral imaging analysis, digital phenotyping, and other non-destructive methods. Novel approaches of hybrid seed production, including apomixis and doubled haploidy, will be integrated further for their commercial application. Multi-dimensional approaches need to be taken for addressing the gaps in understanding the key physiological processes underlying the regulation of dormancy (and pre-harvest sprouting), germination, recalcitrance, loss of viability, and seed priming.

M. Dadlani (✉)
Formerly at ICAR-Indian Agricultural Research Institute, New Delhi, India

© The Author(s) 2023
M. Dadlani, D. K. Yadava (eds.), *Seed Science and Technology*,
https://doi.org/10.1007/978-981-19-5888-5_17

Keywords

AI · Apomixis · Doubled haploidy · Non-destructive testing for viability ·
Multispectral imaging · Regulation of dormancy · Seed longevity · Molecular
markers for seed quality · EDVs

1   Introduction

Seed technology, the vital link between plant breeding and crop husbandry, has
turned out to be a multi-disciplinary subject, relying primarily on the scientific
advancements in the fields of plant biology, genetics, plant breeding, plant physiology, plant pathology, entomology, agronomy, and agricultural engineering, as all of
them would be involved in quality seed production, testing of its genetic purity, seed
germination, vigour and health, and proper processing and storage to retain high
quality till used by the farmers. There is, however, no defining line between the
science and technology of seeds. While seed science refers to the research and
advancements of scientific knowledge of the processes that determine the quality
of seed, seed technology relates to commercial application of this knowledge for
superior crop performance and also increased profitability for the seed industry.
Identifying and understanding the problems, and finding scientific explanations for
these, is the first step towards developing a successful technology. With the rapid
advancements in modern agriculture, be it the genetic improvement through selection, hybridization, molecular breeding, gene editing, or genetic modifications; novel
approaches to hybrid seed production; development of precise seed enhancement
technologies to withstand abiotic and biotic stresses; or automation and AI, scientific
research has led to the technological advancements as is evidenced by the enormous
progress in this field. While significant scientific advancements have been made
almost in every field of seed biology, production, and storage (also conservation),
there are critical gaps in our understanding of several aspects of seed genetics,
physiology, production, quality assessment, and quality improvement, which calls
for more deep delving into these functions. More effective technologies can be built
only on further advancements in the scientific bases of these functions. Recent years
have seen some significant advances in addressing emerging problems in this sector,
some of which are discussed here.

2   Hybrid Seed Production

The discovery of cytoplasmic male sterility (CMS) facilitated the hybrid breeding in
several crop species, e.g. rice, sorghum, pearl millet, brassicas, sunflower, etc.,
enabling the production of high-yielding hybrids. Whereas heterosis breeding has
been more effectively translated into hybrid development in cross-pollinated species,
there is still plenty of scope to utilize heterosis in other economically important crop
species, where pure line are still dominating, and the productivity has nearly reached

a plateau. Identifying new sources of CMS and their restorer genes, broadening the genetic background by utilizing the crop wild relatives (CWR), and molecular breeding are being pursued most vigorously by plant geneticists, breeders, and molecular biologists globally. While these initiatives will undoubtedly lead to further increase in crop productivity, making genetically pure and high-quality hybrid seeds available to the smallholding farmers at an affordable cost remains a challenge in the absence of stable and functional CMS systems in many agriculturally important species. A stable expression of male sterility is of utmost importance not only for cost-effective hybrid seed production but also for the maintenance of the seed parents and breeder and foundation seed production. Effects of erratic environmental conditions on the gene flow and breakdown of male sterility need to be studied more comprehensively, specially in high-value crops, to understand the specific roles of different factors and accordingly plan hybrid seed production under suitable growing conditions.

The potential of genic (or genetic) male sterility (GMS) has found limited use in commercial hybrid seed production in field crops. In this system, the recessive/wild-type (WT) heterozygote is used to produce both the GMS line and the maintainer line. The need for roguing out 50% male fertile plants in seed parent population, and difficulty in differentiating the male fertile plants from the male sterile ones at an early stage makes this a tedious and economically non-viable except in some cucurbits and ornamental marigold. Despite the widespread application of PTGMS in rice, the systems are easily affected by unpredictable environmental conditions, causing intrinsic problems for sterile and maintainer line seed propagation and hybrid seed production (Chang et al. 2016). The *barnase-barstar* system of GMS is successful in hybrid development and seed production in Indian mustard (Mariani et al. 1992). Wu et al. (2016) proposed a next-generation hybrid seed production technology (SPT) by adopting a three-step approach comprising (1) a WT full coding sequence (CDS) of a male fertility gene for fertility restoration, (2) a pollen-inactivating gene that disrupts transgenic pollen development, and (3) a seed colour marker gene for seed sorting. Several similar strategies are being tested in rice (Chang et al. 2016) and maize (Zhang et al. 2018). Qi et al. (2020) proposed a simple next-generation sortable hybrid seed production strategy based on CRISPR/Cas-9 technology that creates a manipulated GMS maintainer (MGM) system via a single transformation. The derived single-copy hemizygous MGM lines bore a mutated MS26 gene, leading to complete male sterility but normal vegetative growth and grain yield. Such approaches will need to be standardized and adopted in other important crops, e.g. cotton.

New strategies for cost-effective hybrid seed production, e.g. blending (Nie et al. 2021), apomixis (Lawit 2012), or doubled haploidy (Pazuki et al. 2018), are being explored for a long time, but have not reached the stage of commercial exploitation for hybrid seed production in many species. Recently, Nie et al. (2021) reported that blends of restorer and male sterile lines (AL-type cytoplasmic male sterile lines), mixed in different proportions, can result in a significant increase in hybrid seed production in wheat, though maintaining the desired genetic purity of the hybrid seed remains an issue to be resolved.

Apomixis, which occurs throughout the plant kingdom (Grossniklaus et al. 2001), may provide one such tool for cost-effective hybrid seed production. The ability of angiosperms to reproduce apomictically is a trait having far-reaching consequences in plant breeding and seed production. Among angiosperms, more than 300 plant species from more than 35 families have been described as apomictic, with a distribution pattern that indicates a polyphyletic origin (Khan et al. 2015). Apomictic plants establish genetically stable, seed-propagating clones, which can produce numerous sporophytic true-to-type generations. This reduces the time to develop a new variety and the cost of hybrid seed production. The stabilization of heterozygous genotypes via apomixis is expected to make breeding programmes faster and cheaper, and hence, it has attracted the attention of seed biologists, plant breeders, and commercial seed producers alike. Strategies for introgression of apomixis from wild relatives into cultivated plant species and transformation of sexual genotypes into apomictically reproducing genotypes have been pursued (Barcaccia and Albertini 2013) for decades, but with limited success. Though from the evolutionary perspective apomixis may be considered a consequence of sexual failure and not advancement (Silvertown 2008), the success in apomictic seed development in agricultural species can have a far-reaching impact, specially in non-grain crops. Hence, molecular regulation of apomixis and its use in hybrid seed production will certainly remain an area of much research interest in the coming years.

Since the early reports of microspore embryogenesis by Guha and Maheshwari (1964), and doubled haploidy in maize (Sarkar and Coe 1971), the potential of doubled haploidy (DH) in hybrid production has been explored in many crops. Recent successes in many crops (Chaikam et al. 2019; Kurimella et al. 2021) have widened its scope as both a tool in molecular mapping and development of hybrids and maintenance of the homozygous parental lines for hybrid development and seed production. Standardization of suitable protocols from induction of DH to the seed production of the derived parental lines has, therefore, attracted much attention of both the public and private research institutions and service providers. Further studies will be needed to make this technology cost-effective and applicable to species that are recalcitrant to such manipulations.

3   Prediction of Seed Longevity

The knowledge of the storage potential of the seed lots is of critical value to the farmers and the seedsmen alike in planning the storage of the seed stock for the next sowing or for managing the seed inventory in a commercial seed programme. Since the early reports on seed longevity of different species by Ewart (1908), many systematic studies have been undertaken on the longevity of seeds of a large number of species under different conditions of commercial stores as well as of the gene banks for short-, medium-, and long-term storage (Cromarty et al. 1982; Steiner and Ruckenbauer 1995; Singh et al. 2016; Lu et al. 2018; Solberg et al. 2020). The viability equations proposed and modified by EH Roberts, RH Ellis, and their colleagues (Roberts and Abdalla 1968; Roberts 1973; Ellis and Roberts 1980;

Ellis et al. 1989; Ellis and Hong 2007) provided a simple, yet valuable, method of estimating seed viability periods of crop species, based on three factors, i.e. temperature, moisture, and a species constant, based on the interception value of the linear survival curve, which, in spite of some lacunae, are in use. However, predicting seed deterioration and longevity of different species requires long-term storage experiments under ambient conditions and identification of the most sensitive physiological parameter that can pinpoint the state of losing viability. As a seed lot represents a population, in which each seed may be at a different physiological state, single seed analyses are receiving greater attention in recent years (Klaus Mummenhoff 2022). This trend, coupled with ambient storage studies, is expected to present a more accurate way of predicting seed longevity during storage.

With more and more studies showing the influence of anoxia in extending the longevity of seeds held in low M.C., validity of accelerated ageing, which predicts longevity by subjecting seeds to high humidity and high temperature, needs to be revisited. Scientists in the University of Wageningen are exploring the possibility of storing seeds in anoxia (for extending longevity) and developing a rapid seed ageing test under low moisture conditions and elevated partial pressure of oxygen to mimic the ambient storage and as an alternative to accelerated ageing test (Groot et al. 2015; Buijs et al. 2020; https://www.wur.nl/en/Research-Results/Projects-and-programmes/Wageningen-Seed-Science-Centre).

4 Seed Germination and Dormancy

The ability of desiccation-tolerant (orthodox) seeds to remain viable in the quiescent state, followed by resumption of metabolic activities culminating into germination upon hydration when other conditions are favourable, or the inability to do so by remaining dormant, and, finally, the loss of viability in the dry state, is still an enigma to the seed scientists.

Despite being an area widely investigated, critical gaps exist in our understanding of the regulation of seed dormancy and germination. In a recent review, Carrera-Castaño et al. (2020) pointed out the critical role of the seed coverings, and the need to study the genetic manipulation of specific cell wall components involving DOG1 and CWRE genes, and cell wall-related enzymes in the endosperm concerning dormancy and germination. Similarly, DOG1 and RDO5 genes appear to play key roles in dormancy regulation both via ABA accumulation and independent of ABA (Xiang et al. 2014; Fedak et al. 2016; Nonogaki 2019), which need to be elucidated. To understand the significance of RNA modifications such as N6 or N1, adenosine methylation and tissue-specific signalling mechanisms, such as pointed out by Carrera-Castaño et al. (2020), will probably need to be studied at the single-cell level.

A large volume of research point out the central role of the reactive oxygen species (ROS), which include free radicals, such as singlet oxygen ($1O_2$), superoxide ($O^{\cdot-2}$) or hydroxyl radical ($\cdot OH$), and non-radical hydrogen peroxide (H_2O_2), in the regulation of seed germination and dormancy (Baalbaki 2021). Bailly et al. (2008)

introduced the concept of the 'oxidative window for germination' to explain the dual role of ROS in either signalling for release from dormancy or oxidative damage, which inhibits germination. However, the mechanism of actions of the ROS is far from fully understood. Of these, hydrogen peroxide, though not a free radical, is considered to be the main ROS involved in cellular signalling, as it is more stable and is capable of crossing biological membranes (Petrov and Van Breusegem 2012). Similarly, even though the involvement of plant hormones in regulating seed germination and dormancy is much investigated, their roles need to be enumerated to understand the modulation of gene expression in the presence of specific hormones (Miransari and Smith 2014). Though the involvement of ROS and hormones in seed germination appears quite evident, the underlying mechanisms of the cross-talk between the two and the regulation of these processes are far from clear (Bailly 2019). Thus, studying the genetic control of the production of hormones (ABA, GA, auxins, cytokinins, brassinosteroids, ethylene, etc.), spatiotemporal accumulation of ROS, and the roles of hormones and ROS in manipulating the processes imposing dormancy or triggering germination will remain an active area of seed research.

This is also expected to influence the breeding programmes both for introducing primary dormancy to control pre-harvest sprouting and for releasing physiological deep dormancy to promote germination.

5   Desiccation-Sensitive (Recalcitrant) and Intermediate Longevity Behaviour

In a recent article, Dickie John (2021) drew attention to the problems in ex situ conservation of wild species, many of which are desiccation sensitive and exhibit recalcitrance or intermediate longevity behaviour. This poses a challenge to the managers of the seed banks because unlike the cultivated plant species, where seed multiplication can be undertaken periodically, in the case of wild species, collecting seeds with high initial germination is aimed at, considering the problem of multiplication. Citing a paper by Ali et al. (2007), who observed an initial increase in desiccation tolerance in *Anemone nemorosa* seeds, followed by a decline, he inferred that continued growth and development of the embryo resulted in loss of desiccation tolerance. He also drew a parallel with the loss of desiccation tolerance in orthodox seeds upon radicle emergence, a benchmark in priming technology, where seed hydration is stopped short of radicle protrusion to arrest the metabolic advancement of seeds without losing the property of desiccation tolerance. This analogy also points out the possibility of learning more about the mechanism of recalcitrance and prolonging their longevity by manipulating the desiccation tolerance trait through investigations of the physiology of primed seeds. Means to protect the longevity of primed seeds might throw more light on the physiological basis of recalcitrance and help find means to extend the longevity of desiccation-sensitive seeds. Recent studies on the re-induction of desiccation tolerance, by either modulating the ROS (Peng et al. 2017) or altering the hormonal regulation (Marques et al. 2019), are promising in understanding the desiccation tolerance and its incorporation in recalcitrant and intermediate seeds.

6 Quantifiable, Universal Vigour Tests

Though good progress has been made in the last 50 years in elucidating the basic mechanisms and development of tests to assess seed vigour in a variety of species, including the large- and small-seeded field crops, vegetables, flowers, forage grasses, and green manuring species, we are still far from having a universal scale which can be used for quantifying seed vigour. Can we predict that wheat seeds recording 85–90% RE (radicle emergence) will certainly result in 85% or more field emergence under normal growing conditions at any location or that the seed lots which recorded 50, 60, and 70% germination after the AA test (or, for that matter, CD test) will result in ~50, ~60, and ~70% germination after 'x' months of storage under ambient conditions? Researchers have generated extensive data on both, vigour tests have been standardized based on the criteria of high r values between these parameters (Powell 2022; Marcos-Filho 2015; and Loeffler 2022), and yet these cannot be used for recommending Minimum Standards for Certification for seed vigour.

Since its constitution in 2001, the ISTA Vigour Committee has introduced several vigour tests (ISTA 2022), e.g. the accelerated ageing and conductivity tests for soybean (*Glycine max*); conductivity test for field bean (*Phaseolus vulgaris* L.), chickpea (*Cicer arietinum* L.), and radish (*Raphanus sativus*); controlled deterioration for *Brassica* spp.; radicle emergence for corn (*Zea mays*), radish (*Raphanus sativus*), and wheat (*Triticum aestivum*); and tetrazolium vigour test for *Glycine max*. The work to standardize many more is in progress. From indirect and lengthy tests of yesteryears, such as the glutamic acid decarboxylase activity (GADA) test (Grabe 1964), paper piercing test, and mean germination time (MGT), which required well-trained seed analysts to perform vigour tests, the focus has shifted to direct, rapid, and more robust tests at present. In some instances, a combination of more than one vigour test has been used for a more reliable vigour estimation (Mavi et al. 2016).

In industrialized countries, customers expect 100% seed germination and field emergence and are ready to pay a premium for that, specially in high-value seeds. Therefore, the seed industry focuses on achieving the highest seed performance and looks for precision in vigour assessment. The future vigour tests are expected to be simpler to perform, automated, and preferably non-destructive, based on measuring the ability of individual seeds in a lot to perform under a given set of storage and/or growing conditions. In a recent study, Ermis et al. (2022) found that RE counts between 30 and 34 h were highly correlated ($P < 0.01$) with normal germination percentage, Ki, and p50. Storing seeds at 75% relative humidity and 35 °C for over 100 days, they constructed seed survival curves based on normal germination using probit analysis and concluded that the RE test can be used to evaluate the seed storage potential of interspecific hybrid crosses of cucurbits. Recent studies using mathematical models for vigour estimation in selected crops based on electrical impedance spectroscopy (Feng et al. 2021), near-infrared spectroscopy (Al-Amery et al. 2018), quantification and profiling of the volatile organic compounds (Umarani et al. 2020; Zhang et al. 2022), and hyperspectral imaging (Yuan et al. 2022) have shown encouraging results, paving the way for further studies under a wide range of seed storage environments and by including more number crops and varieties. Application of mathematical models, combined with the physical, biochemical, and imaging techniques, is foreseen to provide better estimates of seed vigour than simple lab tests.

7   Non-destructive Rapid Assessment of Seed Viability/Germination

Rapid and reliable assessment of the seed viability is required for research purposes, germplasm conservation and regeneration, and management of commercial seed inventory. Research groups both in public research institutions and in the private seed and allied industry are pursuing innovative approaches to achieve this.

Respiration, being the key to all metabolic processes, is most vital in seed germination and seedling development. Thus, it is not surprising that many attempts have been made by the researchers for over a century (Hasegawa 1935; Grabe 1964; Lakon 1928) to identify the most stable and reliable parameter of respiration that can be used for the assessment of seed vigour and viability. While this led to the tetrazolium test for a quick assessment of seed viability (ISTA 1985), continuing efforts are being made to develop an automated, quick, and yet reliable indicator of seed viability and vigour, which can be used commercially for differentiating good seeds from the poor ones.

Oxygen consumption at the beginning of germination is considered an indicator of seed vigour (Reed et al. 2022). Determination of the respiration of individual seeds, thus, is a potential tool for seed quality assessment. A rapid and non-destructive method proposed by Tai-Gi Min and Woo Sik Kang (2011) for assessing the individual seed viability of Brassicaceae species using resazurin reagent (RR) attracted a lot of attention. Recently, a somewhat similar test 'Seedalive™' has been proposed for viability assessment. This rapid, non-destructive, and artificial intelligence (AI)-based test was reported to have been tested on seeds from 15 plant families including cultivated and wild species. Being non-destructive, and showing high correlation with germination rate, the test may have a potential application in germplasm conservation and regeneration planning (Klaus Mummenhoff 2022) besides its use in the commercial seed. 'Fytagoras', a research-based company, recommends seed respiration analyser (SRA), for a fast (10–72 h) and accurate measurement of individual seed respiration, which may be correlated to seed lot quality. They suggest that SRA, among other things, may be useful as a measuring tool to quickly determine seed viability, in the mapping of multiple germination and seed-related parameters, and in predicting the effect and feasibility of priming methods (https://www.fytagoras.com/en). Both the technologies show merit but need further investigation and appropriate validation.

8 Imaging Technologies for Quality Assessment

Potential of X-ray and other imaging technologies has been explored for a long time for various aspects of seed quality, from the morphological characterization of seed and other plant parts for variety characterization and DUS testing to detection of slight seed damages and insect invasion and visualization of embryonic structures and germinating seeds. Optical sensing and imaging technologies being rapid, non-destructive, and simple are also gaining importance for the quality assessment of single seeds, particularly in the management of seed banks (Hilhorst 2020).

ISTA rules already have provisions for using X-ray images for seed testing. AOSA developed a handbook for X-ray testing of seeds (AOSA 1979), and ISTA has included X-ray test for seed testing with the following objectives:

- To provide a quick, non-destructive method of differentiating between filled, empty, insect-damaged, and physically damaged seed from the morphological characteristics evident on an X-radiograph.
- To create a permanent photographic record of the proportions of filled, empty, insect-damaged, and physically damaged seeds in a sample.

Considering the growing importance of innovative technologies in seed programmes, ISTA organized a seed seminar on 'Advancements and innovation in seed testing: from science to robust test' during the Congress, 2022, at Cairo, Egypt, which deliberated on the possible uses of different types of imaging technologies for seed quality assessment. Incotec NL has introduced innovative seed sorting systems using X-ray imaging. This fully automated 2D X-ray upgrading technology uses fourth generation of analysis software (based on deep learning) and was found useful in upgrading seeds of tomato and pepper (based on. Among other similar technologies, infrared thermography seemed quite promising in identifying the differences in seed vigour in soybean seed lots (presentations in the ISTA Congress, Cairo, 2022).

Delayed luminescence (DL) is another imaging approach, which is based on the long-term decay of weak photon emissions from materials following exposure to light with a wavelength of 400–800 nm and provides a comprehensive method for measuring biological systems, including seeds (Vesetova et al. 1985). Adeboye and Börner (2020), based on the information available for over 40 years on DL as a possible tool for seed quality assessment, concluded that the diagnostic potentials of DL, based on its ability to penetrate and analyse molecular structures, could be explored for future use in digital varietal discrimination, determination of seed quality based on membrane integrity, and genetic composition of the seed.

Red-blue-green (RBG) and multispectral imaging showed the potential to differentiate seed quality based on the unique images of individual seeds at different wavelengths, which was found particularly useful in detecting seed-borne fungi in sugar beet seeds (Boelt et al. 2018). The imaging technologies combined with machine learning algorithms have made significant advances in seed science

(Medeiros et al. 2020), which would support and facilitate the decision-making in the seed industry related to the marketing of seed lots.

Though non-invasive, these technologies are mostly single seed-based, require costly instrumentation and other infrastructure, and hence are costly for wider application. The advancements in the field of artificial intelligence (AI) are expected to reduce the cost and expand the scope.

Automatic imaging-based assessment of seeds and seedlings has shown some quantifiable advantages over visual assessment not only in terms of precision or speed but also in a range of parameters, including length, width, area of the seed or seedling, colours, shape, and other morphological parameters of each seed, which can be recorded simultaneously. This not only offers a more precise approach but also provides information for consistent documentation of the samples by storing the original images for comparison and future use (Jansen 2022, https://analytik.co.uk/webinar-artificial-intelligence-enables-new-ways-in-seed-quality-testing/ accessed 20/04/2022).

The industry prefers to follow a protocol, based on advance seed research, that is suitable for rapid and wider application (not crop- or condition-specific). While all the above-mentioned technologies showed promise, more work is required before these methods can be recommended for wide adoption. AOSA as well as ISTA with its Advanced Technologies Committee, Seed Science Advisory Group, and other specific technical committees would need to critically assess the scientific merits of the new technologies and validate their use in phenotyping and seed quality testing.

9   Application of Molecular Markers for Variety Identity and Seed Testing

Application of the protein and DNA markers in variety identification and hybrid purity testing has been long recognized. ISTA has standardized the protocols for protein markers in establishing variety identity in barley, *Pisum* and *Lolium* (SDS-PAGE); wheat (Acid–PAGE); wheat and Triticosecale (SDS-PAGE) and for hybrid purity testing in maize and sunflower (UTLIEF); and SSR markers (using a set of at least eight polymorphic markers) for variety identity in wheat and maize (ISTA Rules, 2022). Jamali et al. (2019), reviewing the applicability of DNA markers in plant variety protection, recognized their efficacy, particularly in case of establishing the status of essentially derived varieties (EDVs). However, the UPOV currently accepts the use of molecular markers only when they perfectly correlate with DUS traits. Knock (2021) questioned the extension of the protection to essentially derived varieties (EDVs) in the UPOV 1991 Convention and expressed concerns that while new breeding technologies (NBT) are the norm of variety development, the inclusion of molecular breeding-derived varieties as EDVs, instead of independent varieties (IV), put these at a serious disadvantage and suggested some modifications in the guidelines so that the innovative breeding is not curtailed.

Notwithstanding the wide application and precision of these markers, the need for using a large number of markers for arriving at a conclusive decision is seen as a deterrent. Hence, the next-generation marker technologies using SNPs (single nucleotide polymorphisms) are being promoted in breeding as well as seed technology (Zhang et al. 2020), which would have to emphasize the precision and stringency of the markers. With cost-effective advanced chip and array technologies consisting of hundreds of thousands of SNP markers, in combination with advanced computing technologies, the use of SNP markers for establishing variety identity and hybrid purity seems only a matter of time before arriving at an acceptable methodology. The advantages of using SNPs for assessing the genetic purity of maize hybrids and parental lines through the use of ddRADseq (double digest restriction-site associated DNA) as the sequencing protocol (Peterson et al. 2012) have been reported. The BMT group of UPOV, in collaboration with ISTA, and after necessary scientific deliberations, will be expected to come out with clear guidelines and protocols for using DNA markers for establishing variety identity, particularly for the essentially derived varieties (EDVs) and gene-edited varieties, which are expected to rise rapidly in the future. Standardization of precise molecular techniques for variety identification, specially for closely related, but independent, varieties, and genetic purity testing will need to be developed parallel to the advances in molecular plant breeding.

With the increasing issues on IPRs of the various varietal technologies and increasing cases of spurious seed in the market, the barcode system based on a set of crop-specific molecular markers needs to be developed for all crop species in commerce. ISTA has recommended protocols for establishing variety identity in wheat and maize using a set of at least eight SSR markers each (ISTA 2022). Making DNA fingerprinting an integral part of crop improvement programme as a regulatory requirement may be considered. A national library of DNA fingerprints needs to be established where a national variety register is maintained. The data of new varieties should be verified with the data available in the national library when a new variety is entered.

Accurate detection of seed-borne viruses, fungi, and bacteria using molecular markers is another important aspect of seed quality that has been pursued vigorously in the last two decades. PCR-based methods, e.g. multiplexed PCR (m-PCR), real-time PCR (RT-PCR), and bio-PCR, have been used effectively in the detection of seed-borne pathogens. However, the cost of equipment and infrastructure and the need for the technical expertise of seed analysts make these somewhat prohibitive. Technologies such as loop-mediated isothermal amplification (LAMP), developed by Notomi et al. (2000), which is a simple, cost-effective, and rapid method for the specific detection of genomic DNA and enables the synthesis of large amounts of DNA in a short time without the use of thermal cycler, may be further refined (see chapter "Molecular Techniques for Testing Genetic Purity and Seed Health" for details).

10 Next-Generation Seed Enhancement Technology

Seed enhancement technologies comprising physical, chemical, and biological treatments, priming, and coating/pelleting are now well recognized as means to improve the planting value of seeds in a wide range of conditions. Seed priming is identified as a potential tool that can provide synchronized, rapid, and on-demand germination, seedling establishment, and resilience against biotic and abiotic stresses improving plant survival in harsh and uncertain environments (Pedrini et al. 2020). Accurate and easy-to-apply seed priming and other enhancement technologies would offer substantial benefits to the rapidly growing business of seedlings (young plants), especially in the cultivation of high-value crops. Despite offering substantial advantages in field emergence under wide-ranging environments, priming is reported to adversely impact the vigour and longevity of seeds, if not sown immediately. Attempts have been made to extend the longevity of primed seeds by storing them in a partial vacuum (Chiu et al. 2003) or low-temperature conditions or induce longevity by exposing them to a hardening treatment (Bruggink et al. 1999). Recently, the advantage of priming for better seedling emergence and vigour has been reported in two desiccation-sensitive tropical rainforest species (Becerra-Vázquez et al. 2020), which not only has a high potential in restoration activities but may also broaden our understanding of the mechanism of recalcitrance.

Another area of growing interest relates to the development of precise seed enhancement technologies using biologicals, specially for organic cultivation, providing protection against pests and pathogens at seedling and early vegetative stages (Singh and Vaishnav 2022). Maintaining the viability and efficacy of fungal inoculum during storage, specially under warm and humid environment, will be a key research area. Nanotechnology is also emerging as a potential technology for supplementing micronutrients, pesticides, tolerants, and seed enhancers, though the standardization of technology and biosafety issues still remain to be addressed.

The future of seed research and technology development appears both exciting and challenging and calls for a closer linkage between the public research institutions, private seed industry, and technology developers cutting the boundaries of specialization. Platforms such as AOSA and ISTA would play important roles in validating their scientific merits and value for application of the new technologies. However, recognizing the fact that the way agriculture continues to be practised in the developing economies with the predominance of smallholders relying on low-cost inputs is quite different from that in the industrialized countries, affordability of quality-enhanced seeds of superior plant varieties will remain a major challenge.

Acknowledgements Sincere gratitude to Dr. H.S. Gupta, Former Director, ICAR-IARI, Dr. D. Vijay and Dr. S.K. Chakrabarty, Principal Scientists, Div. Seed Science & Technology, ICAR-IARI, and Dr. Partha R. Das Gupta, Emeritus Advisor, Syngenta Foundation, India, for their insightful observations and valuable inputs in preparing this chapter.

References

Adeboye K, Börner A (2020) Delayed luminescence of seeds: are shining seeds viable? Seed Sci Technol 48:167. https://doi.org/10.15258/sst.2020.48.2.04

Al-Amery M, Geneve RL, Sanches MF, Armstrong PR, Maghirang EB, Lee C, Vieira RD, Hildebrand DF (2018) Near-infrared spectroscopy used to predict soybean seed germination and vigour. Seed Sci Res 28:245–252

Ali N, Probert RJ, Hay FR, Davies H (2007) Post-dispersal embryo growth and acquisition of desiccation tolerance in Anemone nemorosa L. seeds. Seed Sci Res 17(3):155–163

Association of Official Seed Analysts (1979) Radiographic analysis of agricultural and forest tree seed: handbook no. 31, by E Belcher and J Bozzo

Baalbaki R (2021) Editorial. Seed Sci Technol 49(3):321–330. https://doi.org/10.15258/sst.2021.49.3.11

Bailly C (2019) The signalling role of ROS in the regulation of seed germination and dormancy. Biochem J 476:3019–3032

Bailly C, El-Maarouf-Bouteau H, Corbineau F (2008) From intracellular signaling networks to cell death: the dual role of reactive oxygen species in seed physiology. C R Biol 31(10):806–814

Barcaccia G, Albertini E (2013) Apomixis in plant reproduction: a novel perspective on an old dilemma. Plant Reprod 3:159–179

Becerra-Vázquez ÁG, Coates R, Sánchez-Nieto S et al (2020) Effects of seed priming on germination and seedling growth of desiccation-sensitive seeds from Mexican tropical rainforest. J Plant Res 133:855–872

Boelt B, Shrestha S, Salimi Z, Jørgensen JR, Nicolaisen M, Carstensen JM (2018) Multispectral imaging – a new tool in seed quality assessment? Seed Sci Res 28:222–228

Bruggink G, Ooms J, Toorn PV (1999) Induction of longevity in primed seeds. Seed Sci Res 9:49–53

Buijs G, Willems LAJ, Kodde J, Groot SPC, Bentsink L (2020) Evaluating the EPPO method for seed longevity analyses in Arabidopsis. Plant Sci 301:110644

Carrera-Castaño G, Calleja-Cabrera J, Pernas M, Gómez L, Oñate-Sánchez L (2020) An updated overview on the regulation of seed germination. Plan Theory 9(6):703

Chaikam V, Molenaar W, Melchinger AE, Prasanna BM (2019) Doubled haploid technology for line development in maize: technical advances and prospects. Theor Appl Genet 132:3227–3243

Chang Z, Chen Z, Wang N, Xie G, Lu J, Yan W, Zhou J, Tang X, Deng XW (2016) Construction of a male sterility system for hybrid rice breeding and seed production using a nuclear male sterility gene. Proc Natl Acad Sci U S A 113(2016):14145–14150

Chiu KY, Chen CL, Sung JM (2003) Partial vacuum storage improves the longevity of primed sh-2 sweet corn seeds. Sci Hortic 98(2):99–111

Cromarty A, Ellis RH, Roberts EH (1982) The design of seed storage facilities for genetic conservation. In: Handbooks for genebanks, no 1, 100p. International Board for Plant Genetic Resources, Rome

Dickie JB (2021) Editorial. Seed Sci Technol 49(1):73–80

Ellis RH, Hong TD (2007) Quantitative response of the longevity of seed of twelve crops to temperature and moisture in hermetic storage. Seed Sci Technol 35:432–444

Ellis RH, Roberts EH (1980) Improved equations for the prediction of seed longevity. Ann Bot 45:13–30

Ellis RH, Hong TD, Roberts EH (1989) A comparison of low moisture content limit to the logarithmic relation between seed moisture and longevity in 12 species. Ann Bot 63:601–611

Ermis S, Oktem G, Mavi K, Hay F, Demir I (2022) The radicle emergence test and storage longevity of cucurbit rootstock seed lots. Seed Sci Technol 50(1):1–10

Ewart AJ (1908) On the longevity of seeds. Proc R Soc Victoria 21:1–120

Fedak H, Palusinska M, Krzyczmonik K, Brzezniak L, Yatusevich R, Pietras Z, Kaczanowski S, Swiezewski S (2016) Control of seed dormancy in Arabidopsis by a cis-acting noncoding antisense transcript. Proc Natl Acad Sci U S A 113:E7846–E7855

Feng L, Hou T, Wang B, Zhang B (2021) Assessment of rice seed vigour using selected frequencies of electrical impedance spectroscopy. Biosyst Eng 209:53–63

Grabe DF (1964) Glutamic acid decarboxylase activity as a measure of seedling vigor. Proc Assoc Off Seed Anal 54:100–109

Groot SPC, de Groot L, Kodde J, Treuren R (2015) Prolonging the longevity of ex situ conserved seeds by storage under anoxia. Plant Genet Resour Characterization Util 13(1):18–26

Grossniklaus U, Spillane C, Page DR, Kohler C (2001) Genomic imprinting and seed development: endosperm formation with and without sex. Curr Opin Plant Biol 4:21–27

Guha S, Maheshwari SC (1964) In vitro production of embryos from anthers of Datura. Nature 204: 497

Hasegawa K (1935) On the determination of viability in seed by reagents. Proc Int Seed Testing Assoc 7:148–153

Hilhorst HWM (2020) Editorial, seed science and technology, vol 48, no 2, pp 315–323

International Rules for Seed Testing (2021), Full Issue I–19-8 (300);https://doi.org/10.15258/ista rules

ISTA (1985) Handbook of tetrazolium testing. Zurich, International Seed Testing Association, p 72

ISTA (2022) International rules for seed testing. International Seed Testing Association, Wallisellen, Switzerland. International Rules for Seed Testing (2021), Full Issue I–19-8 (300). https://doi.org/10.15258/istarules

Jamali SH, Cockram J, Lee T (2019) Insights on deployment of DNA markers in plant variety protection and registration. Theor Appl Genet 132(7):1911–1929

Jansen (2022) https://analytik.co.uk/webinar-artificial-intelligence-enables-new-ways-in-seed-quality-testing/. Accessed 20 Apr 2022

Khan YJ, Choudhary R, Tyagi H, Singh AK (2015) Apomixis: the molecular perspectives and its utilization in crop breeding. J AgriSearch 2(3):153–161

Klaus Mummenhoff (2022) Dead or alive: simple, nondestructive, and predictive monitoring of seedbanks. Presented in ISTA Seed Seminar, "Advancements and innovation in seed testing: from science to robust test", 8 May 2022. ISTA Congress, Cairo

Kock Michael A (2021) Essentially derived varieties in view of new breeding technologies – plant breeders' rights at a crossroads. GRUR Int 70(1):11–27

Kurimella RK, Bhatia R, Singh KP, Panwar S (2021) Production of haploids and doubled haploids in Marigold (Tagetes spp.) using anther culture. In: Segui-Simarro JM (ed) Doubled haploid technology. Methods in molecular biology, vol 2289. Humana, New York. https://doi.org/10.1007/978-1-0716-1331-3_18

Lakon G (1928) Ist die Bestimmung der Keimfahigkeit der Samen ohne Keimversuch moglich. Angewandte Botanik (Zeitschrift der Vereinigung fur angewandte Botanik) 10:470

Lawit SJ (2012) Self-reproducing hybrid plants. US20120266324A1 (Google Patents)

Loeffler T (2022) Editorial. Seed Sci Technol 50(1):163–174

Lu X, Chen X, Cui C (2018) Germination ability of seeds of 23 crop plant species after a decade of storage in the National Gene Bank of China. PGR Newslett 139:42–46

Marcos Filho JM (2015) Seed vigour testing: an overview of the past, present and future perspective. Sci Agric 72:363–374

Mariani C, Gossele V, De Beuckeleer M, De Block M, Goldberg RB, De Greef W, Leemans J (1992) A chimeric ribonuclease inhibitor gene restores fertility to male sterile plants. Nature 357:384–387

Marques A, Nijveen H, Somi C, Ligterink W, Hilhorst H (2019) Induction of desiccation tolerance in desiccation sensitive Citrus Limon seeds. J Integr Plant Biol 61:624–638

Mavi K, Powell AA, Matthews S (2016) Rate of radicle emergence and leakage of electrolytes provide quick predictions of percentage normal seedlings in standard germination tests of radish (Raphanus sativus). Seed Sci Technol 44:393–409

Medeiros AD, Silva LJD, Ribeiro JPO, Ferreira KC, Rosas JTF, Santos AA, Silva CBD (2020) Machine learning for seed quality classification: an advanced approach using merger data from FT-NIR spectroscopy and X-ray imaging. Sensors (Basel) 20(15):4319

Min T-G, Kang WS (2011) Simple, quick and nondestructive method for Brassicaceae seed viability measurement with single seed base using resazurin. Hortic Environ Biotechnol 52(3):240–245

Miransari M, Smith DL (2014) Plant hormones and seed germination. Environ Exp Bot 99:110–121

Nie, Kong, Cui, Sang, Mu, Xu, Tian (2021) Blend wheat AL-type hybrid and using SSRs to determine the purity of hybrid seeds. Seed Sci Technol 49(3):275–285

Nonogaki H (2019) Seed germination and dormancy: the classic story, new puzzles, and evolution. J Integr Plant Boil 61:541–563

Notomi T, Okayama H, Masubuchi H, Yonekawa T, Watanabe K, Amino N, Hase T (2000) Loop-mediated isothermal amplification of DNA. Nucleic Acids Res 28(12):E63

Pazuki A, Aflaki F, Gurel S, Ergul A, Ekrem G (2018) Production of doubled haploids in sugar beet (Beta vulgaris): an efficient method by a multivariate experiment. Plant Cell Tissue Org Cult 132:85–97

Pedrini S, Balestrazz A, Mdsen MD, Bhaising K, Hardegree SP, Dixon KW, Kildisheva OA (2020) Seed enhancement: getting seeds restoration-ready. Restor Ecol 28(3):266–275

Peng L, Lang S, Wang Y, Pritchard HW, Wang X (2017) Modulating role of ROS in re-establishing desiccation tolerance in germinating seeds of Caragana korshinskii Kom. J Exp Bot 68:3585–3601

Peterson BK, Weber JN, Kay EH, Fisher HS, Hoekstra HE (2012) Double digest RADseq: an inexpensive method for de novo SNP discovery and genotyping in model and non-model species. PLoS One 7(5):e37135

Petrov VD, Van Breusegem F (2012) Hydrogen peroxide-a central hub for information flow in plant cells. AoB Plants 2012:pls014. https://doi.org/10.1093/aobpla/pls014

Powell AA (2022) Seed vigour in the 21st century. Seed Sci Technol 50(1):45–73

Qi X, Zhang C, Zhu J et al (2020) Genome editing enables next-generation hybrid seed production technology. Mol Plant 13(9):1262–1269

Reed RC, Bradford KJ, Khanday I (2022) Seed germination and vigor: ensuring crop sustainability in a changing climate. Heredity, pp 1–10. https://doi.org/10.1038/s41437-022-00497-2

Roberts EH (1973) Predicting the storage life of seeds. Seed Sci Technol 1:499–514

Roberts EH, Abdalla FH (1968) The influence of temperature, moisture, and oxygen on period of seed viability in barley, broad beans and peas. Ann Bot 32:97–117

Sarkar KR, Coe EH (1971) Origin of parthenogenetic diploids in maize and its implications for the production of homozygous lines. Crop Sci 11(4):543–544

Silvertown J (2008) The evolutionary maintenance of sexual reproduction: evidence from the ecological distribution of asexual reproduction in clonal plants. Int J Plant Sci 169:157–168

Singh HB, Vaishnav A (eds) (2022) New and future developments in microbial biotechnology and bioengineering - sustainable agriculture: revitalization through organic products. Elsevier, Amsterdam, p 366

Singh N, Dash S, Khan YJ (2016) Survival of chickpea, sesame, niger, castor and safflower seeds stored at low and ultra-low moisture contents for 16-18 years. Seed Sci Technol 44:1–14

Solberg SØ, Yndgaard F, Andreasen C, von Bothmer R, Loskutov IG, Asdal Å (2020) Long-term storage and longevity of orthodox seeds: a systematic review. Front Plant Sci 11:1007

Steiner A, Ruckenbauer P (1995) Germination of 110-year-old cereal and weed seeds, the Vienna sample of 1877. Verification of effective ultra-dry storage at ambient temperature. Seed Sci Res 5:195–199

Umarani R, Bhaskaran M, Vanitha C, Tilak M (2020) Fingerprinting of volatile organic compounds for quick assessment of vigour status of seeds. Seed Sci Res 30(2):112–121

Vesetova TV, Veselovsky VA, Rubin AB, Bochvarov PZ (1985) Delayed luminescence of air-dry soybean seeds as a measure of their viability. Physiol Plant 65(4):493–497

Wu Y, Fox TW, Trimnell MR, Wang L, Xu RJ, Cigan AM, Huffman GA, Garnaat CW, Hershey H, Albertsen MC (2016) Development of a novel recessive genetic male sterility system for hybrid seed production in maize and other cross-pollinating crops. Plant Biotechnol J 14:1046–1054

Xiang Y, Nakabayashi K, Ding J, He F, Bentsink L, Soppe WJ (2014) Reduced Dormancy5 encodes a protein phosphatase 2C that is required for seed dormancy in Arabidopsis. Plant Cell 26:4362–4375

Yuan P, Pang L, Wang LM, Yan L (2022) Application of hyperspectral imaging to discriminate waxy corn seed vigour after ageing. Int Food Res J 29(2):397–405

Zhang D, Wu S, An X, Xie K, Dong Z, Zhou Y, Xu L, Fang W, Liu S, Liu S et al (2018) Construction of a multicontrol sterility system for a maize male-sterile line and hybrid seed production based on the ZmMs7 gene encoding a PHD-finger transcription factor. Plant Biotechnol J 16(2018):459–471

Zhang J, Yang J, Zhang L, Luo J, Zhao H, Zhang J, Wen C (2020) A new SNP genotyping technology target SNP-seq and its application in genetic analysis of cucumber varieties. Sci Rep 10(1):5623

Zhang T, Ayed C, Fisk ID, Pan T, Wang J, Yang N, Sun Q (2022) Evaluation of volatile metabolites as potential markers to predict naturally-aged seed vigour by coupling rapid analytical profiling techniques with chemometrics. Food Chem 367:130760

Printed in the United States
by Baker & Taylor Publisher Services